Library of
Davidson College

Developmental Plasticity
*Behavioral and Biological Aspects
of Variations in Development*

DEVELOPMENTAL PSYCHOLOGY SERIES

SERIES EDITOR
Harry Beilin

Developmental Psychology Program
City University of New York Graduate School
New York, New York

LYNN S. LIBEN. *Deaf Children: Developmental Perspectives*

JONAS LANGER. *The Origins of Logic: Six to Twelve Months*

GILBERTE PIERAUT-LE BONNIEC. *The Development of Modal Reasoning: Genesis of Necessity and Possibility Notions*

TIFFANY MARTINI FIELD, SUSAN GOLDBERG, DANIEL STERN, and ANITA MILLER SOSTEK. (Editors). *High-Risk Infants and Children: Adult and Peer Interactions*

BARRY GHOLSON. *The Cognitive-Developmental Basis of Human Learning: Studies in Hypothesis Testing*

ROBERT L. SELMAN. *The Growth of Interpersonal Understanding: Developmental and Clinical Analyses*

RAINER H. KLUWE and HANS SPADA. (Editors). *Developmental Models of Thinking*

HARBEN BOUTOURLINE YOUNG and LUCY RAU FERGUSON. *Puberty to Manhood in Italy and America*

SARAH L. FRIEDMAN and MARIAN SIGMAN. (Editors). *Preterm Birth and Psychological Development*

LYNN S. LIBEN, ARTHUR H. PATTERSON, and NORA NEWCOMBE. (Editors). *Spatial Representation and Behavior Across the Life Span: Theory and Application*

W. PATRICK DICKSON. (Editor). *Children's Oral Communication Skills*

EUGENE S. GOLLIN. (Editor). *Developmental Plasticity: Behavioral and Biological Aspects of Variations in Development*

In Preparation

GEORGE E. FORMAN. (Editor). *Action and Thought: From Sensorimotor Schemes to Symbolic Operations*

Developmental Plasticity

Behavioral and Biological Aspects of Variations in Development

Edited by

EUGENE S. GOLLIN

Department of Psychology
University of Colorado
Boulder, Colorado

1981
ACADEMIC PRESS
A Subsidiary of Harcourt Brace Jovanovich, Publishers
New York   London   Toronto   Sydney   San Francisco

This volume is based, in part, on a continuing program in biobehavioral development conducted by the Developmental Psychology Area of the Psychology Department in the University of Colorado

COPYRIGHT © 1981, BY ACADEMIC PRESS, INC.
ALL RIGHTS RESERVED.
NO PART OF THIS PUBLICATION MAY BE REPRODUCED OR
TRANSMITTED IN ANY FORM OR BY ANY MEANS, ELECTRONIC
OR MECHANICAL, INCLUDING PHOTOCOPY, RECORDING, OR ANY
INFORMATION STORAGE AND RETRIEVAL SYSTEM, WITHOUT
PERMISSION IN WRITING FROM THE PUBLISHER.

ACADEMIC PRESS, INC.
111 Fifth Avenue, New York, New York 10003

United Kingdom Edition published by
ACADEMIC PRESS, INC. (LONDON) LTD.
24/28 Oval Road, London NW1 7DX

Library of Congress Cataloging in Publication Data
Main entry under title:

Developmental plasticity.

(Developmental psychology series)
Includes bibliographies and index.
1. Adaptability (Psychology) 2. Adaptation
(Physiology) 3. Developmental psychology.
4. Learning, Psychology of. 5. Psychology,
Comparative. I. Gollin, Eugene S. II. Series.
[DNLM: 1. Child development. 2. Learning. WS
105 D48917]
BF713.D476   155.2'2   80-2331
ISBN 0-12-289620-3   AACR2

PRINTED IN THE UNITED STATES OF AMERICA

81 82 83 84   9 8 7 6 5 4 3 2 1

Contents

List of Contributors	ix
Preface	xi

I
Evolutionary and Genetic Background 1

1
Evolution and Genetic Variability 3
GERALD E. McCLEARN

Introduction	3
Contemporary Variability between Species	3
Evolutionary Variability	5
Sources of Variability	12
Variability within Species	18
Molecules and Variability	19
Polygenic Systems	21
Developmental Variability within the Organism	26
Summary	29
Bibliography	30
References	30

II
The Sensory Base — 33

2
The Infancy of Human Sensory Systems — 35
JOHN S. WERNER AND LEWIS P. LIPSITT

Introduction	35
Prenatal Origins of Sensory Development	36
Sensory Bases of Infant Perception	37
Conclusions	61
References	61

III
Learning and Ethology — 69

3
Learning Theory, Ethological Theory, and Developmental Plasticity — 71
DAVID CHISZAR

Introduction	71
Problems Raised by the Goal-Directedness of Behavior	73
Learning Theory	77
Ethological Theory	81
Developmental Plasticity in the Perspectives of Learning and Instinct Theory	87
Evolution of Developmental Plasticity: Implications for the Analysis of Immediate Causation	88
The Confluence of Ideas: Emergence of Interest in Animal Perception	93
References	98

4
The Infancy of Human Learning Processes — 101
LEWIS P. LIPSITT AND JOHN S. WERNER

Introduction	101
The Pleasures of Sensation as Incentives for Infant Learning	104

The Infant as Learner	114
The Functions of Early Human Learning	124
Summary	128
References	129

5
Innate Programs for Perceptual Development: An Ethological View 135

PETER MARLER, STEPHEN ZOLOTH, AND ROBERT DOOLING

Introduction	135
Perceptual Adaptations to Local Conditions: Honeybees	137
Sensitive Periods for Perceptual Change: Chick Pecking	138
Imprinting and Perceptual Development: Birds	140
Configurational Features and Sign Stimuli: The Herring Gull	143
Innate Constraints on Vocal Imitation: Birdsong	146
Species-Specific Processing of Vocal Stimuli: Macaques	150
Color Vision, Naming, and Preferences	154
The Ontogeny of Speech Perception	157
Conclusions on the Ethology of Perceptual Development	162
References	166

IV
Asymmetries and Variation 173

6
Lateralization and Its Implications for Variation in Development 175

JERRE LEVY

The Origins of Asymmetry	175
Lateralization of the Brain	180
Development of the Hand-Brain Relationship	196
Sex Differences in Cerebral Asymmetry	209
Why Variation?	220
References	222

V
Epistemology, Theory, and Method 229

7
Development and Plasticity 231
EUGENE S. GOLLIN

Introduction	231
Developmental Perspectives	233
Developmental Perspective and Plasticity	235
The Roles of Experience	237
Multimodal and Polyphasic Development	239
Theoretical and Methodological Implications of a Multimodal-Polyphasic Model	243
Cultural and Biological Instances of Plasticity in Development	247
Conclusions	249
References	249

8
Epistemology and Developmental Psychology 253
STEPHEN TOULMIN

Introduction	253
Does Mental Development Have a Unique Destination?	256
The Grounds for Skepticism about Any Universal Destination	257
The Need for an Alternative Approach	259
Alternative Developmental Destinations and Trajectories	262
References	267

Author Index 269
Subject Index 279

List of Contributors

Numbers in parentheses indicate the pages on which the authors' contributions begin.

DAVID CHISZAR **(71)**, Department of Psychology, University of Colorado, Boulder, Colorado 80309

ROBERT DOOLING **(135)**, Rockefeller University Field Research Center, Millbrook, New York 12545

EUGENE S. GOLLIN **(231)**, Department of Psychology, University of Colorado, Boulder, Colorado 80309

JERRE LEVY **(175)**, Department of Behavioral Sciences, University of Chicago, Chicago, Illinois 60637

LEWIS P. LIPSITT **(35, 101)**, Department of Psychology, Brown University, Providence, Rhode Island 02912

GERALD E. McCLEARN **(3)**, College of Human Development, The Pennsylvania State University, University Park, Pennsylvania 16802

PETER MARLER **(135)**, Rockefeller University Field Research Center, Millbrook, New York 12545

STEPHEN TOULMIN **(253)** Committee on Social Thought, University of Chicago, Chicago, Illinois 60637

JOHN S. WERNER **(35, 101)**, Department of Psychology, University of Colorado, Boulder, Colorado 80309

STEPHEN ZOLOTH **(135)**, Rockefeller University Field Research Center, Millbrook, New York 12545

Preface

Developmental scientists in the behavioral and biological areas are faced with a twofold problem. First, they must strive to comprehend the enormous diversity in form and function that exists among organisms of different species, among members of the same species, and within individuals over the course of ontogenesis. The second problem confronting developmentalists is that against this *background* of diversity there is the *figure* of species and individual integrity. How are the themes of diversity and integrity that characterize living systems to be reconciled? The contributors to this volume are in general agreement that both diversity and integrity are phenotypic expressions of developmental processes, and that the task is to explore the constraints on and opportunities for variation in the course of development.

In this volume these themes are examined from a variety of theoretical viewpoints and research contexts. In Chapter 1 McClearn reviews the broad evolutionary landscape and the specific genetic mechanisms implicated in biological and behavioral development. Next Werner and Lipsitt describe the sensory apparatus available to neonatal human beings. Chiszar, in Chapter 3, details the similarities and differences between ethological theories and learning theories and considers developmental plasticity in interdisciplinary contexts. The acquisition of behavior patterns during early postnatal development is examined by Lipsitt and Werner from a traditional learning theory point of view in Chapter 4, and in the following chapter the same general phenomenon is approached by Marler, Zoloth, and Dooling from ethological and comparative vantage points. The role played by asymmetry in general and by cerebral asymmetry in particular in the generation of individuality is examined by Jerre Levy in Chapter 6. In Chapters 7 and 8 Gollin and Toulmin, respectively, explore epistemological, theoretical, and methodological questions that arise from a consideration of developmental plasticity.

Developmental Plasticity
*Behavioral and Biological Aspects
of Variations in Development*

I

EVOLUTIONARY AND GENETIC BACKGROUND

A recurrent theme in this book is the adaptive significance of organismic change and the adaptive value of morphogenetic stabilization. Neither of these aspects of living systems is understandable without a consideration of the proliferation of life forms during the history of the planet. It is the task of evolutionary theorists to trace that history, to order it, and to render it into a tale that makes sense. In the following chapter, McClearn portrays in dramatic fashion the vastness of the time scale that serves as the evolutionary stage. It is a heroic story fashioned by brilliant, albeit temporary, successes and many, many failures. The factors that contribute to success in the sense of survival and to failure in the sense of extinction are, of course, the subject matter of many scientific disciplines. To understand how selective processes work to favor the vigor and prosperity of some organisms and the waning or demise of others requires knowledge about how hereditary mechanisms coact with environmental forces to produce particular phenotypes in the course of individual development, for it is upon the phenotypic arrangements that the selective pressures are exerted. In this chapter, the ground plan for phenotypic variability is presented. Developmental plasticity must be considered within the structure provided by that ground plan.

1

Evolution and Genetic Variability

GERALD E. McCLEARN

Introduction

People differ from starfish and from squirrels and from elephants. Furthermore, people, starfish, squirrels, and elephants differ one from the other. These observations are so obvious that they might be judged trite, but implications of the genetic perspective on this intra- and interspecific variability are of fundamental importance.

In terms of variance analysis, one might conceive of the total variability of *all* organisms. The within-species term would represent the subject matter of individual differences, and the between-species term would be the province of evolutionary biology. The purpose of this chapter is to provide a framework for thinking about variability in the developmental sciences. The picture is necessarily painted with the broadest of strokes, and the interested reader is directed to the cited references for more fine-grained expositions.

Contemporary Variability between Species

As a beginning, let us regard our species in an evolutionary perspective. Together with the gorillas, chimpanzees, orangutans, and gibbons, collectively of the family *Pongidae*, we Hominidae constitute the superfamily *Hominoidea*. Together with the superfamily *Cercopithecoidea* (the Old World monkeys, including 13 genera) and the superfamily *Ceboidea* (New World monkeys, 10 genera), we Hominoidea compose the suborder *Anthropoidea*. All together, we Anthropoidea number about 140 species. The suborder *Prosimii*—which includes the treeshrews, lemurs, and tarsiers—has about 53 species. The *Anthropoidea* and the *Prosimii* make up the order *Primates*, a group of mammals characterized most notably by mobile digits on hands and feet, a shortened snout, frontally placed eyes, a tendency toward upright posture, and a brain that is large relative to body size (see Jolly, 1972; Le Gros Clark, 1965). Thus, we see that about 193 species are in the immediate phylogenetic neighborhood of humankind.

For a more comprehensive picture, we note that primates belong to the class *Mammalia*, which also includes the orders *Monotremata* (such as the duck-billed platypus), *Marsupialia* (kangaroo, opossum, and anteater), *Lagomorpha* (rabbit), *Dermoptera* (flying lemur), *Chiroptera* (bat), *Insectivora* (shrew), *Fissipeda* (dog, cat), *Rodentia* (mouse, rat), *Pinnipedia* (seal), *Artiodactyla* (pig, deer, hippopotamus), *Perissodactyla* (horse, rhinoceros), *Proboscoidea* (elephant), and several others. All of these mammals belong to the subphylum *Vertebrata*. This is to say, they have backbones. Our species constitutes only a small part of all vertebrates; there are, in fact, about 55,000 vertebrate species. Approximately 4300 are mammalian species, about 3000 are amphibian species, 6000 are reptiles, 11,000 are birds, and the bony fishes come in the enormous variety of 28,000 species. Thus, even though diversity of species is a commonplace observation, common knowledge does not give one an adequate view of the range of diversity of living things. We do not encounter even a small sample of this diversity in the ordinary course of events. Even a trip to the zoo can only whet the intellectual appetite. No zoo can stock all 55,000 vertebrate species!

If this number of vertebrates is awesome, consider that there are about 1,055,000 specific types of creatures without backbones—among others, the jellyfishes, crabs, spiders, insects, oysters, flatworms, roundworms, sponges, and starfishes of the world. These invertebrates surpass us vertebrates in species number almost 20 to 1.

Particularly unlikely to be the object of everyday observation are single-celled organisms, but the fact that they are small does not mean that they are unimportant either qualitatively or quantitatively. It has been estimated that there are about 100 octillion living cells in the world today—that is, 100,000,000,000,000,000,000,000,000,000 (Hockett, 1973). Of these 100 octillion cells, perhaps as many as 99 octillion, and at least as many as 90 octillion, are tied up in single-celled organisms (bacteria, algae, and so on). We, the metazoan, multicellular animals, are in a decided minority. We should note further that the preceding discussion has concerned only the animal kingdom. We have not even considered the plants, which also exist in dazzling diversity.

How different are human beings and squirrels? In view of the enormous diversity of living things, the answer to this question will depend on the measuring stick used to assess variability. If a visitor from Mars were to examine a squirrel and a man, many similarities would be observed. It would be found that both are responsive to similar kinds of energies; our sense organs work according to the same basic principles and have similar sensitivities. If the human being and the squirrel were dissected, the specimens would look remarkably alike except for the difference in size. In both cases, the Martian investigator would find a little pump for blood, an inflatable bellows to transfer oxygen from the surrounding environment into the blood stream, a bean-shaped organ for waste disposal, and so on. These are remarkable similarities. If one's frame of reference were the whole array of over a million living animal species, one would conclude that squirrels and human beings are not very different at all.

However, there *are* differences between human beings and squirrels—differences in size, hair covering, complexity of behavioral processes, and so on. Whether species are alike or different is thus a matter of perspective. There is no absolute yardstick with which to measure interspecies distance. The apparent distance will shrink or expand depending upon our emphasis on the similarities or the differences—whether we are emphasizing the theme of all living beings, or the variety of specific forms the living assume.

Evolutionary Variability

The variability of living beings represents only differences among the survivors of the winnowing process called natural selection. Many more species have become extinct than are alive today. To consider the full range of variability, therefore, we need to turn to the evolutionary sequence that has culminated in the species that are our contemporaries.

The evolutionary scale in Figure 1.1 begins 5 billion years ago when the sun started glowing. Matters of professional interest to biologists or psychologists began about 3.5 billion years ago when life originated. Life evidently began in the oceans when exposure of a particular set of atmospheric ingredients to high temperatures and lightning produced the forerunners of amino acids—the building blocks of protein. Amino acids gradually accumulated in the ocean, making it a dilute organic soup. In this soup, combinations of organic constituents formed, and one of these yielded a molecule that could replicate itself. At that point, life was off and running (see Orgel, 1973).

Shortly thereafter, a matter of only a few million years, another fundamental development occurred. In the presence of the energy from sunlight, photosynthesis permitted the building up of carbohydrates from the carbon dioxide that was accumulating in the atmosphere. Photosynthesizing organisms were self-feeding; moreover, and of profound ultimate importance to us, they produced oxygen as a waste product. Oxygen, which was

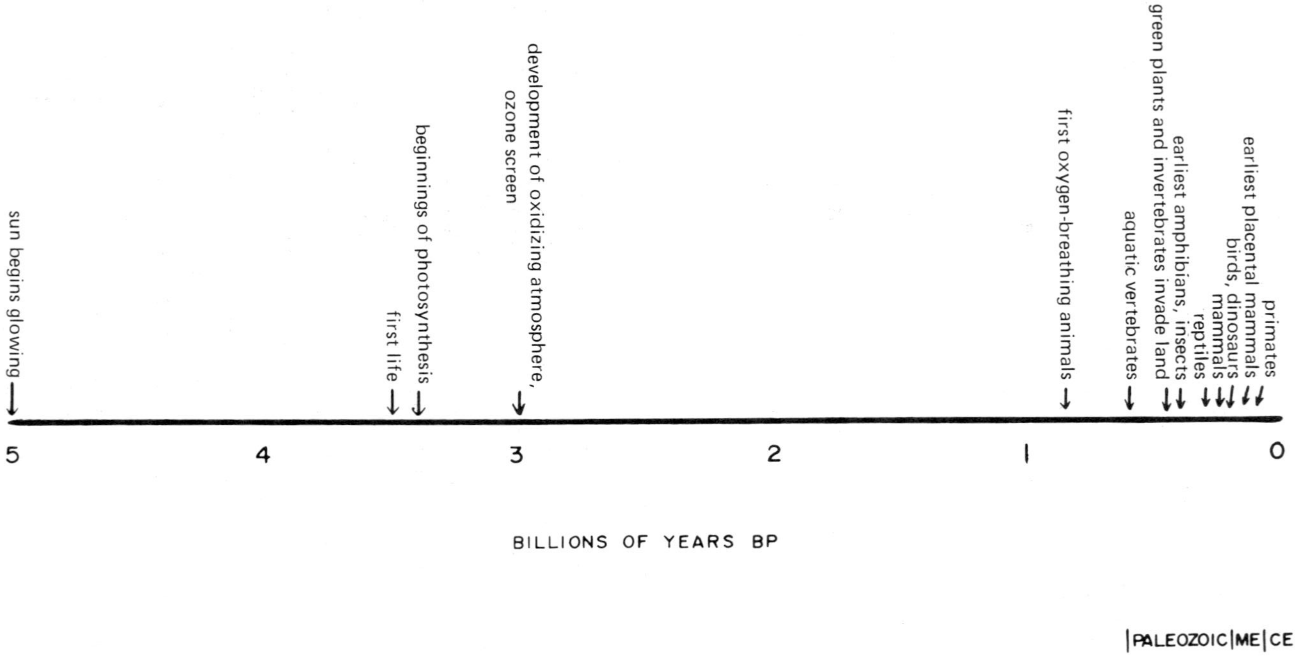

primates →
earliest placental mammals →
birds, dinosaurs →
mammals →
reptiles →
earliest amphibians, insects →
green plants and invertebrates invade land →
aquatic vertebrates →

first oxygen-breathing animals →

development of oxidizing atmosphere, ozone screen →

beginnings of photosynthesis →
first life →

sun begins glowing →

BILLIONS OF YEARS BP

| PALEOZOIC | ME | CE

Figure 1.1. *Evolutionary time scale from 5 billion years ago to the present.*

1. EVOLUTION AND GENETIC VARIABILITY

initially very rare, increased slowly as more and more photosynthesis took place. The first oxygen-breathing animals appeared a little less than a billion years ago. The ability to breathe oxygen was a tremendously successful adaptation that led to an explosion (in slow motion, of course, over millions of years) of various life forms occupying all kinds of different ecological niches.

For hundreds of millions of years, life was restricted to the waters of the earth. Then, about half a billion years ago, green plants and invertebrates led the invasion of the land. Ability to live on the land provided a wealth of ecological niches, and explosive evolutionary radiation again occurred. There were tiny plants and huge plants. Many were fernlike, and some of the ferns reached heights of 100 feet. They came to blanket the land. At the same time, the invertebrates, many with their living tissue encased in exoskeletons to prevent drying out, were adapting to many new ways of life.

Just slightly before the invasion of the land, aquatic vertebrates developed. These were jawless fishes, somewhat reminiscent of today's lamprey. A successful group, they were soon overshadowed by their even more successful jawed descendants. As noted earlier, there are about 28,000 species of these fish alive *today*.

The distinction between predators and prey was important by this time, and thus arose the need for sensory systems. The more successful predators were those with better means of locating their prey; the more successful of the prey had better alarm systems for detecting their predators at a distance. In addition to sensory systems, of course, organisms needed appropriately adapted systems of locomotion. A better detection system would be of small advantage without increasing ability to capture (if a predator) or to evade (if a prey). The advantages of a communication network between sensory input and response output led to development of the nervous system. Nervous tissue accommodating such communication, and becoming increasingly capable of mediating complex behaviors, is another essential ingredient in the evolution of vertebrates.

Figure 1.2 expands the last half of the last subdivision of Figure 1.1. The vertebrate theme was so successful that it too extended to the land. The first such vertebrate exploiters of the land were the amphibians that rose from a particular kind of fish about a half billion years ago. Fish can have either ray fins or lobe fins. Ray fins are possessed by the most commonly encountered fish today. Lobe fins, much fleshier, were more common 500 million years ago. Over millions of years, some of the lobe-finned fish developed the capacity to survive out of water for short periods of time, perhaps allowing them to move from dried-out mud holes to more favorable habitats. The ability to survive on land allowed amphibians to swarm into a wide array of ecological niches. However, their success in earning a living on land was

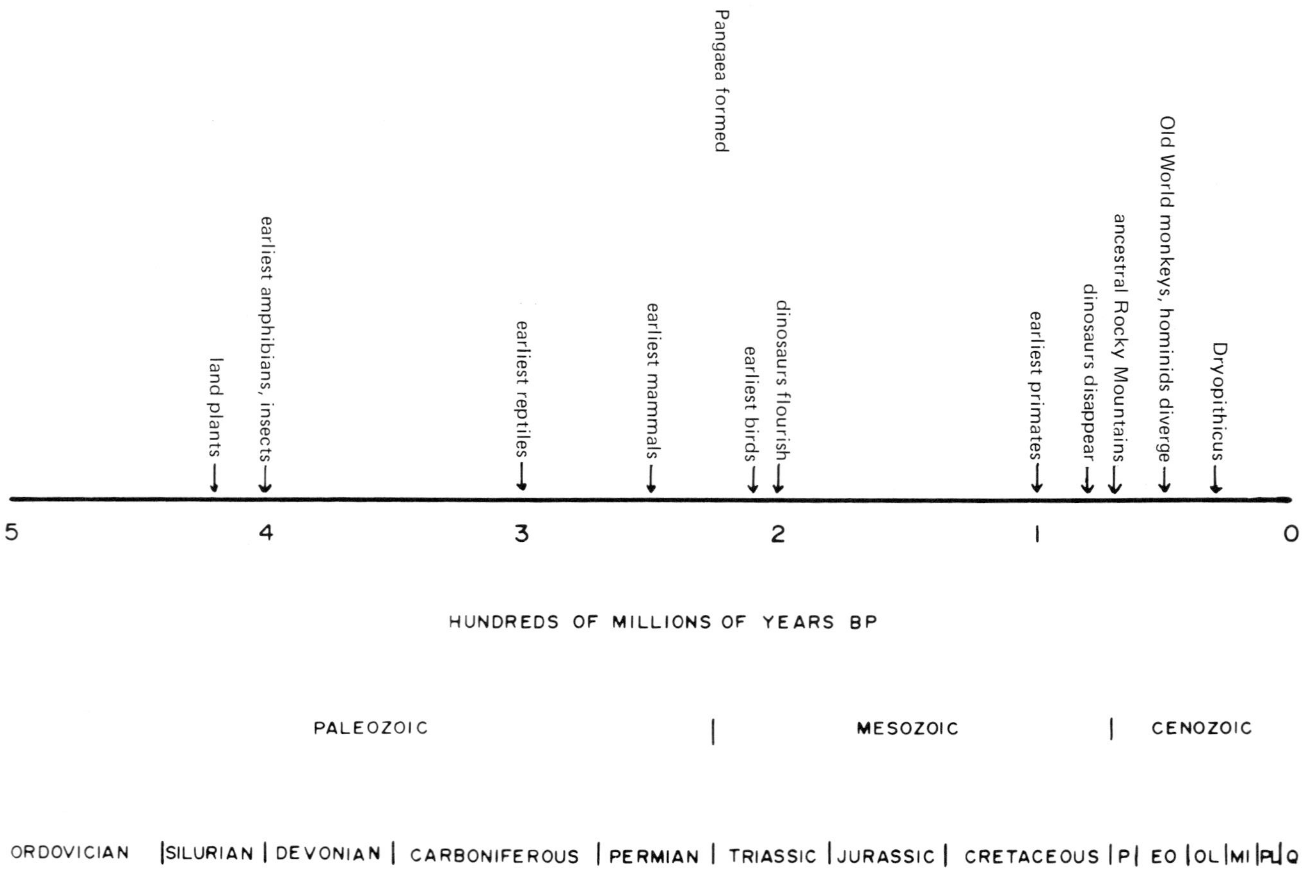

Figure 1.2. Evolutionary time scale from 500 million years ago to the present.

tempered by the necessity of returning to water for reproduction and early development of the next generation.

The next major evolutionary accomplishment, at the taxonomic *class* level, is represented by the reptiles, which produced larger eggs with more room for food storage and with a leathery coat permitting survival without dessication on the land. The reptiles arose 100 million years after the first amphibians and dominated the scene until about 100 million years ago. Reptiles were and are remarkably diverse. Some early ones were quite mammal-like. In one group, called the therapsids, the limbs tended to be parallel with the body axis instead of protruding at right angles to the body, which is typical of reptiles. This development permitted a much more efficient gait (Romer, 1972). Some also probably evolved more efficient mechanisms for regulation of body temperature, freeing them from temperature fluctuations of the environment. Other such "inventions" that led to the class *Mammalia* included a hairy coat, internal fertilization, and the development of the young inside the mother.

The early mammals were competing with a very large array of dinosaurs and other reptiles. These early mammalian ancestors of ours were small, furry creatures and were possibly most active at night, when the reptiles were less active because of the decrease in temperature. (One should, however, consider the lively discussion concerning "hot-blooded dinosaurs" by Desmond, 1976.) Eventually, about 100 million years ago, a specialized mammal with the primate characteristics mentioned earlier came along. When the primates had become fairly well established, dinosaurs disappeared during the "great dying," which is still poorly understood. Within a few million years, the extinction of these lords of the land opened up a whole world of opportunities for mammals, and they diversified greatly.

Figure 1.3 puts this part of the evolutionary sequence under a higher temporal magnification. About 50 million years ago, Old World monkeys and the hominids began to diverge. *Australopithecus*, regarded by some as being at least a collateral relative, and by others as being in our direct line of descent, appeared about 9 million years ago. *Australopithecus* was small, about 4 feet tall, lived in small groups, and *may* have used stones and sticks for tools. *Homo erectus,* whom we are willing to consider as belonging to the same genus, although a different species, appeared about 2 million years ago. Java man and Peking man are representatives of *Homo erectus;* they built fires, made and used tools, and developed what we regard as the rudiments of culture.

Figure 1.4 expands the evolutionary scene from about 5 million years ago. During this period, *Australopithecus* died, *Homo erectus* was established, and *Homo sapiens neanderthalensis* appeared. As indicated by its name, this last-named group is considered to be a subspecies of *Homo sapiens*. That is

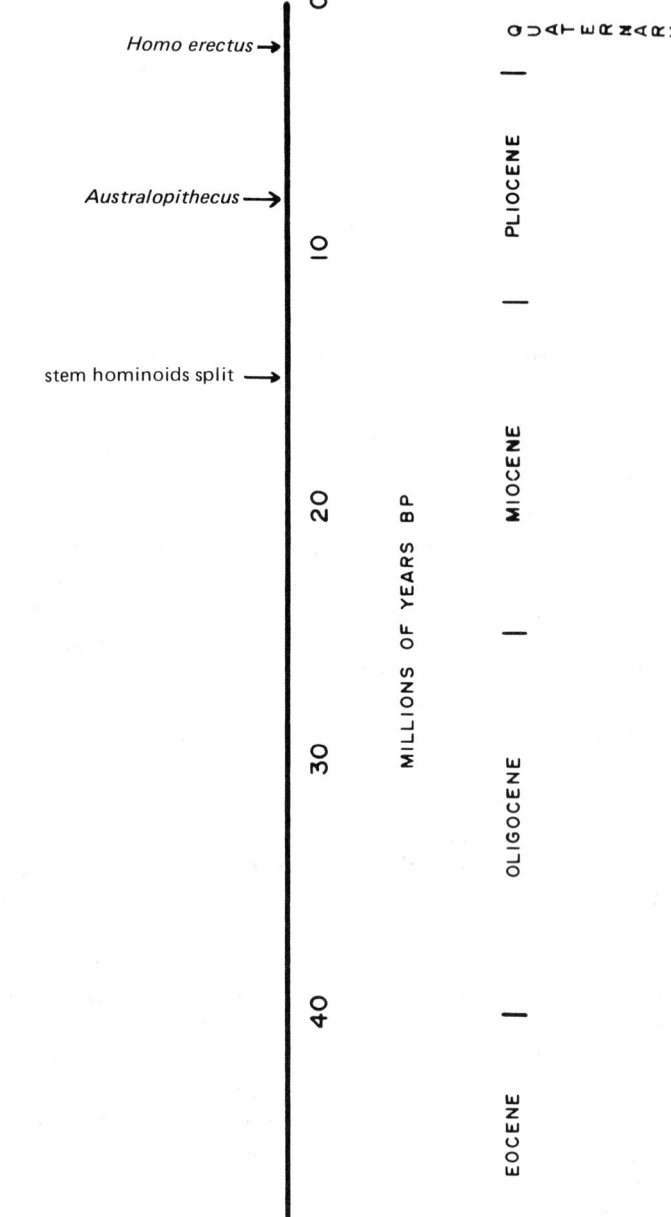

Figure 1.3. Evolutionary time scale from 50 million years ago to the present.

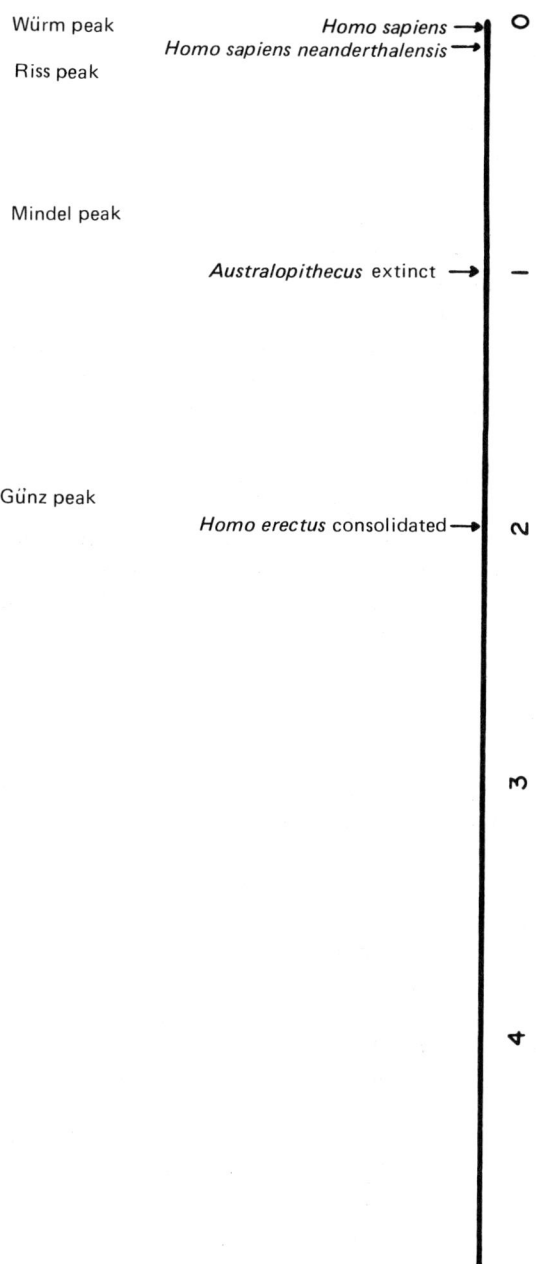

Figure 1.4. Evolutionary time scale from 5 million years ago to the present.

to say, they were human. In addition to their remarkably varied tools, they left evidence of religious beliefs and burial ceremonies.

Figure 1.5 is a further amplification. *Homo erectus* dwindled, the Neanderthal man came into ascendancy. Then, about 40,000 years ago, modern man came upon the scene. Although his brain was no larger than Neanderthal's, he soon prevailed. We do not know whether Neanderthal was outbred, outfought, or absorbed into the population of *H. sapiens*. Soon came cave paintings, spears, bows and arrows, and pottery; then the wheel, writing, iron, Rome, and 1776 (see Figure 1.6).

The purpose of this drastically abbreviated evolutionary review has been to present a temporal scale on which to examine our origins from the beginning of life on earth through the development of amphibians and reptiles to the emergence of the family of man. The variability revealed in this review is dynamic in the sense that it took place over vast periods of time. It is difficult to grasp the expanses of evolutionary time without the aid of an analogy. If we take the time when the sun began to glow as one goal line of a football field and view the other goal line as the present moment, the span of human life on earth, reckoning from the appearance of *Homo Sapiens*, is about .029 of an inch. In contemplating humankind, it is useful to bear in mind how much of the evolutionary process preceded our advent.

Sources of Variability

The tremendous variety of life was explained rationally by Charles Darwin. His theory, like all others, was not completely novel. Lamarck, for example, had speculated about the transformation of species from one to another, and Darwin's theory of natural selection and the mass of evidence he adduced to support it provided the integrative key. As presented in his work, *On the Origin of Species*, his theory became the central dogma of modern biology.

For a while there was some antipathy between the major proponents of evolutionary theory and researchers engaged in expanding on Mendel's discoveries concerning the rules of biological inheritance, but a synthesis was forged in 1930 by population geneticists. Figure 1.7 presents a schematic illustration of some of the important concepts of this synthesis. The circles represent individuals; within each circle, the two small circles represent the alleles that constitute a single gene pair. Heterozygous individuals are those having different alleles (represented here by a solid and an open circle) of a particular gene pair; those with the same allele are called homozygous. We are obliged to Mendel for our understanding of these two basic facts: There are two discrete units of inheritance, and they are inherited independently—one from each parent.

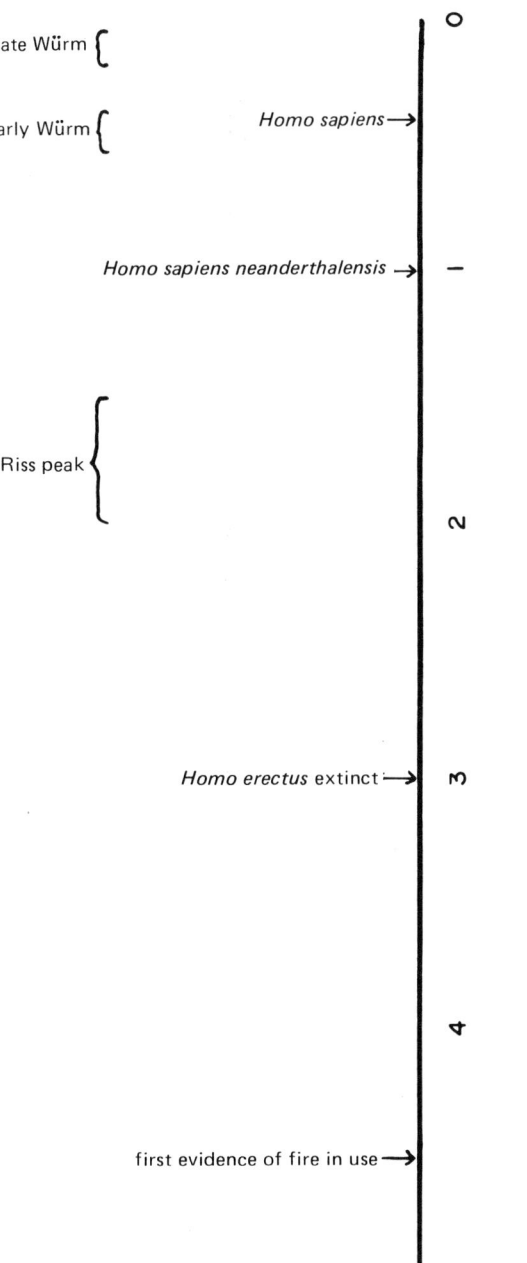

Figure 1.5. Evolutionary time scale from 500,000 years ago to the present.

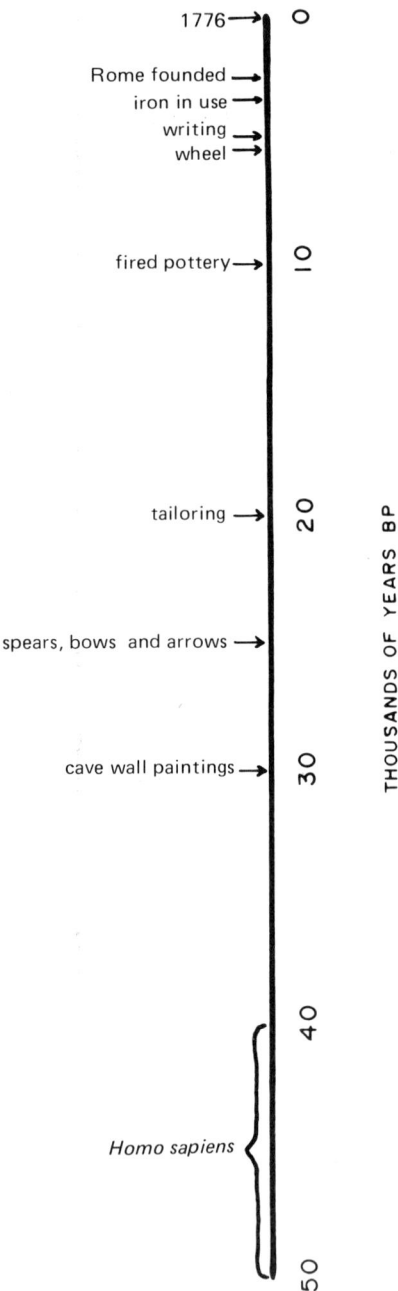

Figure 1.6. *Evolutionary time scale from 50,000 years ago to the present.*

1. EVOLUTION AND GENETIC VARIABILITY   **15**

Population genetics takes the story of variability from there. In the population represented in Figure 1.7, a count would reveal that 80% of the alleles are "solid" and 20% are "open." If the frequency of the solid allele is designated p, and if there are just two forms of this gene, then the frequency of the open allele will be $1 - p$, or q. Further analysis would show that the three possible *genotypes* (pairs of alleles) have the following frequencies: p^2(64%) are homozygous for the solid allele, $2pq$ (32%) are heterozygous, and q^2 (4%) are homozygous for the open allele.

The cornerstone of population genetics is the observation that, in the absence of disturbing factors, genotypic variability remains the same generation after generation. This phenomenon is referred to as the Hardy-Weinberg-Castle equilibrium (see McClearn & DeFries, 1973). Eight out of 10 eggs or sperm produced by individuals in the population will have one of the alleles (the solid one), and 2 out of 10 will have the other (open) allele.

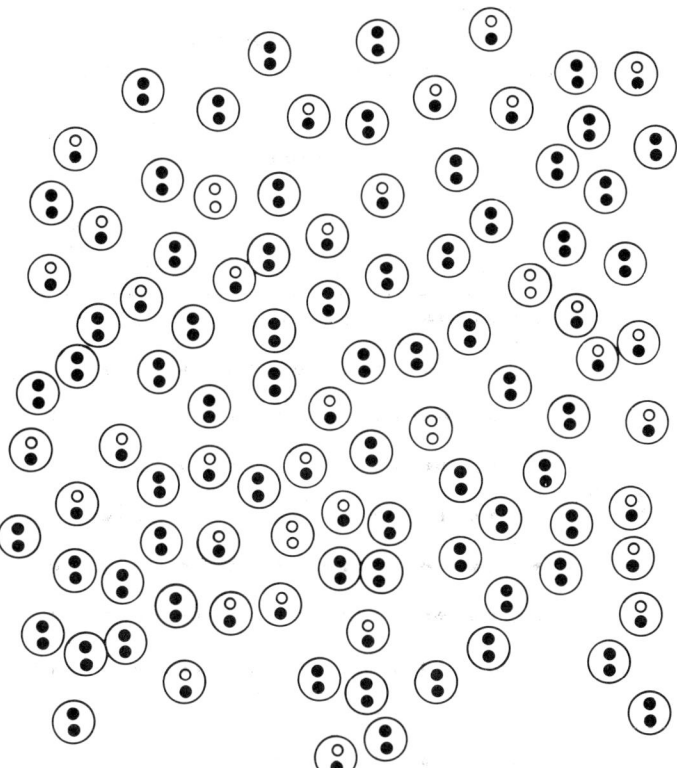

Figure 1.7. *Schematic illustration of a population of individuals segregating for a particular gene pair with allelic frequencies of .8 ("solid" allele) and .2 ("open" allele).*

From Mendel, and from much subsequent research, we know that fertilization of eggs by sperm is independent of the particular allele in the sperm or in the egg. The situation is thus as represented in Figure 1.8, in which 8 solid and 2 open sperm are shown as heads of 10 columns, and 8 solid and 2 open eggs appear at the beginning of 10 rows. The intersection of each row and column is occupied by a pair of alleles constituting the genotype of the offspring that would result from the union of that egg and that sperm. It can be seen that there is a large segment of the table in which the offspring are homozygous for the solid allele. Indeed, there is a literal p^2 (a square with the dimension p) filled with these homozygotes. Furthermore, there are two rectangles, each with dimension $p \times q$, filled with heterozygotes; that is, $2pq$ heterozygotes. Finally, the small square at the bottom right, filled with open homozygotes, has the dimension $q \times q$; in other words, there are q^2 such homozygotes. Thus, the same genotypic frequencies will be observed in each succeeding generation; 64% will be homozygous for one allele, 32% will be heterozygous, and 4% will be homozygous for the other allele.

This stability of genetic variability will occur only if individuals mate randomly and if the frequencies of the solid and open alleles remain the same—

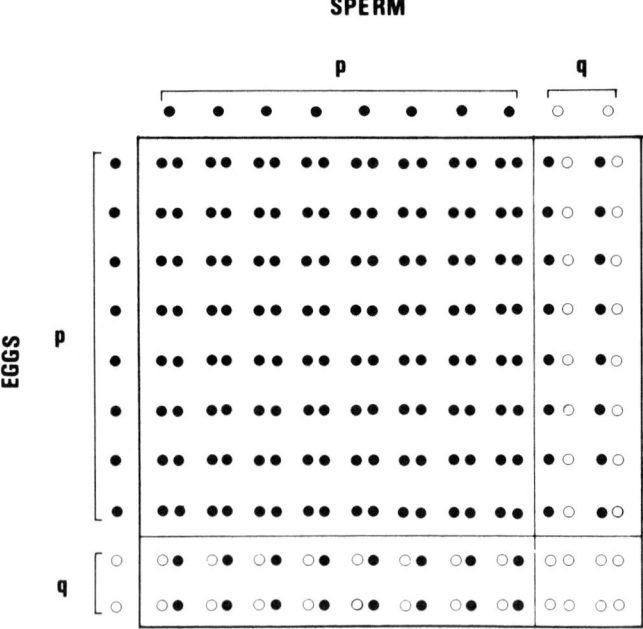

Figure 1.8. *Tabular representation of the Hardy-Weinberg-Castle equilibrium state.*

1. EVOLUTION AND GENETIC VARIABILITY

that is, there is no substantial natural selection that would give a reproductive advantage to one or another of the genotypes, and if there is no mutation. These assumptions may be true for some genes in some populations for substantial periods of time. But when evolution occurs, the population genetics description will be in terms of a change in gene frequency. Suppose that individuals homozygous for the allele initially present in the frequency of q are at a reproductive disadvantage. They will leave proportionally fewer offspring than will the heterozygotes or the other homozygotes, and the frequency of their allele will be diminished below q in the next generation. However, individuals homozygous for the allele may have a slight reproductive edge, in which case the frequency of their allele will increase and that of the other allele will decrease correspondingly. It should be remembered that very slight reproductive differentials may be involved in the process of natural selection. Given the time span of evolution, even a very small reproductive advantage may be quite significant.

A final point related to evolution concerns our tendency to regard *H. sapiens* as being in some way at the top of the evolutionary heap. But we must remember that the criterion of evolutionary success is survival, and that *all* living species are survivors. On these grounds, we human beings have no pride of place. Clearly, we do some things very well—those things that we prize, such as cognitive functioning. We do not fly or swim particularly well, however, nor do we photosynthesize, or digest cellulose, or perceive ultraviolet light, or regenerate limbs, or change skin color for protection, or continuously replace teeth—attributes that are the stock in trade of some living species. Our short span of existence, relative to that of, for instance, alligators or sharks or cockroaches, offers no basis for assurance concerning our long-term prospects. Because far more species have become extinct than exist today, it seems a fairly safe bet that *H. sapiens* in our present configuration will not be here at some point in the future. We shall have either terminated without descendants, or our descendants will have been sufficiently altered by natural selection that they will classify our remains and artifacts as belonging, depending upon their *bonhomie,* to a different species or to a different genus.

Although cultural evolution is important, it has not stopped the process of natural selection. In fact, our cultural inventions themselves may have placed tremendous selection pressures on our species. We are facing these pressures with a gene pool that was established under very different living conditions. There is no reason to think that this gene pool is not responding to a changing environment at the present time as it has in the past. The question is whether there will be enough time for us to adapt to the ever more rapid changes in our environment.

Variability within Species

For human beings, the usual estimates of the number of gene pairs per individual are between 10,000 and 100,000 (Bodmer & Cavalli-Sforza, 1976; Stern, 1973). When we consider more than one gene pair and take into account the mechanisms pertaining to the inheritance of traits in populations, we get a fresh view of variability—variability within species. Consider, for example, the variability that could result if a single male and a single female each had three gene pairs (see Figure 1.9). Both the male and the female are homozygous for one pair and heterozygous for the other two. Each can create four types of gametes and their offspring are of 12 different genotypes, only 4 of which (those encircled) are like either parent. Such

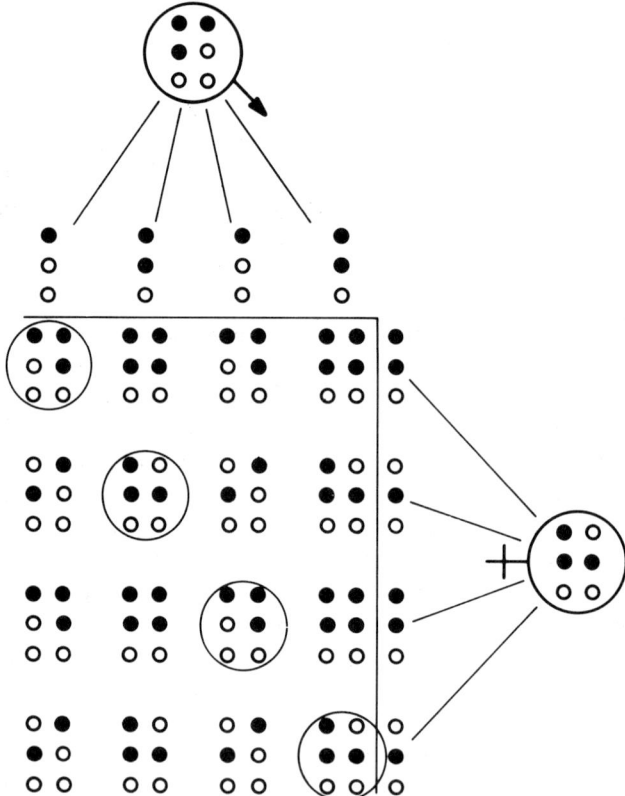

Figure 1.9. *Illustration of genotypic variability of offspring produced by random uniting of male and female gametes if each parent has three gene pairs in the allelic configurations shown. Of the 16 offspring genotypes, only 4 (those encircled on the diagonal) have the same genotypes as one or the other of the parents.*

genotypic variability is produced by the reshuffling process of meiosis. If this same notion is applied to the 100,000 gene pairs, the potential for variability can be seen to be truly enormous. The potential is so great that it is next to impossible that there have ever been two individuals with the same combination of genes. In fact, Bodmer and Cavalli-Sforza (1976) have estimated that each of us has the capacity to generate 10^{3000} different eggs or sperm. As a comparison, they estimated that the number of sperm of *all* men who have *ever* lived is only 10^{24}, so the number 10^{3000} really deserves to be called mind boggling. If we consider 10^{3000} possible eggs being generated by an individual woman and 10^{3000} possible sperm being generated by an individual man, the likelihood of anyone ever—in the past, present, or future—having the same genotype as anyone else (excepting multiple identical births, of course) becomes dismissably small.

Consideration of the results of meiosis provides a picture of variability different from the viewpoint historically characteristic of much of psychology. Individuality has often been treated as an error term in psychological experiments and in the theories derived from or generated by the research. We now see that our biological system is not just tolerant of individual differences; the system generates differences, insists upon them. Individuality is an indispensable ingredient of the evolutionary process. It is the raw stuff of evolution, the quintessence of life.

Molecules and Variability

We have been regarding the origins of variability in evolutionary and Mendelian terms. We shall now look at variability at the molecular level, considering the genetic code and the way in which that code is transcribed and translated into proteins. Figure 1.10 illustrates the double helical molecule called DNA and the sequence of nucleotides that constitute the code. The nucleotide bases, represented by the letters T, A, G, and C, are paired in such a fashion that T is always opposite A, and G is always opposite C. The code itself is a three-letter sequence of nucleotide bases designating a specific amino acid that will be hooked up in a polypeptide chain of protein. In addition to forming structures, proteins form enzymes that act as organic catalysts to speed and facilitate biochemical reactions in our bodies.

Changes in hereditary material, called *mutation,* occur when the genetic code is altered as by exposure to radiation and certain chemicals. Mutation, it may be seen, is the ultimate source of the genetic variability with which evolution can work. With the number of nucleotide bases in the human genome estimated at 6 billion, it is obvious that the possibilities for mutation are great even though the transcription process of DNA is highly reliable.

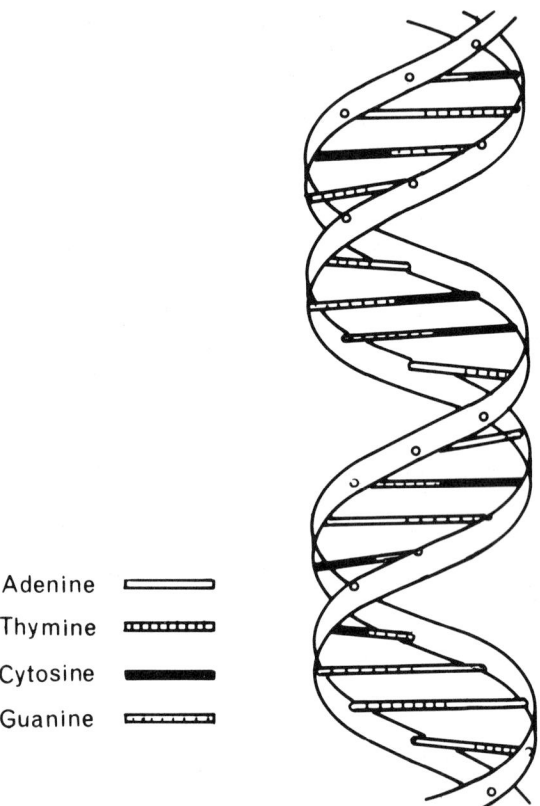

Figure 1.10. *A three-dimensional view of a segment of DNA.* (From Heredity, Evolution, and Society, *Second Edition, by I. Michael Lerner and William J. Libby. W. H. Freeman and Company. Copyright © 1976. Reprinted by permission.*)

Many alterations may be matters of little consequence, so-called neutral mutations. However, some can be very significant. Some may confer an evolutionary advantage, and others can be very deleterious. One of the best understood examples of the latter sort is sickle-cell anemia, a hereditary disorder that involves an abnormal kind of hemoglobin. Normal hemoglobin has four subunits—two chains called alpha chains, and two others called beta chains. Alpha chains are 141 amino acids long; beta chains consist of 146 amino acids. If we started at one end of the beta chain in normal hemoglobin and counted the amino acids, we would find glutamic acid in the sixth position. In sickle-cell anemia, valine occurs in the sixth position, and this one alteration produces the symptomatology of sickle-cell anemia (Bodmer & Cavalli-Sforza, 1976; Sutton, 1975).

Although we understand little about the amino acid sequences involved in another condition, phenylketonuria, it provides perhaps our best example of the path from a gene to a behavioral consequence. The protein involved is the enzyme, phenylalanine hydroxylase. One of the two components of this protein is inactive in individuals homozygous for the "deficient" allele. Because phenylalanine, a common component of our diet, cannot be metabolized by these individuals, there is an accumulation of phenylpyruvic acid and other metabolic by-products, which can be detected in the urine. As the individual develops, these byproducts give rise to behavioral consequences that we identify as mental retardation. Understanding this process had led to the development of techniques for detecting affected individuals in very early infancy and providing rational therapy, and procedures for identifying normal adults who are "carriers" of the deleterious recessive allele and who are therefore potential parents of affected children.

Polygenic Systems

Many traits are not dichotomous, as are the examples just cited, but are continuously and normally distributed. Such *polygenic* traits are subject to the joint effects of a number of genes, each of which may have only a small influence on the trait. The effects of many genes are no more magical than the effects of genes considered one at a time. Genes work only through the biochemical, physiological, and anatomical consequences of the action of their products, the proteins. The eventual effect upon behavior is accomplished by way of the nervous system, the endocrine system, and the other systems of the body.

But genes, of course, do not act alone. Observable attributes (or *phenotypes*) are influenced by both genes and environment. Genes are the sine qua non of a viable organism, and that organism can exist only in an environmental milieu. Nevertheless, it is possible to assess the *relative* contributions of genetic differences and environmental differences to the variability of a phenotype *in a particular population*.

A simple numerical example may elucidate the combined effects of genes and environment in producing a *phenotype*. At the top of Figure 1.11 are shown two gene pairs of each of 16 individuals. The assumption is made that the genetic value (G) conferred upon a phenotype is one for each closed allele. Thus, the first individual has a G score of 0, and the last individual has a G score of 4. We assume also that environmental influences are graded, as indicated by the symbols at the left of the figure. A strong negative influence will subtract 1 point, and so on. These environmental influences were randomly assigned to individuals as shown in the E row. The G and the E scores are combined to provide the P (phenotypic) score. Figure 1.12 illustrates the

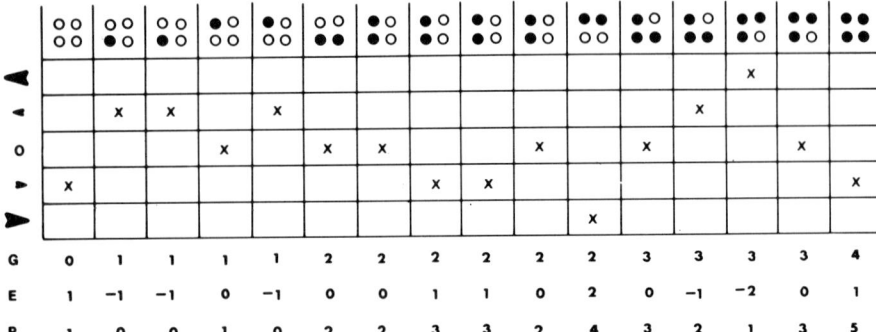

Figure 1.11. *Representation of the combination of genetic effects and environmental influences to produce a phenotypic effect.*

distribution of the genetic factors (top) and the environmental factors (middle), with the resulting phenotypic distribution at the bottom.

Both genetic and environmental influences upon a behavioral phenotype are shown in Figure 1.13, which depicts a distribution of open-field activity scores for mice (McClearn, 1961). The scores range from 0 to more than 180 squares entered in a 3-min period. The individuals contributing to the distribution were members of three groups—two genetically uniform inbred strains, and their hybrid cross. The A strain exhibits low activity, the C57BL strain is quite active, and the activity of the F_1 hybrids is intermediate between the parental strains. Two points emerge from examination of the partitioned distributions: First, the relatively large differences among group means indicate substantial genetic influence. Second, the variability must be environmental in origin. Just as the manipulation of genetic factors by intermating can reveal characteristics of genetic influences upon traits, the systematic study of variability within genetically uniform groups (such as inbred strains) under different environmental conditions can elucidate the mechanism of action of these environmental variables.

In behavioral genetic research, we manipulate variables, both genetic and environmental, to determine if the manipulations affect some behavior of interest. For example, we have investigated sensitivity to the effects of alcohol by measuring sleep time in mice. *Sleep time* is defined as the duration of loss of the righting response subsequent to an injection of alcohol. The top section of Figure 1.14 gives the distribution of sleep-time scores for a genetically diverse population of mice. The range is very substantial—from essentially 0 to over 7500 sec. By selecting males and females from the upper end of the distribution and mating them together, we established a Long-Sleep line and a Short-Sleep line (McClearn & Kakihana, 1973). These selected lines are thereafter maintained by mating the highest of the high (Long-Sleep line)

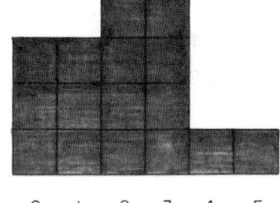

Figure 1.12. *Distributions of genetic effects (top), environmental influences (middle), and the phenotypic effects (bottom) resulting from random assignment of environmental influences to the genetic effects.*

and the lowest of the low (Short-Sleep line). Also shown in Figure 1.14 is the gradual divergence of the lines over the first five generations. By the eighteenth generation of selection, the distributions of sleep-time scores for the two lines did not overlap. In addition to demonstrating that genetic variability underlies differences in alcohol sensitivity, the Long-Sleep and Short-Sleep mice have proved to be valuable tools for investigating the biochemical and physiological mechanisms through which alcohol elicits its behavioral effects.

The results of another behavioral genetic study of the effects of alcohol, in which the activity of mice of various inbred strains was measured after they

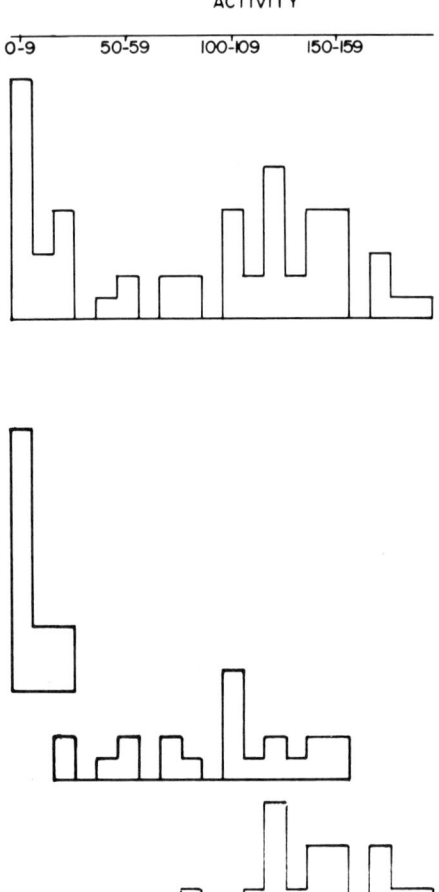

Figure 1.13. *Open-field activity scores of mice. The top distribution is an aggregate of the other three distributions: inbred A strain animals, F_1 hybrids, and inbred C57BL strain animals, respectively.*

had inhaled alcohol vapor for 5 min (McClearn, 1962), will illustrate that genetic and environmental factors may interact to influence behavioral variability. Alcohol markedly reduced the activity of the C57BL mice, whereas As and DBAs showed a marginal effect in the same direction. A prototypic experimental design in psychology is to compare a treated group to a control group to assess the influence of the treatment. The results shown in Figure 1.15 clearly demonstrate that the effect of an environmental variable (treatment) can depend greatly upon the genotype of the individual to whom the treatment is administered. They can also serve as an object lesson for people who want to claim that completely generalizable laws of nature can be based upon studies done with small groups of animals that have not been characterized genetically.

1. EVOLUTION AND GENETIC VARIABILITY

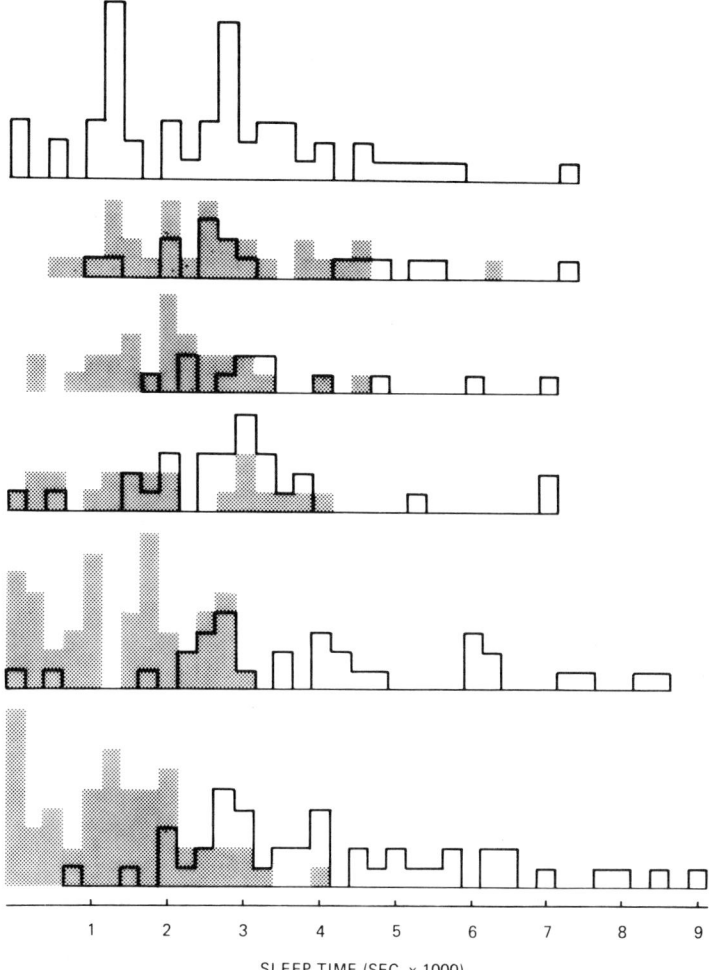

Figure 1.14. *Distributions of sleep-time scores of a genetically diverse population of mice (top) from which selected lines were derived by upward (open bars) and downward (shaded bars) selective breeding. The other distributions illustrate divergence of the lines over five generations of selection.*

Another example of genotype-environment interaction may be taken from research on aggression in mice (Klein, Howard, & DeFries, 1970). In paired encounters between male C57BLs and BALBs under standard laboratory illumination, the C57BL animals were typically the winners. In two separate replications of the experiment, they tallied 27 wins to 3 losses and 22 wins to 3 losses, respectively. However, when the same experiment was conducted

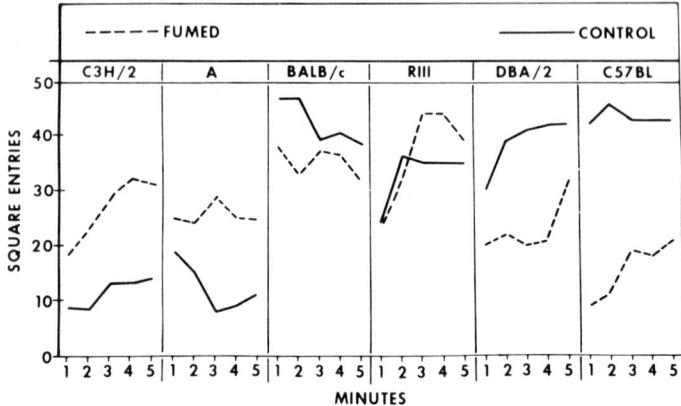

Figure 1.15. *Effects of forced inhalation of ethanol vapor on activity scores in five inbred mouse strains. (From G. E. McClearn, Influence of genetic variables on means, variances, and covariances in behavioral responses to toxicological and pharmacological substances, Journal of Toxicology and Environmental Health, 1979, 5, 145-156. Copyright © 1979 by Hemisphere Publishing Corporation. Reprinted by permission.)*

under low illumination, the BALBs did much better—winning 14 out of 18 bouts in the first replication, and 13 out of 19 in the second. So, if one were asked about the influence of illumination on aggression in mice, it is clear that the answer must specify the strains of the animals used in the experiment.

Developmental Variability within the Organism

The examples of genotype-environment interaction provided to this point have involved manipulation of environmental variables at the time of observation. Examples more pertinent to the theme of this volume are those cases in which the developmental consequences of early environmental manipulations are assessed. In this context, it becomes particularly important to note that "genetic" does not mean "congenital." The total genome is not functioning at fertilization, or at birth, or at any other time of life. Different genes are decoded and come into play at various times during the lifetime of a particular organism. One illustration of this phenomenon is the differential production of certain kinds of hemoglobin during various phases of development. For example, production of the beta chain accelerates at the time of birth and peaks after a few months, whereas production of the alpha chain rises prenatally and maintains a high level.

The operon model of Jacob and Monod (1961), described in Figure 1.16, gives us a hint as to how genes can be turned on and off during develop-

1. EVOLUTION AND GENETIC VARIABILITY

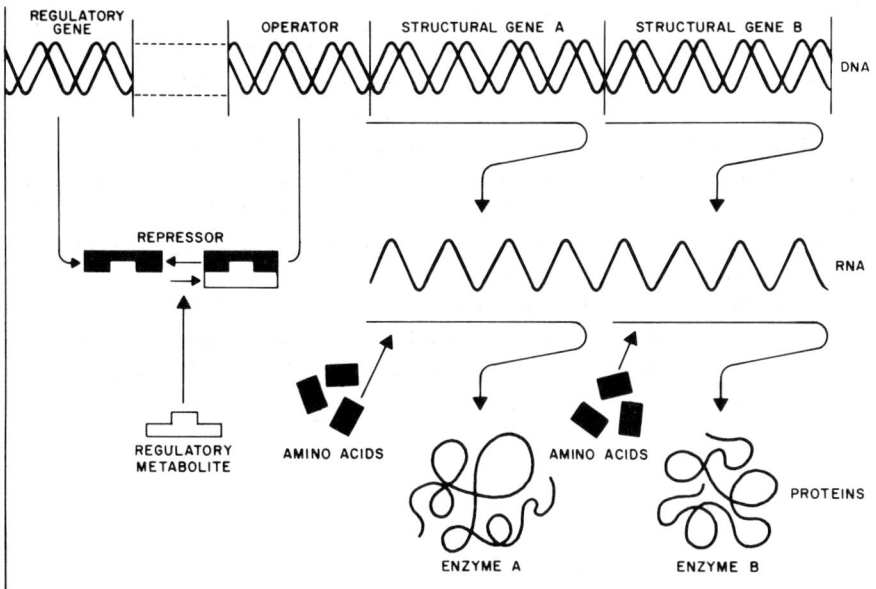

Figure 1.16. The operon model. (From The control of biochemical reactions, by J.-P. Changeux. Copyright © 1965 by Scientific American, Inc. All rights reserved. Reprinted by permission.)

ment. Because most of the research upon which this model is based was conducted on single-celled organisms, it must be regarded as only suggestive for metazoan animals. The model involves regulator genes in addition to the structural genes that actually code for proteins. The regulator genes produce a substance called "repressor," and the repressor binds to a region of DNA that constitutes a structural gene. When the structural gene is bound with the repressor, it is not transcribed. If something produced within the cell or coming into the cell binds with the repressor, then the repressor does not bind with the structural gene and the gene can code for its protein. By cascading models of this sort, schemes can be developed to explain how genes can be turned on and off in delicately timed sequences.

Although huge gaps remain to be filled before we can relate such molecular events to behavioral development, there have been convincing demonstrations of genetic influences on the development of various behaviors (see McClearn, 1970; Schaie, Anderson, McClearn, & Money, 1975). Only two of the many possible examples will be cited here. When Dixon and DeFries (1968) investigated the development of open-field activity in C57BL and BALB mice and their F_1 hybrids, they found that the activity of the BALBs was consistently low and the C57BL activity started at a low level and followed a typical growth curve (see Figure 1.17). The curve for the F_1s is interesting in that it parallels the development of C57BL activity until adulthood and then shows a reduction in activity. This suggests the possiblity that different genes may be turned on at that point.

In human research, Wilson's (1972) analyses of the development of cognitive ability in monozygotic (genetically identical) twins revealed that the spurts and lags in development of the twin pairs are strongly correlated (see Figure 1.18). These data strongly suggest that the development of cognitive abilities is genetically influenced.

Observations of this sort may be the prolegomena of an exciting advance in our understanding of the processes of behavioral development, elucidated by the integrated application of the perspectives of molecular, population, quantitative, and evolutionary genetics.

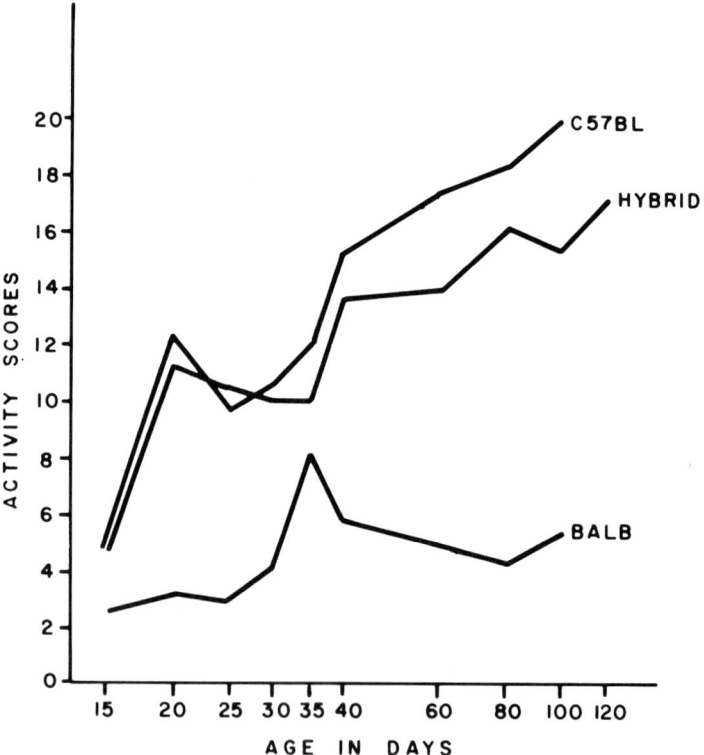

Figure 1.17. *Mean transformed open-field activity scores of two inbred strains of mice (BALB/cJ and C57BL/6J) and their F_1 hybrids in cross-sectional and longitudinal studies. Age is presented on a logarithmic scale.*

1. EVOLUTION AND GENETIC VARIABILITY

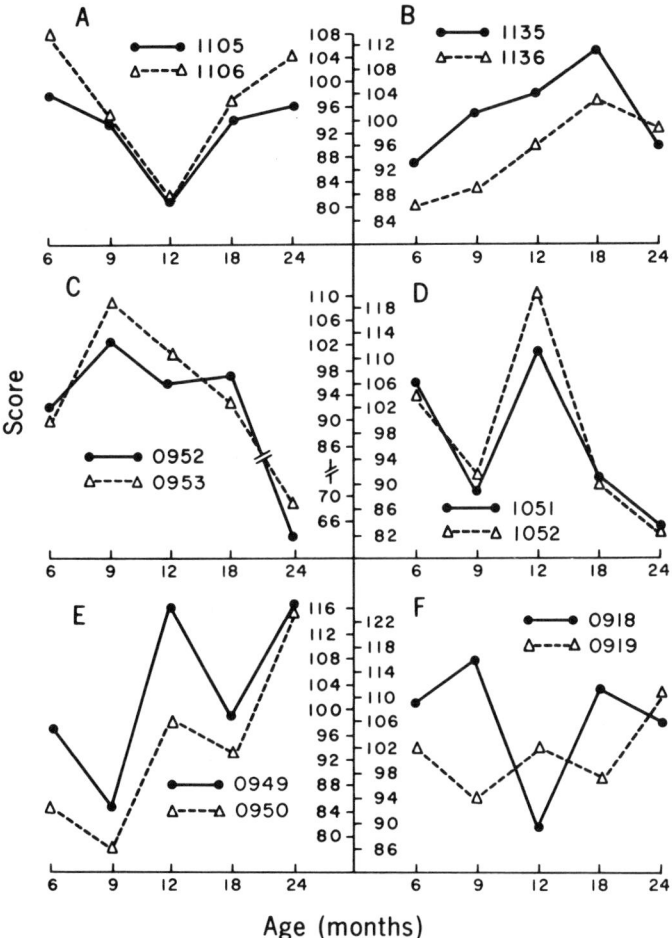

Figure 1.18. *Profiles of mental development scores for MZ twins at ages 6 through 24 months. The pairs in A through E exhibit moderate to high profile congruence; the pair in F is obviously noncongruent. (From R. S. Wilson, Twins: Early mental development, Science, 1972, 175, 914-917. Copyright © 1972 by the American Association for the Advancement of Science. Reprinted by permission.)*

Summary

My intent has been to provide a context in which variability can be examined and evaluated. We have seen that there is a remarkable variability in different forms of life and that the generation and distribution of this variability has a systematic mechanism. This variability is not noise in the system or a deviation from some Platonic ideal. If the world were really put right, with all environmental influences controlled, individuals would still not be as alike as television sets coming off the assembly line, differing only in the channels to be tuned in. Variability exists between species, within species, and within organisms over time. All of this variability can be profitably viewed as having a component that is due to environmental sources and another component that is due to genetic sources. The frame of reference that incorporates these considerations will be a useful one for studying the development of behavioral processes. We need not choose sides in the obsolete nature-nurture dichotomy. We can view behavioral development in a context that acknowledges that both genetic and environmental influences may be important, and that they can interact and coact in very interesting, complicated, and revealing ways.

Acknowledgment

I wish to thank Rebecca G. Miles for her expert editorial assistance.

Bibliography

The following books are recommended as sources of more general information about the evolutionary process.

Dobzhansky, Th. *Genetics of the evolutionary process*. New York and London: Columbia University Press, 1970.

Mayr, E. *Animal species and evolution*. Cambridge, Mass.: The Belknap Press of Harvard University Press, 1963.

Mayr, E. *Evolution and the diversity of life*. Cambridge, Mass., and London: The Belknap Press of Harvard University Press, 1976.

Oakley, K. P., & Muir-Wood, H. *The succession of life through geological time* (7th ed.). London: Trustees of the British Museum, 1967.

Simpson, G. G. *The major features of evolution*. New York: Columbia University Press, 1953.

Simpson, G. G. *The view of life*. New York: Harcourt, Brace and World, 1964.

Stebbins, G. L. *Processes of organic evolution* (2nd ed.). Englewood Cliffs, N.J.: Prentice-Hall, 1971.

References

Bodmer, W. F., & Cavalli-Sforza, L. L. *Genetics, evolution and man*. San Francisco: Freeman, 1976.

Changeux, J.-P. The control of biochemical reactions. *Scientific American*, 1965, *212*, 36-45.

Desmond, A. J. *The hot-blooded dinosaurs*. New York: The Dial Press, 1976.

Dixon, L. K., & DeFries, J. C. Effects of illumination on open-field behavior in mice. *Journal of Comparative and Physiological Psychology*, 1968, *66*, 803-805.
Hockett, C. F. *Man's place in nature*. New York: McGraw-Hill, 1973.
Jacob, F., & Monod, J. On the regulation of gene activity. *Cold Spring Harbor Symposia on Quantitative Biology*, 1961, *26*, 193-209.
Jolly, A. *The evolution of primate behavior*. New York: MacMillan, 1972.
Klein, T. W., Howard, J., & DeFries, J. C. Agonistic behavior in mice: Strain differences as a function of test illumination. *Psychonomic Science*, 1970, *19*, 177-178.
Le Gros Clark, W. E. *History of the primates* (5th ed.). Chicago: The University of Chicago Press, 1965.
McClearn, G. E. Genotype and mouse activity. *Journal of Comparative and Physiological Psychology*, 1961, *54*, 674-676.
McClearn, G. E. Genetic differences in the effect of alcohol upon behavior of mice. In J.D.J. Havard (Ed.), *Proceedings of the Third International Conference of Alcohol and Road Traffic*. London: British Medical Association, 1962. Pp. 153-155.
McClearn, G. E. Genetic influences on behavior and development. In P. Mussen (Ed.), *Carmichael's manual of child psychology* (Vol. 1). New York: Wiley, 1970. Pp. 39-76.
McClearn, G. E., & DeFries, J. C. *An introduction to behavioral genetics*. San Francisco: Freeman, 1973.
McClearn, G. E., & Kakihana, R. Selective breeding for ethanol sensitivity in mice. *Behavior Genetics*, 1973, *3*, 409-410. (Abstract)
Orgel, L. E. *The origins of life*. New York: Wiley, 1973.
Romer, A. S. *The procession of life*. New York: Doubleday, 1972.
Schaie, K. W., Anderson, V. E., McClearn, G. E., & Money, J. (Eds.). *Developmental human behavior genetics*. Lexington, Mass.: Heath, 1975.
Stern, C. *Principles of human genetics* (3rd ed.). San Francisco: Freeman, 1973.
Sutton, H. E. *An introduction to human genetics* (2nd ed.). New York: Holt, Rinehart & Winston, 1975.
Wilson, R. S. Twins: Early mental development. *Science*, 1972, *175*, 914-917.

II

THE SENSORY BASE

The senses are not windows open to the world admitting "pure" copies of outer reality to the sensorium. They are dynamic systems with species-specific characteristics modifiable within certain partially understood limits. In the following chapter, Werner and Lipsitt present a stochastic model of human sensory development that takes account of experiential impacts on maturing sensory systems. These systems are demonstrated to be capable of function prior to normal term indicating the precocial nature of human sensory development. When considered together with altriciality in human motor development, the scenario is established for the prolonged infant-mother relationship typical of primates in general and human primates in particular. The significance of this pattern of sensory-motor organization for human socialization is clear. The developing human experiences a prolonged period during which it is a captive audience exquisitely equipped to capitalize on events occurring in its surround.

At least three aspects of sensory systems play a role in developmental variation. First, there are individual differences of both qualitative and quantitative character attributable to the hereditary materials. Second, differences in experience will introduce additional factors that contribute to variations in individual sensory systems, and last, these two sets of factors acting conjointly will affect the totality of experience. Toulmin's (1967) statement that "We see, not through our optical system, but by its use [p. 829]" underscores the importance of knowing as much as we can about how sensory systems develop and how they affect transactions with the surround.

Reference

Toulmin, S. Neuroscience and human understanding. In G. C. Quarton, T. Melnechuk, & F. O. Schmitt (Eds.), *The neurosciences: A study program.* New York: The Rockefeller University Press, 1967.

2

The Infancy of Human Sensory Systems

JOHN S. WERNER
LEWIS P. LIPSITT

Introduction

Knowledge of the world has only one route to the brain—by way of the senses. Although it has only been recognized in the last 15 or 20 years, our species has evolved a *precocial* pattern of sensory systems. That is, even though none of the human sensory systems is fully mature at birth, it seems likely that they are all mature enough to sustain some adaptive behavior. In this regard, primate ontogenesis follows a developmental pattern that is different from most other animals, as the precocial pattern of sensory systems is usually not associated with the altricial pattern of motor systems (Gottlieb, 1971). This pattern is, no doubt, of significance from an evolutionary as well as from a developmental perspective. Infant sensory systems are capable of abstracting information from the environment, and this information can in turn be used to modify the course of development as required for adaptation. Thus, to understand the constraints on early experience, it is necessary to understand the functional capabilities of the sensory systems.

Precise mapping and organization of sensory input onto the brain has played an important part in the evolution of primate behavior. This mapping, or neurogenesis, takes place during prenatal and postnatal human development and provides an avenue for experiential modifications and, ultimately, evolutionary refinements. It is nature's modification and selection of ontogenetic processes that make evolution possible (deBeer, 1958). The precision required in human sensory neurogenesis may be too high to entrust only to a genetic code, for such a system would be easily defeated by unexpected developmental changes in physical growth. A precise system must be plastic to some extent if it is to accommodate other modifications during development. Thus, even though the genetic code controlling sensory systems may canalize experiences in particular directions (Scarr-Salapatek, 1976), experiences may also modify the sensory abilities. A full understanding of infant development must, therefore, consider the reciprocating relations between experience and the sensory-neural mechanisms. At least for the sense of vision, recent experiments are beginning to make such an analysis possible. This conceptualization of human infancy stands in marked contrast to those developmental perspectives that insert the word *versus* between the words *nature* and *nurture* (e.g., Gesell, 1954; McDougall, 1908; Watson, 1924). Rather, it is assumed here that the course of epigenesis is probabilistic in nature, and not predetermined (see Gottlieb, 1976).

Prenatal Origins of Sensory Development

In reviewing data from diverse vertebrate species, Gottlieb (1971) found an invariant sequence in the onset of function of the sensory systems. Using behavioral, histological, and electrophysiological criteria, he found that the sensory systems develop in the following order: tactile, vestibular, auditory, and visual. (The proprioceptive and chemical senses were not considered in this analysis.) Although this sequence appears to hold for man, the onset of function in all sensory systems probably occurs before birth. For example, Humphrey (1964) observed responses to cutaneous stimulation of the perioral region of a 2-month-old fetus. She also reported that tactile responsivity could be demonstrated for all parts of the fetal body before the thirty-second week of gestation. Taste receptors, identifiable as early as the second month of gestation (Bradley, 1972), are probably functional in the fetus, as a stereotyped gustofacial response—lip pursing—may be elicited in preterm infants by a stimulus that was described by adults as sour, .12 M citric acid (Steiner, 1979). It has also been reported that fetal swallowing may vary with chemical injections in the amniotic fluid, although it is not certain that the response is mediated only by the taste of the chemicals (see Mistretta & Bradley, 1977). Approximately 2 months prior to birth, changes in fetal heart rate can be detected in response to auditory stimuli (Bernard & Sontag, 1947; Johansson, Wedenberg, & Westin, 1964). Some, albeit limited, prenatal visual function can also be inferred from tests of preterm infants. Visually evoked cortical potentials have been recorded from infants at 28 weeks conceptional age (Ellingson, 1960), and discrimination of novel color-pattern stimuli was observed for a group of infants with an average gestational age of 35 weeks (Werner & Siqueland, 1978).

These data illustrate that the human sensory systems are capable of some function before birth. Sufficient data do not yet exist to allow conclusions about the role of experience in the development of these functions. On the basis of experiments with other species (see Gottlieb, 1973), however, it appears likely that prenatal experience can be important in the establishment and maintenance of some neurobehavioral structures and functions that are more clearly identifiable in the newborn.

Sensory Bases of Infant Perception

The bands of energy that we perceive as odor, touch, sound, and light are limited. The limitations of our sensory systems during development thus define some of the constraints on perceptual development. We are thereby led to the assumption that understanding sensory development through infant psychophysics and psychophysiology is a sine qua non for understanding perceptual development. Although this perspective is not unique to this chapter, it is certainly not the universally accepted direction in the study of perceptual development. For example, Gibson (1976) writes:

> Here is where the new experimental psychology of infant perception is in danger of going astray. The experimental methods of sensory physiology do not provide good models. The measurement of stimuli is important for psychophysics, but it becomes a useless exercise in experiments with babies. It can lead to as much confusion as the tachistoscope [p. 64].

Gibson's (1966) concern with the higher-order invariants of perception is a worthwhile approach for research in infant perception, as has already been demonstrated for infant speech perception (Eimas, Siqueland, Jusczyk, & Vigorito, 1971). However, to use this approach without reference to more basic sensory abilities is to ignore a large body of useful information. The literature of adult psychophysics and physiology has provided substantial insight into the mechanisms of adult perception. It is assumed here that a similar statement will be made in the future for psychophysical and psychophysiological studies of infant sensory development.

Vestibular System

The vestibular apparatus, or labyrinth, contained in the inner ear is sensitive to linear and rotatory acceleration, and thereby conveys information about the position of the body in relation to the environment. Together with the other sense organs, but especially proprioceptors, it makes possible the maintenance of body posture. Its early maturity and its importance in the development of sensory-motor skills might make the vestibular system an important avenue for early stimulation (Korner & Thoman, 1970). Unfortunately, some basic features of the vestibular system have not yet been elucidated.

The vestibular system is functional at birth (Galebsky, 1927; Heck, 1952; Lawrence & Feind, 1953; McGraw, 1941), but significant postnatal development probably also occurs. Galebsky elicited nystagmus by rotating newborns to demonstrate that the three semicircular canals are sensitive to their appropriate direction of acceleration (horizontal, vertical, and rotatory). Nystagmus is a correctional response to rotation that is characterized by an involuntary deviation of the head and eyes. During rotation the eyes move in two distinct phases: a slow sweeping movement of the eyes in a direction that is opposite to the rotation, followed by a fast phase in which the eyes are swept back in the direction of rotation. If the rotation is suddenly terminated, the sequence of head and eye movements is observed in the opposite direction. Nystagmus can also be elicited by visual stimuli such as a vertical grating that is swept in front of the eyes. This is known as optokinetic nystagmus. Galebsky (1927) was unsuccessful in eliciting optokinetic nystagmus in the newborn human infant, but several subsequent investigations have demonstrated it clearly (Dayton, Jones, Aiu, Rawson, Steele, & Rose, 1964; Gorman, Cogan, & Gellis, 1957).

McGraw studied rotatory nystagmus in 67 subjects ranging in age from newborns to about 2 years. Three developmental phases were identified: (a) neonatal responses to rotation that include lateral deviation of the head and eyes (slow phase of nystagmus), but an absence of the return eye movement during rotation (fast phase); (b) a "gross oscillatory phase" beginning around 4 months of age in which slow horizontal eye movements are observed during and after rotation; and (c) a "fine oscillatory phase" in which the horizontal eye movements during rotation occur with a shorter latency and in which the fast phase of nystagmus appears more like the nystagmus of an adult. In contrast to these findings, Tibbling (1969) found that the fast component of nystagmus was not only present at an early age in infancy but that it decreased, rather than increased, in velocity with age. Goodkin (1979) was unable to substantiate Tibbling's results. Indeed, like McGraw, Goodkin showed that the fast return eye movement phase in nystagmus is much slower in infants under 3 months of age than it is in adults. Thus, "the saccadic motor driving the high velocity movements is not yet developed at three months [Goodkin, p. 89]."

Evaluation of the results of these studies is difficult because they included few comparisons with older children and adults tested under identical conditions to those used with infants. It seems likely that further progress in understanding the development of vestibular function and its contribution to postural adjustments and motor skills may also be enhanced by investigations that attempt to facilitate development through early vestibular stimulation.

Chemical Senses

For most sensory systems the major emphasis in research has been on the limits of sensitivity. Sensitivity and acuity have also been a concern in research on olfaction and gustation, but in addition work on the chemical senses has attended to the hedonic qualities of stimulation, or what Pfaffmann (1960) has called "the pleasures of sensation." The hedonics of chemical sensitivity are probably essential in the regulation of eating and drinking. How these hedonic properties might be modified by experience is of considerable practical, as well as theoretical, interest (see also Lipsitt & Werner, Chapter 4, this volume).

OLFACTORY SENSITIVITY

Newborn infants are capable of discriminating and remembering odorous substances, and it is conceivable that the hedonic qualities of olfactory experiences play a role in the formation of maternal-infant bonds. For example, Macfarlane (1975) placed two breast pads above supinely positioned infants. One of the pads was previously used by the infant's nursing mother, and one was unused. Infants between 2 and 7 days of age showed a significant orientation preference toward their mother's breast pad relative to the unused pad. Moreover, when the new breast pad was replaced by one that was used by an unfamiliar nursing mother, the infants showed a significant orientation preference toward the pad containing the odor of their own mother. This discrimination is subtle, and it is possible that acuity for such ethologically significant olfactory stimuli is greater in infants than in adults. Furthermore, a preference for the mother's breast pad was not observed for infants tested at 2 days of age, but was observed when tests were conducted at either 6 or 8-10 days of age. Thus, experience may be required for infants to recognize their mothers through olfactory cues. These observations are consistent with Engen's (1974) more general hypothesis that experience is an important determinate of odor preferences.

As neonates are capable of responding positively to odors that may be presumed to be hedonically rewarding, so too they are able to respond to odors that may be presumed to be aversive. Rieser, Yonas, and Wikner (1976) presented a cotton swab with a small drop of ammonium hydroxide on either the left or the right side of newborns who were from 1 to 6 days of age. On 64% of the trials the infants turned their heads away from the stimulus, and on 30% of the trials they turned toward the stimulus. The neonates were thus able to discriminate the spatial location of the odor and to respond by avoidance. The facial expressions of neonates also illustrate their ability to respond differentially to odors that adults describe as pleasant and odors that adults describe as aversive and offensive. Steiner (1977, 1979) demonstrated this clearly when he had a panel of observers rate the facial expressions of newborn infants when a cotton swab was placed under each baby's nose. Ratings of acceptance and satisfaction were associated with the odor of banana, vanilla, and strawberry, whereas high ratings on expressions of rejection were associated with the artificial odors of fish and rotten eggs.[1] Unlike the odor preference described by Macfarlane (1975) that seemed to accrue through experience with the olfactory stimulus, the facial expressions reported by Steiner were evident in neonates less than 12 hours old, and apparently without previous experience with food odors.

In addition to developmental changes in olfactory preference within the first few days of life (Macfarlane, 1975), there also appear to be changes in olfactory sensitivity. Indeed, it is possible that the substrate for the preference change observed by Macfarlane is a sensitivity change that occurs shortly after birth. For example, Lipsitt, Engen, and Kaye (1963) determined the minimum concentration of an olfactory stimulus, tincture asafoetida, required to produce a criterion response on four consecutive daily tests. They found that newborn infants' thresholds progressively decreased over the 4 days of testing.

Olfactory sensitivity has also been assessed by use of a habituation paradigm, in which response decrement is observed to the repeated presentations of a stimulus, whereupon a second stimulus is presented. Dishabituation, or recovery of the habituated response, constitutes evidence of discrimination between the two stimuli. In addition, differential response to the familiar (stimulus to which habituation occurred) and novel stimuli implies some memory of the habituation stimulus. Engen, Lipsitt, and Kaye (1963) used this procedure to test olfactory discriminations in newborn infants. The dependent measures were derived from polygraphic records of activity and of respiration during stimulus presentations and during interspersed non-odorous control trials. A decrement in response to the olfactory stimulus (either asafoetida or anise oil) was observed over trials, and dishabituation occurred when a novel olfactory stimulus was presented. Thus, the newborns were able to discriminate between the two stimuli.

Subsequent experiments by the same investigators (Engen & Lipsitt, 1965) used odor mixtures as habituation stimuli, followed by dishabituation tests with one component of the odor mixture. For example, the top panel of Figure 2.1 illustrates the decrease in mean response over the course of five trial blocks to the presentation of a 50-50% mixture of undiluted anise oil and asafoetida.

[1] We have found that Steiner's (1977) pictures of infant facial expressions in response to olfactory stimuli provide excellent demonstrations for introductory child psychology courses. On several occasions we have used the class as blind judges who rated the affective responses of the babies (i.e., they had no knowledge of the stimulus associated with a particular set of facial expressions). In each case, Steiner's original ratings were easily replicated.

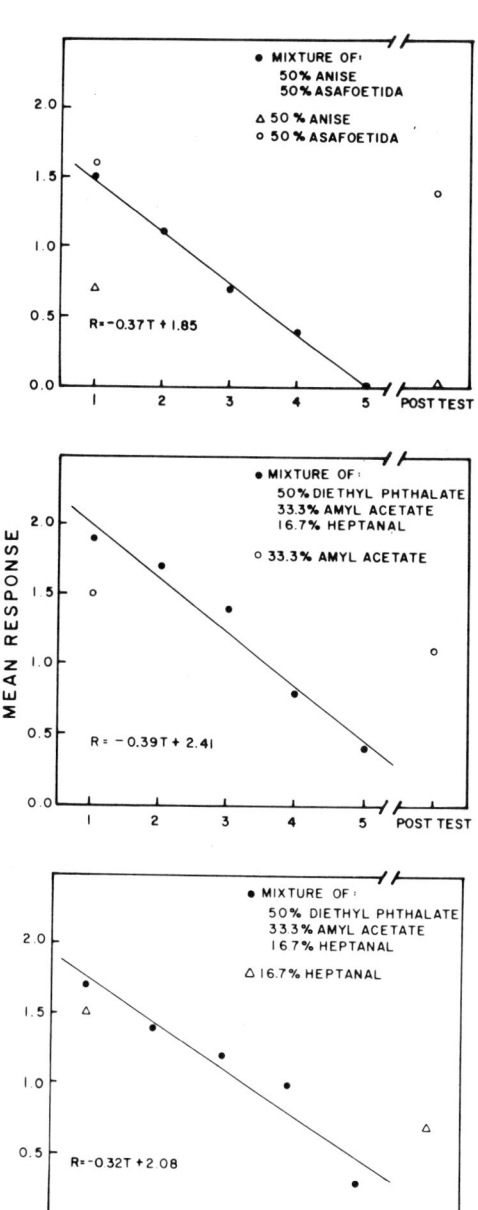

Figure 2.1. Mean number of responses plotted as a function of trial blocks and posttest stimulus presentations. Each panel represents different stimulus conditions. Open symbols on top block show response to odor components prior to odor mixture presentations. (From T. Engen and L. P. Lipsitt, Decrement and recovery of responses to olfactory stimuli in the human infant, Journal of Comparative and Physiological Psychology, 1965, 59, 312-316. Copyright © 1965 by the American Psychological Association. Reprinted by permission.)

Following habituation, half of the subjects were tested with a 50% concentration solution of anise oil, and half were tested with a 50% concentration solution of asafoetida. Recovery of responding on the posttest was observed for the 50% asafoetida, but not for the 50% anise oil. This pattern of responding was consistent with adult similarity judgments, which showed that the 50% anise oil solution was more similar to the anise-asafoetida combination than was the 50% asafoetida solution.

To demonstrate discrimination of component stimuli of an odor mixture independently of intensity, Engen and Lipsitt (1965) tested for dishabituation with components that were approximately equated in intensity to the odor mixture that was used as a habituation stimulus. The odor intensity equation was made in terms of infant responses and adult similarity judgments. Two component stimuli, 33.3% amyl acetate and 16.7% heptanol, were each of approximately equal subjective intensity (even though the concentrations differed) to a mixture of 33.3% amyl acetate, 16.7% heptanol, and a 50% dilutent (diethyl phthalate). As seen in the middle and bottom panels of Figure 2.1, after infants habituated to the odor mixture, their responses increased upon presentation of each of the component stimuli. Thus, in the first days of life, infants are capable of discriminating specific features of olfactory stimuli. The primary odor qualities for both infants and adults are, however, still unknown.

GUSTATORY SENSITIVITY

The four basic gustatory qualities are generally considered to be sweet, bitter, sour, and salt. Different regions of the tongue are relatively more or less sensitive to each of these qualities. Early studies, summarized in several reviews (Lipsitt, 1977; Peiper, 1963; Pratt, Nelson, & Sun, 1930), attempted to determine whether these basic taste qualities can be discriminated in infancy. In general, most of the early studies demonstrated positive responses to sweet fluids, but they failed to make clear whether infants can discriminate bitter, sour, and salt.

Included with their review, Pratt et al. reported data on their own taste experiments with 28 infants. They recorded bodily movement, sucking, and facial reactions when an applicator that had been dipped in a solution was inserted in the infant's mouth. The solutions included varying concentrations of sugar, quinine, citric acid, and salt, with distilled water serving as a control stimulus. Relative to the water, for which 32% of the reactions were sucking responses, sugar (16.66% solution) and salt (8.33% solution) appeared to potentiate sucking, whereas quinine (0.25% solution) and citric acid (2.15% solution) decreased sucking. These latter two solutions also elicited the largest percentage of facial reactions. By parametric variation of the concentration of the taste solutions, they discovered that infants and adults differ in their reactions to different tastes. That is: "The final strength of citric acid solution used, 2.14 per cent, seemed rather weak for the adult experimenters. Yet the infants reacted strongly. To quinine 0.27 per cent the adult experimenters reacted strongly, while the infant reactions were relatively weak [p. 123]." These data, coupled with the olfactory data of Macfarlane, raise the possibility that some dimensions of the chemical senses are more acute in infants than in adults, whereas other dimensions of chemical sensitivity are more acute in adults.

Several studies have generally confirmed infants' ability to discriminate sweet solutions. Usually a positive preference is obtained for sweet fluids, although two studies (Dubignon & Campbell, 1969; Kron, Stein, Goddard, & Phoenix, 1967) reported that milk formula elicited more sucking than did solutions of 5% sugar (dextrose or corn syrup). In contrast, Desor, Maller, and Turner (1973) observed that the volume of sugar solutions ingested by neonates was greater than the volume of water ingested in the same time period. Moreover, the infants discriminated between various concentrations and types of sugars as evidenced by greater ingestion of .20 and .30 M solutions than of .05 and .10 M solutions and their preference for sucrose and fructose over glucose and lactose. The authors pointed out that these consummatory preferences on the part of the infants parallel adult ratings of the sweetness of the sugar solutions. A subsequent investigation by Nowlis and Kessen (1976) provided similar results with two concentrations of sucrose (.058 M and .117 M) and glucose (.277 M and .555 M) and anterior tongue pressure changes during sucking as the dependent measure.

Lipsitt and his colleagues (Crook & Lipsitt, 1976; Lipsitt, 1977; Lipsitt, Reilly, Butcher, & Greenwood, 1976) have conducted a series of studies to examine neonatal gustatory discriminations and the relations between sucking, respiration, and heart rate. In these experiments, the infant sucks on a nipple that contains several small tubes so that different taste stimuli may be presented without removing the nipple during the tests. The infant controls the amount of fluid ingested because the stimuli are presented contingent on sucks of a criterion amplitude. However, the amount of fluid presented following a criterion response is only .02 ml, rendering it unlikely that differential response to the taste stimuli is affected by postingestional consequences. These experiments show that sucking patterns for different fluids are closely related to the tastes of the fluids. The pattern of sucking under nonnutritive conditions is characterized by bursts of sucks that are separated by short pauses. These pauses between bursts become shorter when the infant is reinforced with water, and still shorter when the infant is reinforced with sucrose. Additionally, there is an increase in the number of sucks invested in the average burst, whereas the rate at which the sucks are emitted decreases (i.e., the interresponse time (IRT) increases). Lipsitt (1977) also showed that changes in heart rate accompany these changes in sucking. That is, while the sucking rate decreases with sweet fluids, the heart rate increases. Because the change in heart rate may be observed with the first burst of sucking with sucrose, even within the first few sucks of the burst, the effect cannot be attributed to a change in blood sugar level. In addition, the energy expenditure associated with sucking for sweet fluids might be expected to be less than under nonnutritive conditions because the rate of activity decreases. Therefore, the associated increase in heart rate has been considered paradoxical. When the sucrose condition is investigated parametrically by manipulating either the quantity of sucrose provided per suck (Crook, 1976) or the concentration of sucrose in the fluid (Crook & Lipsitt, 1976), a correlation is observed between sweetness and these paradoxical effects of sucking rate deceleration and heart rate acceleration.

Burke (1977) showed that swallowing of fluids may be partly responsible for changing sucking patterns with sucrose solutions. Both sucking and swallowing may be, at least in part, under hedonic control of the infant. Indeed, Lipsitt (1977) has suggested that the decreasing sucking rate and increasing heart rate may occur when the infant savors the sweeter fluids: "The temptation is to conclude anthropomorphically that when sucrose is introduced, rate slows down to enable savoring of the tastant, and the heart rate goes up to signal the joy that the infant thus derives from this 'pleasure of sensation' [p. 138]." Steiner's (1977, 1979) observations of infants' facial expressions while they are ingesting sweet fluids are consistent with this hypothesis. Steiner (1977) noted that: "Adult observers, viewing pictures of these expressions without knowledge of the stimulus applied interpret the facial expressions elicited by sweet in terms of satisfaction, appreciation, enjoyment or liking [p. 175]." Examples of newborn facial expressions while ingesting sucrose (.73 M) are depicted in Column 3 of Figure 2.2.

Infants are clearly sensitive to the quality and concentration of sweet fluids. Studies of other taste qualities have been less well documented. Early work by Jensen (1932) suggested that facial expressions were not a very sensitive measure for testing infants' discrimination of taste, although Steiner (1979) found differential expressions in newborns to sour (.12 M citric acid) and bitter (.0003 M quinine sulfate) stimuli. Illustrations of the expressions of infants tested within the first 20 hours of life are presented in Figure 2.2. According to Steiner, these expressions are similar to the expressions obtained in adults who were tested with the same substances. For example, sour elicited a puckering of the lips, and bitter evoked a retching or spitting reaction. In contrast, Maller and Desor (1973), on the basis of fluid volume consumed, failed to show discrimination of salt (.003-.2 M NaCl), bitter (.03-.18 M urea), or sour (.001-.012 M citric acid) solutions from water. Recently, however, Crook (1978) compared sucking responses to salt solu-

2. THE INFANCY OF HUMAN SENSORY SYSTEMS 45

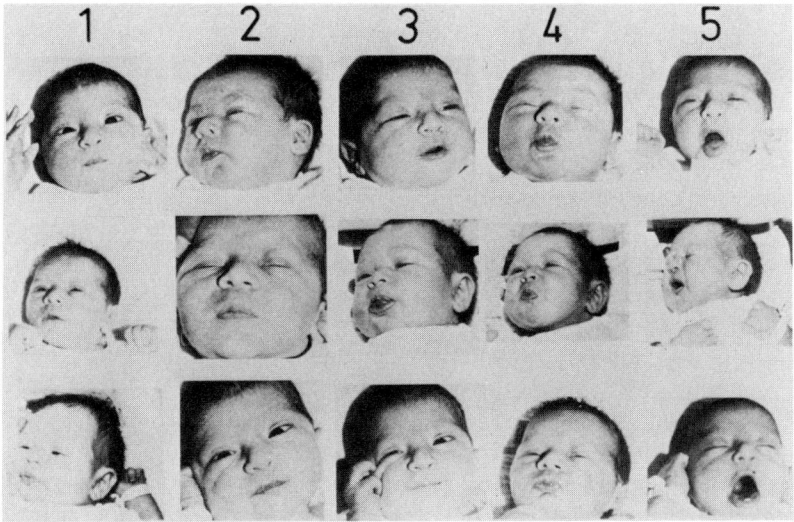

Figure 2.2. Facial expressions of newborn infants to different taste stimuli. Tests were made before the first postnatal feeding. Each of the numbered columns represents a different stimulus condition: (1) no taste stimulus; (2) distilled water presentation; (3) sweet stimulus; (4) sour stimulus; and (5) bitter stimulus. (From J. E. Steiner, Human facial expressions in response to taste and smell stimulation, in H. W. Reese and L. P. Lipsitt (Eds.), Advances in child development and behavior (Vol. 13). New York: Academic Press, 1979. Reprinted by permission.)

tions (.1 M, .3 M, or .6 M NaCl) and water when they were presented contingent on criterion amplitude sucking. The length of sucking bursts was significantly shorter for NaCl solutions than for water. Thus, newborn infants were able to discriminate the taste of salt, even when the concentration was as low as .1 M.

It may be concluded that infants are capable of responding differentially to some taste qualities. Responses to taste qualities are perhaps also modified by experience, and thereby exert significant control over the taste preferences of the developing infant. Thus, even though newborns are capable of discriminating between the tastes of different fluids, they may ingest a variety of sometimes harmful fluids as readily as they ingest water. The tragic and fatal incidents in which newborns ingested solutions that were mistakenly prepared with salt instead of sucrose is a case in point.

Cutaneous Sensitivity

Intellectual development, according to Piaget (1952), and social development, according to Freud (1938), have their foundations in the sucking response of the infant. Tactile stimulation is reflexively linked to the newborn's sucking. Light touch around the corners of the mouth often elicits a rooting reflex, an ipsilateral head turn to the tactile stimulus with accompanying mouth opening. Pressure on the palms of the hands elicits head movement to the midline and opening of the mouth, the Babkin reflex. These responses increase the probability of the infant finding an object to suckle, with concomitant oral-tactile feedback. When the inside of the mouth of an infant is stimulated, there is an immediate closing of the mouth followed by sucking. Although these responses are readily observed and can be shown to be affected by experience (see Lipsitt & Werner, Chapter 4, this volume), they and other aspects of cutaneous sensitivity have received little systematic psychophysical study, even though the infant's responsiveness to tactile stimuli plays a central role in two major theories of development.

Sherman and Sherman (1925; Sherman, Sherman, & Flory, 1936) noted the number of pin pricks over various regions of the body that were required to produce a withdrawal or avoidance response. They found that sensitivity increased rapidly during the first 3 days after birth. Subsequent investigations have confirmed this observation using more precisely controlled stimuli, usually an electrical stimulus that is gradually increased until a criterion response is elicited. Graham (1956; Graham, Matarazzo, & Caldwell, 1956) used small voltages to determine sensitivity around the knee, whereas Lipsitt and Levy (1959) determined sensitivity with an electrode attached to a toe. Both of these studies showed that tactile sensitivity increases over the first 4-5 days of life. Moreover, by use of both a longitudinal and cross-sectional design, Lipsitt and Levy (1959) were able to establish that changes in threshold with age were indeed sensitivity changes, and effects of sensitization and/or conditioning could be ruled out.

Lipsitt and Levy (1959) suggested several variables that might account for an increase in cutaneous sensitivity during the first few days after birth: increased wakefulness or alertness, recovery from the effects of anesthesia, and recovery from anoxia that may have been experienced by some of the infants. Another possibility follows from Richter's (1930) electrical resistance measures obtained from the palm and back of the hands of 50 newborns. The skin resistance of the newborn was substantially higher than that found in adults. A similar result was reported by Kaye (1964) for skin conductance measures that were made on the palmar, plantar, and calf regions. Electrical conductance increased throughout the daily tests that were carried out for the first 4 days of life. This variable was assessed by Kaye and Lipsitt (1964) who measured basal skin conductance and electrotactual threshold over the first 4 postnatal days. They found that basal skin conductance increased and tactile threshold decreased over the 4 days. However, an analysis of covariance, controlling for the effects of skin conductance, indicated that there was still a significant increase in sensitivity with age. Thus, although skin conductance change may contribute to age-correlated increases in cutaneous sensitivity, it cannot fully account for the complete developmental effects.

Many basic cutaneous phenomena such as spatial and temporal summation, and receptive field organization, have yet to be studied in human infants. Investigations of these phenomena over a broad age range could be informative, not only in describing developmental changes in cutaneous sensitivity, but also in contributing to our understanding of basic tactile mechanisms. For example, attempts to relate sensory codes to specific types of cutaneous receptors have had a long, but not very successful history (Kenshalo, 1971). However, recent advances through adult psychophysical studies have led to a duplex model of mechanoreception (Verrillo, 1968). According to this model, mechanoreception is mediated by two systems that differ in temporal and spatial summation. High frequency cutaneous sensitivity is mediated by Pacinian corpuscles, whereas low frequency responses are mediated by other receptors, probably hair follicle endings and Meissner corpuscles. The relative distribution of these receptor types changes from birth through old age (Cauna, 1965) so that a developmental analysis of frequency sensitivity led to advances in the evaluation of this model. Verrillo (1977) obtained threshold versus frequency functions that indicated that children (8-12 years of age) were more sensitive than adults to most frequencies of sinusoidal vibration. Children and adults do not, however, appear to differ in sensitivity at low frequencies (18-25 Hz), which are thought to be mediated by receptors other than Pacinian corpuscles (Frisina & Gescheider, 1977). This effect follows from Verrillo's model on the basis of differences between the two age groups in the structure of the Pacinian corpuscle. Because the size and shape of the Pacinian corpuscle also differs between infants and children, it would be interesting to compare infant sensitivity on a comparable task to that used with children and adults. Developmental analyses of this sort may have much to contribute to a better understanding of sensory receptor processes.

Audition

Sound waves traveling through the external ear create a pattern of vibration on the ear drum, or tympanic membrane. Faithful transference of the pattern of stimulation in the external ear to the fluid-filled cochlea of the internal ear, which contains the neural receptors, is made possible by the middle ear. The ossicles of the middle ear serve as an impedance-matching device between the air and fluid media of the external and internal ear. The inner ear is well developed at birth. Indeed, Bredberg (1968) reported that the organ of Corti is fully differentiated, and the cochlear duct is innervated, by the sixth month of fetal development. The ossicles in the middle ear of the newborn are well developed, but their efficiency of function may be impaired by the presence of trapped fluid and mesenchymal tissue. Most of this fluid disappears in the first few days of postnatal life, although it is not completely dissipated until several months after birth (Spears & Hohle, 1967). The effect of this fluid might be to increase auditory threshold, but it does not preclude auditory function in either the fetus or the newborn.

Newborn and young infants' cardiac acceleration (Bartoshuk, 1964; Steinschneider, Lipton, & Richmond, 1966) and changes in activity level (Hoversten & Moncur, 1969) are a function of sound intensity. Bartoshuk reported a linear relation between cardiac response (log beats min $^{-1}$) and intensity for 1000 Hz tones between 49 and 78 dB (resting sound level was 10.5 dB relative to 2×10^{-4} dynes cm^{-2}). Thus, over this range, equal stimulus ratios produced approximately equal response ratios. The exponent for the power function fit to the data was .53 which, as Bartoshuk pointed out, is close to that obtained in psychophysical experiments with adults (Stevens, 1961).

The absolute threshold for infants has not been determined, but it is probably higher than for the average adult (Hecox, 1975; Robertson, Peterson, & Lamb, 1968). The lowest intensity for which Eisenberg (1965) obtained reliable cardiac changes was about 40 dB (1 sec pulses of white noise, relative to 2×10^{-4} dynes cm^{-2}). Whether other dependent measures would yield a lower absolute threshold estimate is not clear. In any case, existing data clearly indicate that many sounds in the neonate's environment are within the audible range. In addition, thresholds have been observed to decrease during the first 8 postnatal months (Hoversten & Moncur, 1969).

Newborn infants are capable of temporal integration of sound energy. Eisenberg (1965) reported that stimuli that were shorter than 300 msec were not associated with changes in cardiac response, but longer stimuli did produce reliable changes (see also, Clifton, Graham, & Hatton, 1968). The critical duration for a response would be expected to be dependent upon the intensity such that, as the sound intensity is decreased, an increased duration would be required to meet a criterion response.

In addition to discriminating intensity and duration changes, it can be inferred from their eye movements that newborns can localize the spatial position of a sound source (Wertheimer, 1961). Sound localization has also been demonstrated by presenting a stimulus at a specified location for repeated trials until a response is habituated, whereupon the stimulus is presented in a new location. Dishabituation to the novel stimulus position implied that the infants could localize the sound (Leventhal & Lipsitt, 1964). Sound localization is accomplished by adults primarily on the basis of binaural differences in time of arrival and intensity (Thurlow, 1971). It is still not clear whether the newborn infant can use both of these cues.

Adults are maximally sensitive to the frequency range between approximately 2000 and 4000 Hz, although the range in recognition may be from 20 to 20,000 Hz (Scharf, 1975). The audible frequency range for infants has not been determined, although several independent investigators have shown that infants do respond differentially for some frequencies. For example, two studies (Eisenberg, Griffin, Coursin, & Hunter, 1964; Hoversten & Moncur, 1969) found that stimuli below 1000 Hz elicit greater responding than do stimuli above 4000 Hz. Using a habituation-dishabituation paradigm, Bartoshuk (1962) presented data suggesting that neonates could discriminate changes in tonal patterns (intensities and frequencies were held constant). Thus, these data show that newborn infants can discriminate frequency, and perhaps the psychological correlate of frequency, which is pitch. Unfortunately, with the exception of the Bartoshuk study, the possibility that the reported frequency discriminations were based upon loudness cues cannot be ruled out. To rule out loudness cues, it is first necessary to determine the intensity versus frequency function required for an equal-loudness contour for infants. Such a function has not been measured for infants, and because of possible developmental changes in early life, an adult or young child's equal-loudness function cannot be assumed for the infant.

Although it seems likely that infants discriminate pure tones, such auditory stimuli are probably not as effective in eliciting responses as are broad-band stimuli (Eisenberg, 1965). In addition, patterned stimuli are probably not only more effective than pure tones (ignoring possible loudness cues), but both of these types of stimuli are most effective in the frequency range corresponding to the human voice (Hutt, Hutt, Lenard, Bernuth, & Muntjewerff, 1968). Hutt et al. concluded that "the structure of the human auditory apparatus at birth ensures . . . that the voice at normal intensities is nonaversive and prepotent [p. 890]." Indeed, this suggestion is consistent with other recent data showing that neonates respond to and perhaps prefer the human voice over other auditory stimuli (Ashmead & Lipsitt, in preparation), and also with the discovery that within the first months of life, infants perceive the spectrum of synthetic speech consonants in a manner that is similar to adults (Eimas, Siqueland, Jusczyk, & Vigorito, 1971). This discriminatory capacity appears to be based on phonetic rather than on acoustic differences (Eimas, 1975).

Although much work remains for the study of infant auditory processing, it is safe to conclude that newborns are sensitive to the spatial location, intensity, duration, and probably frequency of sounds in their normal environments.

Vision

The axial length of the human eyeball increases rapidly during the first 2 years of life (by about 4 mm), and then more slowly until it reaches the adult size during adolescence (Larsen, 1971). Paralleling this change in the globe, there are other changes in the size and relative position of the refractive media (Scammon & Wilmer, 1950). On the basis of the optics of the eye alone, developmental changes in vision might be expected.

Rod and cone photoreceptors begin to develop prior to birth, but their full elaboration in number, length, and relative distribution, especially around the fovea (the center of gaze), does not resemble the adult retina until at least the fourth postnatal month (Barber, 1955; Mann, 1928). Functional considerations and the limited histological data on the infant retina have engendered some controversy concerning its relative maturity during the first few months of life.

The order of myelination of the optic nerve and optic tract occurs from the brain toward the eye. According to Magoon and Robb (1980) some lamellae of myelin are seen in the optic nerve and the tract at the time of birth, although the optic nerve is less well myelinated, especially near the globe. Substantial thickening of all parts of the optic nerve and tract were observed during the first 2 years of life, with subsequent slower age-correlated increases in myelin thickness.

The central neural mechanisms of the human infant visual system have received little systematic study, although it is clear that significant postnatal development does occur. Hickey (1978) has identified two overlapping phases of cell growth in the lateral geniculate nucleus (of the thalamus): a rapid growth phase in the parvocellular layers ending about 6 months postpartum, and a longer phase of growth in the magnocellular layers that ends about 2 years postpartum. Postnatal increases in cell size, axonal length, and dendritic arborization have also been identified in the visual cortex (Conel, 1939-1963).

Prenatal and postnatal neurogenesis of the visual cortex of rhesus monkeys appears to be similar to human neurogenesis, except for a shifted timetable (Sidman & Rakic, 1973). The ontogeny and organization of the visual cortex in monkey fetuses and infants has been mapped out in substantial detail, particularly through the elegant methods of Rakic (1975, 1976, 1977).

Visual development has probably been the focus of more intensive investigation than any other aspect of human infancy. The material considered here will be illustrative, rather than comprehensive. It is divided into three sections: wavelength sensitivity, contrast sensitivity, and possible roles of experience in visual development. Some important areas of human visual development that are not discussed here have been reviewed elsewhere; for example, eye movements (Maurer, 1975; Salapatek, 1975; Salapatek & Banks, 1978), depth perception (Bower, 1974; Walk, 1978; Yonas & Pick, 1975), attentional preferences (Fantz, Fagan, & Miranda, 1975; Karmel &

Maisel, 1975), recognition memory (Cohen & Gelber, 1975; Werner & Perlmutter, 1979), and models (Bronson, 1974; Haith, 1978).

WAVELENGTH SENSITIVITY

Sensitivity to different regions of the spectrum is a function of quantal absorption by the photopigment in the receptors, and the adaptive state of the visual system as a whole. Scotopic, or twilight, vision is mediated by a single class of photoreceptor, the rods, whereas photopic, or daylight, sensitivity is mediated by three classes of cone photoreceptor. Spectral sensitivity differs for the scotopic and photopic systems, as evidenced by changes in spectral sensitivity from dark-adapted to light-adapted states, a change known as the Purkinje shift. Early attempts to demonstrate a Purkinje shift in infants, and to measure the spectral sensitivity of their scotopic and photopic systems, have been flawed by technical problems (see Werner & Wooten, 1979). Recent evidence has, however, confirmed the conclusion drawn from early research that both the rods and the cones are functional in early life.

Spectral sensitivity in infants and adults has been measured with the visually evoked cortical potential (VECP) under conditions favoring the scotopic system (i.e., dark adaptation and large homogeneous fields; Werner, 1979). The resultant functions from 400 to 650 nm for two subjects, an infant and an adult, are presented in separate panels in Figure 2.3, where log quantal sensitivity is plotted as a function of wavelength, or wavenumber. The functions for these observers are both biphasic, suggesting that sensitivity was mediated by more than one class of receptor. Additional evidence showed that the main branch of the curve from 400 to 600 nm was attributable to rods, whereas long-wave cone intrusion can be seen in the branch from about 600 to 650 nm. The data in Figure 2.3 indicate that both rods and cones are functional in these subjects. The scotopic portion of the curve was subjected to systematic analysis with 50 observers, ranging in age from 1 month to 70 years. It was found that shortwave sensitivity decreased as a monotonic function of age. Although there were substantial individual differences at each age, the average infant was about a factor of four more sensitive at 400 nm than the average 30-year-old adult. Furthermore, analyses of these data showed that this developmental effect cannot be attributed to developmental changes in the optical density of the rod photopigment, rhodopsin. Rather, it is due to more efficient transmission of short-wavelength light on the part of the refractive media of younger observers. These data allowed in vivo determinations of ocular media density that were consistent with direct spectrophotometric measures with a small number of excised eyes (Boettner & Wolter, 1962).

Spectral sensitivity has also been measured under photopic conditions with

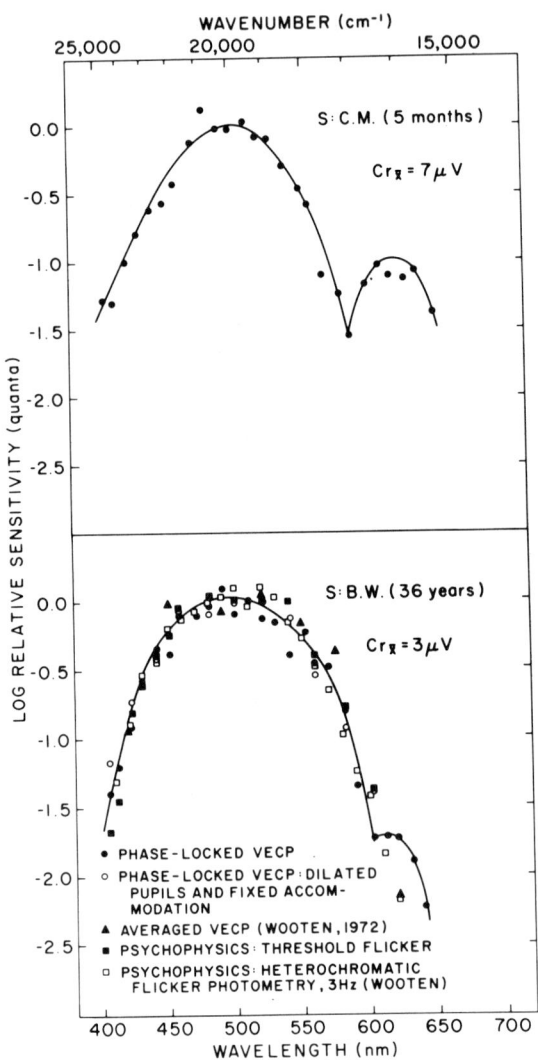

Figure 2.3. Log relative sensitivity of an infant (C.M.) and an adult (B.W.) observer expressed in terms of the relative number of quanta incident on the cornea at the criterion VECP amplitude voltage indicated in each panel. Additional VECP and psychophysical measurements for B.W. are indicated in the figure key and described in the original report. (From Werner, 1979.)

the VECP (Dobson, 1976; Moskowitz-Cook, 1979) and with behavioral methods (Peeples & Teller, 1978). Dobson compared adult and 2-month-old infant sensitivity at eight spectral bands. Her averaged data are replotted in Figure 2.4 on a quantal sensitivity basis. As shown by the bottom curves, the infants were more sensitive than the adults at short wavelengths, but similar in relative sensitivity at mid and long wavelengths. This finding is con-

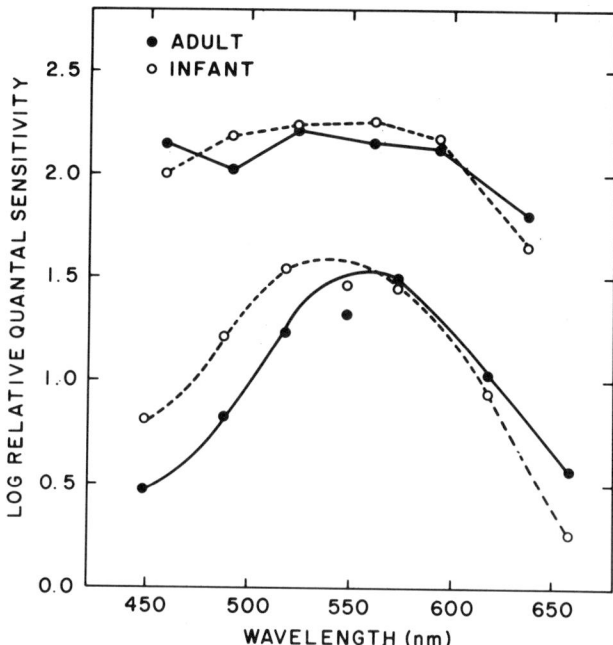

Figure 2.4. Averaged infant and adult log relative sensitivity expressed in terms of the relative number of quanta incident on the cornea at the response criterion. Top curves show data obtained by behavioral methods under white light adaptation (data from Peeples and Teller, 1978). Bottom curves represent sensitivity determined by the VECP (data from Dobson, 1976). (Figure originally appeared in J. S. Werner and B. R. Wooten, Human color vision and color perception. Infant Behavior and Development, 1979, 2, 241-274. Norwood, N.J.: Ablex Publishing Corporation. Reprinted by permission.)

sistent with infants' less dense refractive media described earlier (Boettner & Wolter, 1962; Werner, 1979), as well as their less dense macular pigmentation (Hering, 1885).

Similar results to those obtained by Dobson were recently reported by Moskowitz-Cook (1979), who tested infants between 3 and 22 weeks of age with the VECP. Her data for the younger infants indicated an elevation in relative sensitivity at short wavelengths, but this elevation diminished to adult levels by about 19-22 weeks of age.

Peeples and Teller (1978) measured photopic sensitivity using behavioral methods. Test flashes from one of six different spectral bands were presented on a white adaptation field. This adaptation field resulted in relatively flat sensitivity functions. As shown by the top curves in Figure 2.4, the adults and infants were similar in relative sensitivity throughout the range of their spectral measurements.

If the infant photopic system is more sensitive than that of adults at short wavelengths, as suggested by the 2-month-old infant data of Dobson and

Moskowitz-Cook, it should have been even more apparent in the data of Peeples and Teller because the latter measures were made against a slightly yellowish (3500° K) background. Lights of nearly the same color temperature (2800°K) have been shown to increase adult short-wavelength sensitivity by selective depression of the mid- and long-wavelength cone systems (Wooten, Fuld, & Spillmann, 1975). Thus, although there is general agreement that infants and adults are similar in relative photopic sensitivity at mid and long wavelengths, further determinations are needed to evaluate possible differences between infants and adults at short wavelengths.

Spectral sensitivity functions describe a fundamental property of the infant visual system, but they do not allow specific inferences about the infant's chromatic mechanisms. Color space has three dimensions: hue, saturation, and brightness. Early attempts to study chromaticity (hue and/or saturation) in infants have been confounded by a failure to equate the stimuli in terms of brightness for the infants. Acute brightness discrimination has been reported for infants in their first months of life (Doris & Cooper, 1966; Hershenson, 1964; Peeples & Teller, 1975). The infant brightness-equation problem has also barred conclusions from some recent studies because of an apparent confusion between equal-luminance spectra and equal-brightness spectra (see Werner & Wooten, 1979). The only safe way to obtain lights of equal brightness for infants is to equate them for individual infants under the specific conditions of an experiment.

Peeples and Teller (1975) studied chromatic discrimination in 2-month-old infants with a procedure that clearly ruled out brightness artifacts. They first showed that infants could discriminate a white bar that was embedded in a white screen, while the intensity of the white bar was varied in small steps over a wide range. There was, however, a small intensity interval over which the bar was not discriminated from the field in which it was embedded (i.e., at the brightness-match point). They showed that the intensity interval over which the lights were matched in brightness was small even for infants. To demonstrate chromatic discrimination, Peeples and Teller measured infants' discrimination of a red bar that was embedded in a white field. The intensity of the red bar was systematically varied in small steps (about .085 log unit) so that the white and red components of the field must have been matched in brightness at some intensity level. Yet, the infants discriminated the chromatic stimulus from the background at all intensities. Thus, Peeples and Teller's results indicated that the infants must have at least dichromatic vision (i.e., at least two spectrally selective mechanisms). This finding has been verified by other work in the same laboratory (Teller, Peeples, & Sekel, 1978), as well as in a different experiment in which an operant conditioning paradigm was used to rule out brightness cues in a chromatic discrimination (Schaller, 1975).

Normal adult color vision is trichromatic. That is, it is mediated by three cone photoreceptor types of differing spectral sensitivity. A trichromat can discriminate all wavelengths of the visible spectrum (about 400-700 nm) from a broad-band white light. Dichromats, however, cannot discriminate all wavelengths from white. The spectral region of their discrimination failure is known as their neutral point (see Hurvich, 1972). The wavelength locus of the neutral point depends upon the color temperature of the white light against which discrimination is tested, but in general it is between about 490 and 510 nm for adult protanopes and deuteranopes (lacking the long- and middle-wavelength cones, respectively) and around 570 nm for tritanopes (lacking the short-wavelengths cones).

Bornstein (1976) conducted an experiment to determine whether 3-month-old infants have a neutral point, because, if they do not, it can be concluded that they are trichromatic. Spectral lights between 490 and 500 nm were presented until visual fixation was habituated, whereupon a white light was presented. The habituated response increased when the white light was introduced and subsequently declined when the spectral lights were presented. On the basis of this discrimination, Bornstein concluded that the infants did not have a neutral point. In a second experiment, Bornstein demonsrated that infants can discriminate spectral lights in the yellow-green region of the spectrum (the stimuli used were 560, 570, and 580 nm), as would be expected if they had trichromatic vision. From these experiments, Bornstein concluded that the 3-month-old infants were indeed trichromatic. This conclusion was, however, challenged on several grounds (Werner & Wooten, 1979). The most salient objections were: (a) inadequate brightness controls; (b) an unsafe assumption that the spectral range from 490 to 500 nm would include the neutral point locus for infant protanopes and deuteranopes; and (c) a failure to test for tritanopia, the third form of dichromacy.

Teller, Peeples, and Sekel (1978) have shown that the color vision of 2-month-old infants may depart considerably from what would be expected of trichromats. When brightness cues were controlled, infants discriminated several chromatic stimuli from a white field; however, they did not discriminate between a white and a yellowish green stimulus. Also, there was an apparent neutral point that could be inferred from discrimination tests with a purple chromatic stimulus. An interesting, and probably significant, feature of these results is that the infant discrimination failures do not parallel any known form of adult color vision deficiency. It is thus possible that infants' color space is substantially different from the color space of trichromatic adults.

CONTRAST SENSITIVITY

As with spectral sensitivity, sensitivity to the spatial distribution of light might differ between infants and adults on the basis of developmental changes in the refractive media, not to mention age-correlated differences in neural processes. Adults and children are capable of visual accommodation. That is, they have the ability to alter the refractive power of the lens to focus near objects. Accommodation usually occurs on a reflexive basis, but it requires form vision for its control. Newborn and 1-month old infants are not capable of accommodation to the range of stimulus distances to which adults normally accommodate. The accommodation seen within the first month of life is most readily elicited, and appropriate, for targets that are at distances of about 75 cm or less (Braddick, Atkinson, French , & Howland, 1979). This ability increases rapidly in the early months, as evidenced by the increased range of accommodation in the 2-and 3-month-old infant and the adultlike capacity that is reached by approximately 6 months of age (Banks, 1980; Braddick et al., 1979; Haynes, White, & Held, 1965). Thus, on first glance, it might be expected that visual resolution before about 3 months of age would be distance dependent. However, acuity estimates for infants in this age range have not been found to change with changes in stimulus distance (Atkinson, Braddick, & Moar, 1977a; Fantz, Ordy, & Udelf, 1962; Salapatek, Bechtold, & Bushnell, 1976). These latter reports are not necessarily inconsistent with severe limitations in accommodation of young infants, given their poorly developed spatial sensitivity at this age (discussed in what follows). Indeed, the most likely explanation for limited accommodation below 2-3 months of age is that infants' neural tuning to high spatial frequencies limits the detection of poor image quality (Braddick, et al., 1979). In addition, calculations by Green, Powers, and Banks (1980) indicate an inverse relation between the depth of focus and eye size and visual acuity. The smaller eye size and reduced visual acuity of young infants is related to a greater depth of focus in infants relative to adults such that accommodation would not significantly improve their vision. As depth of focus decreases in the first months of life, there is a correlated increase in the range of visual accommodation.

Form sensitivity is usually described in terms of the smallest detail that can be resolved (i.e., visual acuity). This measure is often useful, but it describes the performance of the visual system only at one extreme of its operating range. A more general measure of visual performance is provided by the function that relates the contrast[2] of spatial patterns required for detection and spatial frequency (specified in terms of the number of cycles per degee of visual angle). This function, referred to as a contrast sensitivity function, is usually measured with achromatic gratings in which the luminance distribution is sinusoidally modulated. A contrast sensitivity function of an adult is shown by the top curve in Figure 2.5. This observer's peak sensitivity is at

---

[2] Contrast, C, is often defined as: $C = (L_{max} - L_{min}) / (L_{max} + L_{min})$, where L_{max} is the luminance maximum in a cycle, and L_{min} is the luminance minimum in a cycle.

2. THE INFANCY OF HUMAN SENSORY SYSTEMS

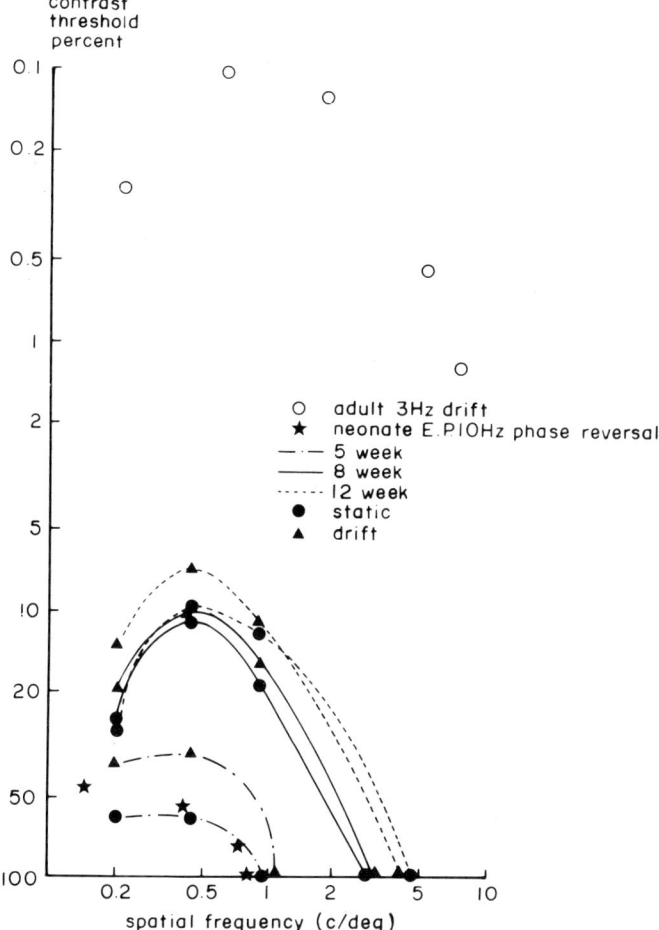

Figure 2.5. *Contrast sensitivity as a function of spatial frequency with static, phase reversal, and drifting grating patterns plotted for an adult and infants. The different symbols denote different stimulus conditions; the mean age of the observers is denoted by different line patterns. (From O. Braddick and J. Atkinson, Accommodation and acuity in the human infant, in R. D. Freeman (Ed.), Developmental neurobiology of vision. New York: Plenum Publishing Corporation. Reprinted by permission.)*

about 2 cy deg^{-1}. Sensitivity progressively declines at higher and lower spatial frequencies.

Contrast sensitivity functions have a number of interesting properties that have been discussed by Cornsweet (1970). Because any two-dimensional stimulus can be described by a set of sine-wave gratings (Fourier's theorem), a contrast sensitivity function should allow predictions of visual performance with any stimulus, provided that certain assumptions (which, for human vision, are probably true only over a limited range) can be made.

Infant contrast sensitivity functions have been determined in a number of independent investigations (Atkinson, Braddick, & Braddick, 1974; Atkinson, Braddick, & French, 1979; Atkinson, Braddick, & Moar, 1977a, 1977b; Banks & Salapatek, 1976, 1978; Pirchio, Spinelli, Fiorentini, & Maffei, 1978). Although the specific conditions of these studies differed, the results in each case agree reasonably well with the infant functions of Braddick and Atkinson (1979) that are presented in Figure 2.5. The functions for subjects of different ages do not overlap much. Rather, they increase progressively along both the sensitivity and spatial frequency coordinates with increasing age. Such developmental changes in contrast sensitivity have been found with both behavioral and evoked cortical potential recordings; indeed, at least under some conditions (but obviously not under all conditions), it can be shown that results from a single infant may be nearly identical for both types of measurement (Atkinson, Braddick, & French, 1979).

Visual acuity can be estimated from a contrast sensitivity function by extrapolating from the descending slope at high frequencies to a contrast of unity. It is apparent from Figure 2.5 that the highest resolvable (static) spatial frequency increases with age, from about .9 cy deg^{-1} at 5 weeks (Banks & Salapatek obtained values of 2.4 cy deg^{-1} at this age[3]) to approximately 4 cy deg^{-1} at 12 weeks. For adults, the high-frequency cutoff under comparable stimulus conditions is usually between about 35 and 50 cy deg^{-1}. These estimates are not dependent on whether the high-frequency cutoff is determined from contrast sensitivity functions for static or drifting gratings because, as is shown in the infant data of Figure 2.5, drifting gratings affected sensitivity to low spatial frequencies, but not to high spatial frequencies. Similar findings with adults have been reported by Van Nes, Koenderick, Nas, & Bouman (1967) for temporal modulations at 0 and 1 Hz, and to a lesser extent at 10 Hz. It is worth noting that the acuity values obtained from the contrast sensitivity functions shown in Figure 2.5 agree fairly well with other behavioral measures of infant acuity such as optokinetic nystagmus, and the preferential-looking technique, whereas acuity estimates based on the visually evoked cortical potential tend to be higher (see review by Dobson & Teller, 1978). The numerous infant acuity measures, covering a more ex-

---

[3] There are at least two important aspects of Banks and Salapatek's (1978) methods that may account for their higher spatial frequency cutoffs for 5-week-old infants than were obtained by Braddick and Atkinson (1979). First, Braddick and Atkinson presented their stimuli on oscilloscope screens (15° each), whereas Banks and Salapatek used a projection system that produced much larger stimuli (total field size of 96° × 40°). Second, although both studies used a preferential looking procedure, Braddick and Atkinson separated the grating and homogeneous fields by 9° whereas the grating and homogeneous field used by Banks and Salapatek were joined at the midline. Therefore, the grating in the Braddick and Atkinson studies would initially have appeared in parafoveal vision.

tensive age range than the measures of contrast sensitivity, suggest that acuity increases rapidly from birth to about 6 months of age (e.g., Marg, Freeman, Peltzman, & Goldstein, 1976; Sokol, 1978).

The decline in contrast sensitivity at low spatial frequencies has also been of interest because it may be indicative of lateral inhibitory processes (Cornsweet, 1970; cf. Estévez & Cavonius, 1976). Low-frequency attenuation in infant contrast sensitivity was present at all ages except at 1 month, the lowest age group tested (Atkinson, Braddick, & Moar, 1977a; Banks & Salapatek, 1978). Of course, it is always possible that a decline in sensitivity would be found with 1-month-old infants if contrast sensitivity were measured at still lower spatial frequencies.

It may be concluded that adults and infants below 6 months of age differ in contrast sensitivity throughout the spatial frequency domain. This difference must be of profound importance in the development of infant form perception.

EXPERIENCE AND VISUAL DEVELOPMENT

Electrophysiological studies with the rhesus monkey, which has color and pattern vision that closely approximates that of humans, have established that a high percentage (50-80%) of the cells in the visual cortex will respond to stimuli that are presented to either eye, although such binocular units are usually biased in their responses to one eye. The anatomical organization of the visual cortex is such that the cells that are dominated by input from the left eye tend to lie in columns that alternate with columns of cells that are dominated by input from the right eye. Primate binocular organization is discrete and can be clearly demonstrated by electrophysiological and histological methods (Hubel & Wiesel, 1972; LeVay, Hubel, & Wiesel, 1975). Of special interest is a group of binocular cells that respond maximally when corresponding images are focused on slightly different horizontal positions in the two eyes. These cells are selectively turned for binocular disparity (Baker, Grigg, & von Noorden, 1974; Hubel & Wiesel, 1970), the stimulus correlate of the depth sensation known as stereopsis.

Monkey ocular dominance columns are present before birth (Rakic, 1976), and although the receptive field tuning in cells of the newborn is quantitatively different from that in adults, they appear to be qualitatively similar (Wiesel & Hubel, 1974). To determine the role of visual experience in establishing or maintaining binocular organization, Wiesel and Hubel conducted a series of deprivation experiments. They found that binocular occlusion for the first month after birth drastically reduced the number of cells receiving input from both eyes. These effects were not observed if the deprivation occurred later in development. Thus, it appears that visual experience during a critical period is required to maintain binocular connections that are present from birth. The type of experience required during the critical, or sensitive, period may be quite specific (i.e., that the two eyes are used simultaneously). When infant monkeys are made exotropic by severing the appropriate eye muscles (one eye is turned outward), both eyes remain functional and develop normal monocular acuity. However, the animals resolve the dilemma of discordant images in the two eyes by using only one eye at a time. Under this condition, the number of cortical binocular cells is permanently reduced (Baker et al., 1974).

The importance of visual stimulation in binocular development has been shown anatomically and physiologically with nonhuman primates, but similar stimulation is probably necessary for human development as well. Unfortunately, instances of interocular differences in retinal input and permanent deficits in humans are probably not rare (e.g., esotropia—crossed eyes), aniseikonia (images in the two eyes are of different size), and anisometropia (the two eyes differ in refractive power). Discordant stimulation of the two eyes in infants and children often results in suppression of input from one eye. This in turn may permanently disrupt binocular cortical organization and its important behavioral functons in the perception of depth. Approximately 2-4% of the population is stereoblind (Richards, 1970; Julesz, 1971). Sometimes these binocular anomalies are associated with amblyopia, which is a reduction in acuity that cannot be optically corrected.

Movshon, Chambers, and Blakemore (1972) conducted an experiment to show that ocular misalignment during early development may indeed be associated with stereoblindness. Binocular connections in the cortex could be inferred by adapting one eye to a visual stimulus and observing whether the adaptation effect was present in responses to stimuli presented in the opposite eye. Specifically, they studied the interocular transfer of the tilt aftereffect in which adaptation to a grating that is tilted in a clockwise orientation leads to the impression that a vertical grating is tilted in a counterclockwise orientation. The magnitude of transfer between the eyes is about 70% for normal subjects, but Movshon et al. found little interocular transfer (about 12%) in subjects with a history of strabismus that was not corrected before they were 2 years of age.

In two recent experiments (Banks, Aslin, & Letson, 1975; Hohmann & Creutzfeldt, 1975), interocular transfer of the tilt aftereffect was determined for observers with known histories of strabismus. Hohmann and Creutzfeldt tested 12 individuals whose squint was surgically corrected between 3 and 5 years of age. They found that the older the child was when the squint began, up to 2.6 years, the higher the interocular transfer of the tilt aftereffect. Normal interocular transfer, and by inference normal binocular cortical connections, was observed for subjects who developed strabisumus after about 2.6 years of age. They therefore suggested that the critical period for human binocular development ends by about 2.0-2.6 years of age. Banks et al. tested 24 strabismic subjects who were surgically corrected and for whom the onset of strabismus was well defined. On the basis of their interocular transfer data, they concluded that there is a sensitive period for binocular development that peaks between about 1 and 3 years of age, and then gradually declines. Abnormal binocular experience during this sensitive period may permanently impair binocular organization.

Binocularity is not the only aspect of visual development that requires early experience. Individuals having an astigmatic error that leaves the images in one orientation defocused develop a permanent and uncorrectable loss of acuity for images in that orientation (Mitchell, Freeman, Millodot, & Haegerstrom, 1973). This effect must be neural because it is not eliminated by laser-generated interference fringes that bypass the optics of the eye (Freeman & Thibos, 1975). The sensitive period for orientation development may differ from the sensitive period in the development of binocularity. Still other aspects and experiments on experiential determinants of human visual development have been discussed by Mitchell (1978).

Conclusions

Newborns enter the world with all sensory systems capable of sustaining some adaptive behavior. The prenatal neurogenesis of these capacities may be regarded as an instance of Carmichael's (1970) *law of anticipatory function*. According to this law, "it is biologically essential to have the structures that make later adaptive responses ready at a period somewhat prior to the time when such reactions must work if the animal is to survive and lead a life that is characteristic of its species [p. 449]." Thus, at birth, the sensory systems make it possible for the infant to learn and to profit from experience. At the same time, it may be noted that the sensory systems limit the stimulus energies to which the newborn may respond, and thereby, may canalize infant experiences in particular directions. It seems likely that further sensory epigenesis, aided and abetted by experiences supplied by the usual environment in which the infant is reared, assure that the sensory modalities will increase in their capabilities. The orchestration of sensory and central neural pathways with learning processes will then make possible increased adaptation, with age, that is required to cope with the developmental tasks of later childhood.

References

Atkinson, J., Braddick, O., & Braddick, F. Acuity and contrast sensitivity of infant vision. *Nature*, 1974, *247*, 403-404.

Atkinson, J., Braddick, O., & French, J. Contrast sensitivity of the human neonate measured by the visual evoked potential. *Investigative Ophtalmology & Visual Science*, 1979, *18*, 210-213.

Atkinson, J., Braddick, O., & Moar, K. Development of contrast sensitivity over the first 3 months of life in the human infant. *Vision Research*, 1977, *17*, 1037-1044. (a)

Atkinson, J., Braddick, O., & Moar, K. Contrast sensitivity of the human infant for moving and static patterns. *Vision Research*, 1977, *17*, 1045-1047. (b)

Baker, F. H., Grigg, P., & von Noorden, G. K. Effects of visual deprivation and strabismus on the response of neurons in the visual cortex of the monkey, including studies on the striate and prestriate cortex in the normal animal. *Brain Research*, 1974, *66*, 185-208.

Banks, M. S. The development of visual accommodation during early infancy. *Child Development*, 1980, *51*, 646-666.

Banks, M. S., Aslin, R. N., & Letson, R. D. Sensitive period for the development of human binocular vision. *Science*, 1975, *190*, 675-677.

Banks, M. S., & Salapatek, P. Contrast sensitivity function of the infant visual system. *Vision Research*, 1976, *16*, 867-869.

Banks, M. S., & Salapatek, P. Acuity and contrast sensitivity in 1-, 2-, and 3-month-old human infants. *Investigative Ophthalmology & Visual Science*, 1978, *17*, 361-365.

Barber, A. *Embryology of the human eye.* St. Louis: C. V. Mosby, 1955.

Bartoshuk, A. K. Human neonatal cardiac acceleration to sound: Habituation and dishabituation. *Perceptual and Motor Skills*, 1962, *15*, 15-27.

Bartoshuk, A. K. Human neonatal cardiac responses to sound: A power function. *Psychonomic Science*, 1964, *1*, 151-152.

Bernard, J., & Sontag, L. W. Fetal reactivity to tonal stimulation: A preliminary report. *Journal of Genetic Psychology*, 1947, *70*, 205-210.

Boettner, E. A., & Wolter, J. R. Transmission of the ocular media. *Investigative Ophthalmology*, 1962, *1*, 776-783.

Bornstein, M. H. Infants are trichromats. *Journal of Experimental Child Psychology*, 1976, *21*, 425-445.

Bower, T. G. R. *Development in infancy.* San Francisco: W. H. Freeman, 1974.

Braddick, O., & Atkinson, J. Accommodation and acuity in the human infant. In R. D. Freeman (Ed.), *Developmental neurobiology of vision.* New York: Plenum, 1979.

Braddick, O., Atkinson, J., French, J., & Howland, H. C. A photorefractive study of infant accommodation. *Vision Research*, 1979, *19*, 1319-1330.

Bradley, R. M. Development of the taste bud and gustatory papillae in human fetuses. In J. F. Bosma (Ed.), *The third symposium on oral sensation and perception: The mouth of the infant.* Springfield, Ill.: Charles C. Thomas, 1972.

Bredberg, G. Cellular pattern and nerve supply of the human organ of corti. *Acta Oto-Laryngologica, Supplementum*, 1968, (Whole No. 236).

Bronson, G. The postnatal growth of visual capacity. *Child Development*, 1974, *45*, 873-890.

Burke, P. M. Swallowing and the organization of sucking in the human newborn. *Child Development*, 1977, *48*, 523-531.

Carmichael, L. The onset and early development of behavior. In P. H. Mussen (Ed.), *Carmichael's manual of child psychology* (3rd ed.). New York: Wiley, 1970.

Cauna, N. The effects of aging on the receptor organs of the human dermis. In W. Montagna (Ed.), *Advances in biology of the skin* (Vol. 6) *Aging*. New York: Pergamon, 1965.

Clifton, R. K., Graham, F. K., & Hatton, H. M. Newborn heart-rate response and response habituation as a function of stimulus duration. *Journal of Experimental Child Psychology*, 1968, *6*, 265-278.

Cohen, L. B., & Gelber, E. R. Infant visual memory. In L. B. Cohen & P. Salapatek (Eds.), *Infant perception: From sensation to cognition.* New York: Academic Press, 1975.

Conel, J. L. *The postnatal development of the human cerebral cortex* (Vols. 1-7). Cambridge, Mass.: Harvard University Press, 1939-1963.

Cornsweet, T. N. *Visual perception.* New York: Academic Press, 1970.

Crook, C. K. Neonatal sucking: Effects of quantity of the response-contingent fluid upon sucking rhythm and heart rate. *Journal of Experimental Child Psychology*, 1976, *21*, 539-548.

Crook, C. K. Taste perception in the newborn infant. *Infant Behavior and Development*, 1978, *1*, 52-69.

Crook, C. K., & Lipsitt, L. P. Neonatal nutritive sucking: Effects of taste stimulation upon sucking rhythm and heart rate. *Child Development*, 1976, *47*, 518-522.

Dayton, G. O., Jones, M. H., Aiu, P., Rawson, R. A., Steele, B. & Rose, M. Developmental study of coordinated eye movements in the human infant. I. Visual acuity in the newborn human: A study based on induced optokinetic nystagmus recorded by electro-oculography. *Archives of Ophthalmology*, 1964, *71*, 865-870.

De Beer, G. R. *Embryos and ancestors* (3rd ed.). Oxford: Clarendon Press, 1958.

Desor, J. A., Maller, O., & Turner, R. G. Taste in acceptance of sugars by human infants. *Journal of Comparative and Physiological Psychology*, 1973, *84*, 496-501.

Dobson, V. Spectral sensitivity of the 2-month infant as measured by the visually evoked cortical potential. *Vision Research*, 1976, *16*, 367-374.

Dobson, V., & Teller, D. Y. Visual acuity in human infants: A review and comparison of behavioral and electrophysiological studies. *Vision Research*, 1978, *18*, 1469-1483.

Doris, J., & Cooper, L. Brightness discrimination in infancy. *Journal of Experimental Child Psychology*, 1966, *3*, 31-39.

Dubignon, J., & Campbell, D. Discrimination between nutriments by the human neonate. *Psychonomic Science*, 1969, *16*, 186-187.

Eimas, P. D. Speech perception in early infancy. In L. B. Cohen & P. Salapatek (Eds.), *Infant perception: From sensation to cognition.* New York: Academic Press, 1975.

Eimas, P. D., Siqueland, E. R., Jusczyk, P., & Vigorito, J. Speech perception in infants. *Science*, 1971, *171*, 303-306.

Eisenberg, R. B. Auditory behavior in the human neonate: I. Methodologic problems and the logical design of research procedures. *The Journal of Auditory Research*, 1965, *5*, 159-177.

Eisenberg, R. B., Griffin, E. J., Coursin, D. B., & Hunter, M. A. Auditory behavior in the human neonate: A preliminary report. *Journal of Speech and Hearing Research*, 1964, *7*, 245-269.

Ellingson, R. J. Cortical electrical responses to visual stimulation in the human infant. *Electroencephalography and Clinical Neurophysiology*, 1960, *12*, 663-677.

Engen, T. Method and theory in the study of odor preferences. In J. W. Johnston, D. G. Moulton, & A. Turk (Eds.), *Human responses to environmental odors.* New York: Academic Press, 1974.

Engen, T., & Lipsitt, L. P. Decrement and recovery of responses to olfactory stimuli in the human neonate. *Journal of Comparative and Physiological Psychology*, 1965, *59*, 312-316.

Engen, T., Lipsitt, L. P., & Kaye, H. Olfactory responses and adaptation in the human neonate. *Journal of Comparative and Physiological Psychology*, 1963, *56*, 73-77.

Estévez, O., & Cavonius, C. R. Low-frequency attenuation in the detection of gratings: Sorting out the artifacts. *Vision Research*, 1976, *16*, 497-500.

Fantz, R. L., Fagan, J. F., & Miranda, S. B. Early visual selectivity as a function of pattern variables, previous exposure, age from birth and conception, and expected cognitive deficit. In L. B. Cohen & P. Salapatek (Eds.), *Infant perception: From sensation to cognition.* New York: Academic Press, 1975.

Fantz, R. L., Ordy, J. M., & Udelf, M. S. Maturation of pattern vision in infants during the first six months. *Journal of Comparative and Physiological Psychology*, 1962, *55*, 907-917.

Freeman, R. D., & Thibos, L. N. Contrast sensitivity in humans with abnormal visual experience. *Journal of Physiology*, 1975, *247*, 687-710.

Freud, S. Infantile sexuality. In A. A. Brill (Ed. and trans.), *The basic writings of Sigmund Freud*. New York: Random House, 1938.

Frisina, R. D., & Gescheider, G. A. Comparison of child and adult vibrotactile thresholds as a function of frequency and duration. *Perception & Psychophysics*, 1977, *22*, 100-103.

Galebsky, A. Vestibular nystagmus in new-born infants. *Acta Oto-Laryngologica*, 1927, *11*, 409-423.

Gesell, A. The ontogenesis of infant behavior. In L. Carmichael (Ed.), *Manual of child psychology*. New York: Wiley, 1954.

Gibson, J. J. *The senses considered as perceptual systems*. Boston: Houghton-Mifflin, 1966.

Gibson, J. J. Commentary. *Monographs of the Society for Research in Child Development*, 1976, *41*,(4), 62-66.

Goodkin, F. The development of mature patterns of head-eye coordination. (Doctoral dissertation, Brown University, 1979).

Gorman, J. J., Cogan, D. G., & Gellis, S. S. An apparatus for grading the visual acuity of infants on the basis of opticokinetic nystagmus. *Pediatrics*, 1957, *19*, 1088-1092.

Gottlieb, G. Ontogenesis of sensory function in birds and mammals. In E. Tobach, L. R. Aronson, and E. Shaw (Eds.), *The biopsychology of development*. New York: Academic Press, 1971.

Gottlieb, G. (Ed.). *Studies on the development of behavior and the nervous system* (Vol. 1), *Behavioral embryology*. New York: Academic Press, 1973.

Gottlieb, G. The roles of experience in the development of behavior and the nervous system. In G. Gottlieb (Ed.), *Studies on the development of behavior and the nervous system* (Vol. 3), *Neural and behavioral specificity*. New York: Academic Press, 1976.

Graham, F. K. Behavioral differences between normal and traumatized newborns: I. The test procedures. *Psychological Monographs: General and Applied*, 1956, *70*(427).

Graham, F. K., Matarazzo, R. G., & Caldwell, B. M. Behavioral differences between normal and traumatized newborns: II. Standardization, reliability, and validity. *Psychological Monographs: General and Applied*, 1956, *70*(428).

Green, D. G., Powers, M. K. & Banks, M. S. Depth of focus, eye size, and visual acuity. *Vision Research, Vision Research*, 1980, *20*, 827-835.

Haith, M. M. Visual competence in early infancy. In R. Held, H. W. Leibowitz, & H-L. Teuber (Eds.), *Handbook of sensory physiology* (Vol. VIII). Berlin: Springer, 1978.

Haynes, H., White, B. L., & Held, R. Visual accommodation in human infants. *Science*, 1965, *148*, 528-530.

Heck, W. E. Vestibular responses in the newborn. *A.M.A. Archives of Otolaryngology*, 1952, *56*, 573.

Hecox, K. Electrophysiological correlates of human auditory development. In L. B. Cohen & P. Salapatek (Eds.), *Infant perception: From sensation to cognition*. New York: Academic Press, 1975.

Hering, E. Ueber individuelle verschienheiten des farbensinnes. *Lotos, Naturwissenschaftliche Zeitschrift (n.F.)*, 1885, *6*, 142-198.

Hershenson, M. Visual discrimination in the human newborn. *Journal of Comparative and Physiological Psychology*, 1964, *58*, 270-276.

Hickey, T. L. Postnatal development of the human lateral geniculate nucleus. In S. J. Cool & E. L. Smith (Eds.), *Springer series in optical sciences* (Vol. 8): *Frontiers in visual science*. Berlin: Springer-Verlag, 1978.

Hohmann, A., & Creutzfeldt, O. D. Squint and the development of binocularity in humans. *Nature*, 1975, *254*, 613-614.

Hoversten, G. H., & Moncur, J. P. Stimuli and intensity factors in testing infants. *Journal of Speech and Hearing Research*, 1969, *12*, 687-702.

Hubel, D. H., & Wiesel, T. N. Cells sensitive to binocular depth in area 18 of the macaque monkey cortex. *Nature*, 1970, *225*, 41-42.

Hubel, D. H., & Wiesel, T. N. Laminar and columnar distribution of geniculocortical fibers in the macaque monkey. *Journal of Comparative Neurology*, 1972, *146*, 421-450.

Humphrey, T. Some correlations between the appearance of human fetal reflexes and the development of the nervous system. *Progress in Brain Research*, 1964, *4*, 93-133.

Hurvich, L. M. Color vision deficiencies. In D. Jameson & L. M. Hurvich (Eds.), *Handbook of sensory physiology* (Vol. VII/4). Berlin: Springer, 1972.

Hutt, S. J., Hutt, C., Lenard, H. G., Bernuth, H. V., & Muntjewerff, W. J. Auditory responsivity in the human neonate. *Nature*, 1968, *218*, 888-890.

Jensen, K. Differential reactions to taste and temperature stimuli in newborn infants. *Genetic Psychology Monographs*, 1932, *12*, 363-479.

Johansson, B., Wedenberg, E., & Westin, B. Measurement of tone response by the human foetus: A preliminary report. *Acta Oto-Laryngologica*, 1964, *57*, 188-192.

Julesz, B. *Foundations of cyclopean perception*. Chicago: University of Chicago Press, 1971.

Karmel, B. Z., & Maisel, E. B. A neuronal activity model for infant visual attention. In L. B. Cohen & P. Salapatek (Eds.), *Infant perception: From sensation to cognition*. New York: Academic Press, 1975.

Kaye, H. Skin conductance in the human neonate. *Child Development*, 1964, *35*, 1297-1305.

Kaye, H., & Lipsitt, L. P. Relation of electrotactual threshold to basal skin conductance. *Child Development*, 1964, *35*, 1307-1312.

Kenshalo, D. R. The cutaneous senses. In J. W. Kling & L. A. Riggs (Eds.), *Experimental psychology*. New York: Holt, Rinehart & Winston, 1971.

Kobre, K. R. A negative contrast effect in newborns. (Master's thesis, Brown University, 1971).

Korner, A. F., & Thoman, E. Visual alertness in neonates as evoked by maternal care. *Journal of Experimental Child Psychology*, 1970, *10*, 67-78.

Kron, R. E., Stein, M., Goddard, K. E., & Phoenix, M. D. Effect of nutrient upon the sucking behavior of newborn infants. *Psychosomatic Medicine*, 1967, *29*, 24-32.

Larsen, J. S. The sagittal growth of the eye. IV. Ultrasonic measurement of the axial length of the eye from birth to puberty. *Acta Ophthalmologica*, 1971, *49*, 873-886.

Lawrence, M. M., & Feind, C. R. Vestibular responses to rotation in the newborn infant. *Pediatrics*, 1953, *12*, 300-306.

LeVay, S., Hubel, D. H., & Wiesel, T. N. The pattern of ocular dominance columns in macaque visual cortex revealed by a reduced silver stain. *Journal of Comparative Neurology*, 1975, *159*, 559-575.

Leventhal, A. S., & Lipsitt, L. P. Adaptation, pitch discrimination, and sound localization in the neonate. *Child Development*, 1964, *35*, 759-767.

Lipsitt, L. P. Taste in human neonates: Its effects on sucking and heart rate. In J. M. Weiffenbach (Ed.), *Taste and development: The genesis of sweet preference*. Washington, D.C.: U.S. Government Printing Office, 1977.

Lipsitt, L. P., Engen, T., & Kaye, H. Developmental changes in the olfactory threshold of the neonate. *Child Development*, 1963, *34*, 371-376.

Lipsitt, L. P., & Levy, N. Electrotactual threshold in the neonate. *Child Development*, 1959, *30*, 547-554.

Lipsitt, L. P., Reilly, B. M., Butcher, M. J., & Greenwood. M. M. The stability and interrelationships of newborn sucking and heart rate. *Developmental Psychobiology*, 1976, *9*, 305-310.

Macfarlane, A. Olfaction in the development of social preferences in the human neonate. In *Parent-infant Interaction* (CIBA Foundation Symposium, #33). Amsterdam: Elsevier, 1975.

Magoon, E. H., & Robb, R. M. Development of myelin in human optic nerve and tract. Abstract in Supplement to *Investigative Ophthalmology and Visual Science*, 1980, p. 3.

Maller, O., & Desor, J. A. Effect of taste on ingestion by human newborns. In J. F. Bosma (Ed.), *Oral sensation and perception: Development in the fetus and infant*. Washington, D.C.: U.S. Government Printing Office, 1973.

Mann, I. C. *The development of the human eye*. Cambridge: Cambridge University Press, 1928.

Marg, E., Freeman, D. N., Peltzman, P., & Goldstein, P. J. Visual acuity development in human infants: evoked potential measurments. *Investigative Ophthalmology*, 1976, *15*, 150-153.

Maurer, D. Infant visual perception: Methods of study. In L. B. Cohen & P. Salapatek (Eds.), *Infant perception: From sensation to cognition*. New York: Academic Press, 1975.

McDougall, W. *An introduction to social psychology*. Boston: Luce, 1908.

McGraw, M. B. Development of rotary-vestibular reactions of the human infant. *Child Development*, 1941, *12*, 17-19.

Mistretta, C. M., & Bradley, R. M. Taste in utero: Theoretical considerations. In J. M. Weiffenbach (Ed.), *Taste and development: The genesis of sweet preference*. Washington, D.C.: U.S. Government Printing Office, 1977.

Mitchell, D. E. Effect of early visual experience on the development of certain perceptual abilities in animals and man. In R. D. Walk & H. L. Pick (Eds.), *Perception and experience*. New York: Plenum, 1978.

Mitchell, D. E., Freeman, R. D., Millodot, M., & Haegerstrom, G. Meridional amblyopia: evidence for modification of the human visual system by early visual experience. *Vision Research*, 1973, *13*, 535-558.

Moskowitz-Cook, A. The development of photopic spectral sensitivity in human infants. *Vision Research*, 1979, *19*, 1133-1142.

Movshon, J. A., Chambers, B.E.I., & Blakemore, C. Interocular transfer in normal humans, and those who lack stereopsis. *Perception*, 1972, *1*, 483-490.

Nowlis, G. H., & Kessen, W. Human newborns differentiate differing concentrations of sucrose and glucose. *Science*, 1976, *191*, 865-866.

Peeples, D. R., & Teller, D. Y. Color vision and brightness discrimination in two-month-old human infants. *Science*, 1975, *189*, 1102-1103.

Peeples, D. R., & Teller, D. Y. White-adapted photopic spectral sensitivity in human infants. *Vision Research*, 1978, *18*, 49-53.

Peiper, A. *Cerebral function in infancy and childhood*. New York: Consultants Bureau, 1963.

Pfaffmann, C. The pleasures of sensation. *Psychological Review*, 1960, *67*, 253-268.

Piaget, J. *The origins of intelligence in children*. (M. Cook, trans.). New York; International Universities Press, 1952.

Pirchio, M., Spinelli, D., Fiorentini, A., & Maffei, L. Infant contrast sensitivity evaluated by evoked potentials. *Brain Research*, 1978, *141*, 179-184.

Pratt, K. C., Nelson, A. K., & Sun, K. H. *The behavior of the newborn infant*. Columbus: The Ohio State University Press, 1930.

Rakic, P. Timing of major ontogenetic events in the visual cortex of the rhesus monkey. In N. A. Buchwald & M. Brazier (Eds.), *Brain mechanisms in mental retardation*. New York: Academic Press, 1975.

Rakic, P. Prenatal genesis of connections subserving ocular dominance in the rhesus monkey. *Nature*, 1976, *261*, 467-471.

Rakic, P. Prenatal development of the visual system in rhesus monkeys. *Philosophical Transactions of the Royal Society of London*, 1977, *278*, 245-260.

Richards, W. Stereopsis and stereoblindness. *Experimental Brain Research*, 1970 *10*, 380-388.
Richter, C. P. High electrical resistance of the skin of new-born infants and its significance. *American Journal of Diseases of Children*, 1930, *40*, 18-26.
Rieser, J., Yonas, A., & Wikner, K. Radial localization of odors by human newborns. *Child Development*, 1976, *47*, 856-859.
Robertson, E. O., Peterson, J. L., & Lamb, L. E. Relative impedance measurements in young children. *Archives of Otolaryngology*, 1968, *88*, 162-168.
Salapatek, P. Pattern perception in early infancy. In L. B. Cohen and P. Salapatek (Eds.), *Infant perception: From sensation to cognition*. New York: Academic Press, 1975.
Salapatek, P., & Banks, M. S. Infant sensory assessment: Vision. In F. D. Minifie & L. L. Loyd (Eds.), *Communicative and cognitive abilities: Early behavioral assessment*. Baltimore: University Park Press, 1978.
Salapatek, P., Bechtold, A. G., & Bushnell, E. W. Infant visual acuity as a function of viewing distance. *Child Development*, 1976, *47*, 860-863.
Scammon, R. E., & Wilmer, H. A. Growth of the components of the human eyeball. II. Comparison of the calculated volumes of the eyes of the newborn and of adults, and their components. *Archives of Ophthalmology*, 1950, *43*, 620-637.
Scarr-Salapatek, S. Genetic determinants of infant development: An overstated case. In L. P. Lipsitt (Ed.), *Developmental psychobiology: The significance of infancy*. Hillsdale, N.J.: Erlbaum, 1976.
Schaller, M. J. Chromatic vision in human infants: Conditioned operant fixation to "hues" of varying intensity. *Bulletin of the Psychonomic Society*, 1975, *6*, 39-42.
Scharf, B. Audition. In B. Scharf (Ed.), *Experimental sensory psychology*. Glenview, Illinois: Scott, Foresman, 1975.
Sherman, M., & Sherman, I. C. Sensori-motor responses in infants. *The Journal of Comparative Psychology*, 1925, *5*, 53-68.
Sherman, M., Sherman, I. C., & Flory, C. D. Infant behavior. *Comparative Psychology Monographs*, 1936, *12*,(4).
Sidman, R. L., & Rakic, P. Neuronal migration, with special reference to developing human brain: A review. *Brain Research*, 1973, *62*, 1-35.
Sokol, S. Measurement of infant visual acuity from pattern reversal evoked potentials. *Vision Research*, 1978, *18*, 33-39.
Spears, W. C., & Hohle, R. H. Sensory and perceptual processes in infants. In Y. Brackbill (Ed.), *Infancy and early childhood*. New York: Free Press, 1967.
Steiner, J. E. Facial expressions of the neonate infant indicating the hedonics of food-related chemical stimuli. In J. M. Weiffenbach (Ed.), *Taste and development: The genesis of sweet preference*. Washington, D.C.: U.S. Government Printing Office, 1977.
Steiner, J. E. Human facial expressions in response to taste and smell stimulation. In H. W. Reese & L. P. Lipsitt (Eds.), *Advances in child development and behavior* (Vol. 13). New York: Academic Press, 1979.
Steinschneider, A., Lipton, E. L., & Richmond, J. B. Auditory sensitivity in the infant: Effect of intensity on cardiac and motor responsivity. *Child Development*, 1966, *37*, 233-252.
Stevens, S. S. To honor Fechner and repeal his law. *Science*, 1961, *133*, 80-86.
Teller, D. Y., Peeples, D. R., & Sekel, M. Discrimination of chromatic from white light by two-month-old human infants. *Vision Research*, 1978, *18*, 41-48.
Thurlow, W. R. Audition. In J. W. Kling & L. A. Riggs (Eds.), *Experimental psychology*. New York: Holt, Rinehart & Winston, 1971.
Tibbling, L. The rotatory nystagmus response in children. *Acta Oto-Laryngologica*, 1969, *68*, 459-467.

Van Nes, F. L. Koenderink, J. J., Nas, H., & Bouman, M. A. Spatiotemporal modulation transfer in the human eye. *Journal of the Optical Society of America,* 1967, *57,* 1082-1088.

Verrillo, R. T. A duplex mechanism of mechanoreception. In D. R. Kenshalo (Ed.), *The skin senses.* Springfield, Ill.: Charles C. Thomas, 1968.

Verrillo, R. T. Comparison of child and adult vibrotactile thresholds. *Bulletin of the Psychonomic Society,* 1977, *9,* 197-200.

Walk, R. D. Depth perception and experience. In R. D. Walk & H. L. Pick (Eds.), *Perception and experience.* New York: Plenum, 1978.

Watson, J. B. *Behaviorism.* New York: W. W. Norton, 1924.

Werner, J. S. Developmental change in scotopic sensitivity and the absorption spectrum of the human ocular media. (Doctoral dissertation, Brown University, 1979).

Werner, J. S., & Perlmutter, M. Development of visual memory in infants. In L. P. Lipsitt & H. W. Reese (Eds.), *Advances in child development and behavior* (Vol. 14). New York: Academic Press, 1979.

Werner, J. S., & Siqueland, E. R. Visual recognition memory in the preterm infant. *Infant Behavior and Development,* 1978, *1,* 79-94.

Werner, J. S., & Wooten, B. R. Human infant color vision and color perception. *Infant Behavior and Development,* 1979, *2,* 241-273.

Wertheimer, M. Psychomotor coordination of auditory and visual space at birth. *Science,* 1961, *134,* 1692.

Wiesel, T. N., & Hubel, D. H. Ordered arrangement of orientation columns in monkeys lacking visual experience. *Journal of Comparative Neurology,* 1974, *158,* 307-318.

Wooten, B. R., Fuld, K., & Spillmann, L. Photopic spectral sensitivity of the peripheral retina. *Journal of the Optical Society of America,* 1975, *65,* 334-342.

Yonas, A., & Pick, H. L. An approach to the study of infant space perception. In L. B. Cohen and P. Salapatek (Eds.). *Infant perception: From sensation to cognition.* New York: Academic Press, 1975.

III

LEARNING AND ETHOLOGY

Are learning theory approaches to development and ethological developmental approaches reconcilable? This is one of Chiszar's themes in Chapter 3. He concludes that there are many more similarities between the two viewpoints than there are differences. The reader may judge whether the union between learning theory and ethology proposed by Chiszar is likely to be fruitful or sterile.

Chiszar then reviews the history and current status of ethological theory and suggests that the controversies surrounding instinct and learning theories have been overstated. He also provides a valuable critical review of the evolution of developmental plasticity.

Following Chiszar's chapter are contributions that exemplify the learning theory approach and the ethological approach. In Chapter 4, Lipsitt and Werner treat development and individual differences in the classical tradition of hedonic behaviorism. They provide repeated demonstrations of the conditionability of human infants and suggest that individual differences in behavior may be due to differential reinforcement of response components as well as to congenital differences in general physiological well-being and the health of particular organ systems.

Marler, Zoloth, and Dooling, in Chapter 5, examine the contribution of innate factors to the organization of complex behavioral systems, and the relationship between such innate specifications and the nonrandom structure of natural environments. They review the emergence of perceptual systems in organisms ranging from insects to primates. They argue that ethologists have overplayed the role of the innate determinants of behavioral development, whereas psychologists have displayed a "complementary bias" in underestimating the role of genetic factors in human behavioral development. They recommend a new synthesis of animal and human research.

3

Learning Theory, Ethological Theory, and Developmental Plasticity

DAVID CHISZAR

Introduction

This chapter consists of two arguments. The similarities and differences between ethological theory and learning theory are discussed in the first four sections of this chapter, with the conclusion that the similarities far outweigh the differences (at least from a metatheoretical perspective), and modern ideas regarding the evolution of developmental plasticity are discussed in the remaining sections. In this connection, the interface among sociobiology, ethology, psychology, and several other disciplines is explored to show that a confluence of all of these fields is required to understand behavioral phenomena theoretically as well as empirically. A preview of this argument is afforded by the recognition that there are four kinds of questions to be asked about any example of adaptive behavior:

1. How is it organized and/or released? (i.e., the question of *behavioral physiology*)
2. How does it develop? (the *ontogenetic* question)
3. What function does it serve? (the *ecological* question)
4. How did it evolve? (the *phylogenetic* question)

It should not be assumed that the four questions (and the disciplines they represent) are independent, and that research in one area can proceed intelligibly without interdisciplinary communication and consequent cross-validation. This is perhaps the greatest error that is made by contemporary behavioral scientists. For example, it is possible that several conspecifics exhibit a common adaptive behavior pattern for *different* reasons, often because of different experiences during formative ontogenetic periods. It happens that nestling birds peck or gape at food-laden parental bills and, thereby, induce the parent to release food. However, the pecking of some nestlings may be released specifically by features associated with the parent's bill, whereas the same behavior in older nestlings can be released by entirely different aspects of the parent's body and/or behavior. Without a knowledge of ontogenetic interactions leading to differential stimulus control, an investigator dealing with nestlings of these two sorts might be lured into making an erroneous generalization. Accordingly, it is appropriate to keep in mind that ontogeny may permit different individuals to arrive at a common behavioral expression via a variety of alternative routes. Therefore, experiments aimed at an analysis of releaser mechanisms must be interpreted against the background of Remy Chauvin's (1977) admonition: "They can do it in a hundred different ways [Chiszar, 1978]."

In this context, it is instructive to remember that the functional consequences of food begging by nestling birds are critical for their survival. The fact that food begging is underdetermined by any particular releasing stimulus makes sense in that we might expect nestlings to generalize this response to any cue(s) associated with parents laden with food, including parental behavior. Under certain conditions, such nestlings may enjoy a considerable advantage compared to those whose begging is under more specific stimulus control. Thus, we observe a scenario wherein ecological functions of behavior lead to the evolution of developmental plasticity with respect to the ability of initially neutral stimuli arising from parental bodies and/or behavior to acquire control over a fixed action pattern (FAP), which was at first released only by a specific aspect of the parent (see Chapter 5 by Marler, Zoloth, & Dooling, this volume, for another treatment of this point).

Another example from an entirely different area of research will make these same points equally well. Harlow, Blomquist, Thompson, Schiltz, and Harlow (1968) showed that the frontal cortex of adult rhesus monkeys was critical for effective performance in delayed-response tasks. Animals sustaining massive lesions in this area exhibited dramatic performance deficits compared to controls sustaining sham operations. However, the same surgery was administered to infant monkeys who were then tested at 10-24 months of age; these subjects performed at a level comparable to controls, indicating nearly complete recovery of function. Histological examinations of the brains of these animals revealed extensive destruction of the frontal cortex. In fact, their brains were not noticeably different from those of monkeys sustaining cortical destruction in adulthood. Accordingly, it was not the case that the infants experienced less cortical damage or that they somehow regenerated damaged brain tissue. We are left to conclude that processes subserving delayed-response performance are normally associated with the frontal cortex. But if damage to this area occurs early in life (i.e., before these processes develop), other parts of the brain can assume these responsibilities and the lesioned animal is able to perform in adulthood as if it were normal. This research clearly shows that two rhesus monkeys can perform identically in a delayed-response test for very different reasons and that it is inappropriate to conclude that the performance is based on the frontal cortex alone. Furthermore, we are at once forced to acknowledge both the plasticity inherent in the primate central nervous system (CNS) and the ontogenetic *behavioral* plasticity made possible by this property of the CNS. Again we sense the need for physiological research to be supplemented by ontogenetic analyses, and again we might wonder about the selection pressures that lead to the evolution of such remarkable plasticity.

Problems Raised by the Goal-Directedness of Behavior

Teleology

Most behaviors are adaptive because they accomplish important things for organisms. This is true whether we are considering instinctive behavior or learned behavior. In both cases the animal executes a response or a series of responses that have some functional consequences. However, the goal-directedness of behavior has given rise to fundamental difficulties in the conceptualization of behavior and, thereby, to equally fundamental challenges for the construction of theoretical frameworks to aid in the organization and explanation of behavioral phenomena. The problems go back in history to the ancient Greeks who inaugurated the traditional rationalistic approach to the explanation of human behavior, which assumes (a) the existence of mental purposes; and (b) the ability of mental purposes to cause or otherwise influence the occurrence of behavior. It is important to recognize that rationalism readily spreads to other areas of experience. Some rationalists see mental purposes everywhere that regularity exists, including the behavior of animals, and thus are provided to such thinkers the means by which to explain these regularities. It is easy to understand how this happens. Because much animal behavior is clearly goal directed and, therefore, clearly functional, it is an easy matter to invest the organism with mental purposes, however vague and imperfect, which can stand as causal antecedents for their adaptive behavior. Generations of prescientific writers illustrate the application of traditional rationalism (in one form or another) to animal behavior. There are two main reasons why such application is problematic. First, the imputation of a mental purpose as a cause for an animal's behavior tells us little or nothing about how the animal came to have that purpose. If our job is to explain behavior, we must recognize that hypothesizing a mental purpose merely restates the problem. The job should be to explain how the purpose developed or where it came from. Clearly, we are a long way from being done with our work simply because we propose the existence of some mental purpose. In essence, then, we have accomplished nothing substantial. The second major difficulty with this way of explaining behavior is that it results in assertions that are untestable. There exist no rules about how purposes are translated into behavior or about how they influence behavior. Accordingly, even if we grant that purposes arise out of experience, perception, or motivation, we have an insurmountable difficulty in predicting how they will be expressed (i.e., in generating testable hypotheses about behavior).

Traditional rationalism is adequate as the basis for an ethical system. Human purposes are objects of evaluation by persons as well as by institutions. Purposes serve very well in this capacity; this is probably their greatest utility and the reason why we cling to them. However, they do not perform nearly as well in the role of causal processes. In fact, they have the worst possible vice in this regard: They are absolutely invulnerable to refutation or to any empirical verification. We must, therefore, look elsewhere for a framework to guide our thinking as we attempt to explain behavior.

Retroactive Causation

I suspect that some form of mechanistic determinism (also called physicalism) was the first major alternative to traditional rationalism to arise in the history of western philosophy. It was alive in some Greek minds but wider acceptance had to wait until the seventeenth-century accomplishments of Galileo and Newton, and eventually until the successful application of this view to biological phenomena. The essential presuppositions of all forms of physicalism are that all natural events have physical causes (i.e., physical determinism), discoverable through observation and experimentation (i.e., finite causation). Also, a slight extention of this reasoning says that all natural events occur according to laws relating those events to their causes. Laws, in turn, may be grouped hierarchically into sets such that higher-order laws can subsume large numbers of lower-order ones.

This early form of the mechanistic doctrine is still guiding the thought of many people, scientists as well as nonscientists. It has, however, undergone several important modifications in certain philosophical quarters, largely as a consequence of the work of Hume and Kant. These developments need not concern us here. Rather, it is necessary only to look at a particular kind of error that can be made in the context of physicalism (or in any of its contemporary forms). We always assume that causes must precede their effects; even the most primitive versions of mechanistic determinism are clear on this issue. But consider a temptation that imposes itself upon the observer of adaptive behavior. Specifically, consider the case of lovebirds building a nest. If a relatively naive observer asks why the birds behave the way they do, there is first the temptation to respond by saying, "Because they want to mate" (or something else containing a teleological postulation). Feeling uncomfortable with the teleological aspect of the explanation, the observer might then restate the explanandum leaving out the troublesome word(s). "Because they want to mate" becomes "Because they are going to mate." The change in words is subtle but important; emphasis is removed from teleological terms and placed on the adaptive consequences of the behavior in question. In one sense, this was a major advance in thinking in that attention came to be squarely directed at the adaptedness of behavior. In another sense, however, statements such as this run the risk of making a fundamental logical error; namely, they tend to imply retroactive causation. Whenever an attempt is made to explain the immediate causation of behavior on the basis of its adaptive consequences, the explanandum will contain either teleological concepts or the implication of retroactive causation. This is true whether we are dealing with learned or instinctive behavior. Several examples will make these points clear.

Food-deprived rats quickly learn to traverse alleyways when food rewards are offered for performance. Upon seeing such an animal exhibit its learned response, an observer can ask, "Why does the rat traverse the alley?" One kind of answer might impute purposes or intentions to the rat (e.g., the rat intends to get the food; or, the rat wants the food), and thus a teleological position would be taken. Another kind of answer might center on the food rather than upon any desire the rat may have for it. In this case, food itself, eating the food, satisfying hunger drives, or satisfying physiological needs created by deprivation might be held up as causes for running down the alley. A moment of reflection will reveal, however, that none of these conditions occur until *after* the response of traversing the alley has been executed. Great care is taken by experimenters to insure that the rat can neither see nor smell the food until it has entered the goal box at the end of the alley. Similarly, eating, satisfaction of hunger, and so on, cannot occur until the animal has arrived at the appropriate goal place (i.e., the traversing response must occur prior to this). Hence, any attempt to explain the behavior of traversing the alley on the basis of food or eating must necessarily blunder into postulation of retroactive causation. This is clearly a problem. How can events A, B, or C be causes for event D if events A, B, and C do not occur until after D?

Instinctive behaviors lead us to a similar situation. A male bluegill sunfish will chase other males from his nest-territory; only females will be permitted to enter. We can hypothesize that the resident male chases other males from his territory because they might interfere in some way with his reproductive success (i.e., by stealing the territory, interfering with the female). But, if we then hypothesize the existence of such a mental purpose in the CNS of the territorial male and argue that the purpose is a cause of this male's aggressivity toward other males, we have constructed a teleological explanation. Unless we can simultaneously show how the purpose influences behavior, we have done little more than restate the problem. If we argue instead that the safety of the territory and/or the safety of eggs (which might be deposited in it by a female at some future time) are the causative factors, then we have constructed an argument containing an element of retroactive causation. Benefits that *eventually* accrue to a territory-holding male (e.g., attracting females, mating) cannot logically be involved in the immediate causation of territory-holding behavior because such benefits do not occur until *after* the execution of the behavior under consideration.[1]

Arguments that imply retroactive causation are rarely clear. The same is true for arguments involving teleological concepts. These are two forms of fuzzy thinking that usually occur together and that initially appear to make sense. They are often alluring arguments, so much so that speculation about the causation of animal behavior proceeded along these lines for decades without arousing the critical sensibilities of thinkers who would never use such arguments in other contexts.

Recognizing the problems inherent in teleological arguments and in statements involving retroactive causation, we are forced to pay careful attention to those phenomena that we shall attempt to hold as potential causal agents in animal behavior. At the very least, they must occur *prior* to the behavior that is to be explained. Then, too, they should have specifiable relationships

---

[1] It is important to distinguish between at least two kinds or levels of causation. Immediate or proximate causation refers to those factors that release or elicit behavior. Ultimate or distal causation refers to those factors that are responsible for the evolution of behavior. The greatest difficulties arise when these kinds of causation are not distinguished. The ecological adaptedness of a particular behavior is the basis for its evolution by natural selection. Thus, adaptedness and selection are ultimate causes in the sense that they are responsible for the continued existence of the species and/or behavior under consideration. However, the occurrence of that behavior in a particular individual at a particular moment or in a particular situation is another matter entirely. To understand specific *instances* of behavior, we must discover immediate proximate causes; to understand the *capacity* of a species to exhibit a behavior, we must focus upon ultimate or distal causes. These levels of causality are not independent, but it is not true that knowledge of one level automatically provides knowledge about the other. It is clear, for example, that the territorial behavior of male bluegill sunfish has adaptive consequences that allow us to understand its evolution, and, therefore, its presence in contemporary individuals. However, these statements do not inform us about the specific conditions under which this behavior will occur. That is, we do not yet understand the stimulus factors that release the behavior. Conversely, the most detailed experiments revealing the releasing factors for a particular behavior may not simultaneously tell us anything about the adaptive significance of that behavior. Again, behavioral physiology must be supplemented by ecology to arrive at a complete understanding of behavior.

to the behavior they are supposed to explain. Finally, it would be particularly useful if we could specify the manner in which these causal phenomena came into existence and the ways in which they can be manipulated. It is now possible to envision the entire history of learning theory as an attempt to identify such causal phenomena. All forms of learning theory, from the most mechanistic stimulus-response (S-R) connectionist formulations to the most cognitive modern views, are alike in that each attempts to identify processes that can stand in the correct logical position to serve as causes for behavior. The next section will support this statement with a series of examples. We shall also see that instinct theory, beginning with Konrad Lorenz's (1937) views and extending to contemporary revisions and extentions of those ideas, does exactly the same thing. Surely the specific kinds of mechanisms and processes proposed by learning theorists and ethologists differ in many ways, especially in the extent to which experience influences their organization and operation. However, from the present metatheoretical perspective, it should be clear that the logical roles of these mechanisms and/or processes are quite analogous within their respective domains.

Learning Theory

The first half of the twentieth century was a classical period for the psychology of learning, especially in the United States. These decades witnessed the construction of the great systems, the battles between them (fought through the vehicle of the "critical experiment"), and the eventual resolution of these tensions as the new age of minitheories came explicitly to recognize the necessity of scientific humility by acknowledging boundary conditions, biological constraints, and genetic predispositions. The contemporary scene is difficult to embrace in a single picture, but the history is relatively easy to trace (Hill, 1971).

Classical and Instrumental Conditioning

The most important early developments were the formalization of classical conditioning by Pavlov (1927) and of instrumental conditioning by Thorndike (1898, 1913). Both of these processes, however mysterious and ill-defined their neurological mediations may be, had the effect of pointing to factors that had the proper logical characteristics to serve as causes for at least certain kinds of behavior. Dogs salivate prior to the presentation of meat powder (US) because there exists in the environment stimuli that have been conditioned to elicit such behavior (CSs). The causes of anticipatory salivation (CR) are not to be found in mental purposes or intentions the dog may have regarding the US. Rather, we must look to the stimulus environment surrounding the dog to identify those aspects of it that have become CSs; these are the immediate or proximate causes of the CR. Pavlov thus helped immensely to clear the air of teleological assertions and retroactive causational blunders. The dog's environment contains stimuli, some of which have been paired with the US and have therefore acquired some control over the salivary response. On any given trial, some of these stimuli are presented to the dog, either by accident or by the experimenter's design, and some amount of salivation occurs. This salivation is anticipatory only with respect to the US; it clearly occurs *after* the presentation of the CSs. Hence, CSs are in the appropriate temporal relationship to the behavior to be explained. At one and the same time, Pavlov showed where causes (at least for respondent behavior) are to be found and how formerly neutral stimuli can come to acquire causal connections with behavior. To be sure, Pavlov demonstrated that pairing a CS with a US was the critical operation required for acquisition of stimulus control of the CR by the CS. He also offered a theory about how such pairings accomplish their effects, but his theory has not enjoyed the same longevity as has the empirical demonstrations. Nonetheless, it was the latter that was necessary to get thinkers out of teleological and retroactive causational traps and to initiate an era of productive theoretical speculation aimed at understanding how conditioning works.

Much the same can be said about Thorndike's research dealing with instrumental conditioning. This time, in the realm of operant behavior, we see another set of manipulations leading to another adaptive change in behavior by which certain formerly neutral or ineffective stimuli come to acquire control over certain responses. Again there were two major consequences of the demonstration: (a) it pointed to stimuli that reliably precede certain responses, and it suggested that the proximate causes of the responses are to be found here; and (b) it gave rise to important speculation regarding the processes by which such stimuli acquire their control over behavior. Often psychologists are so drawn to this speculation and to the research it inspires that they forget how important the first consideration is from a logical and metatheoretical perspective. In a sense, it justifies all the speculation about the operational and neurological bases of classical and instrumental conditioning. As was true for Pavlov, Thorndike offered a theory about how his conditioning phenomena came about, and these ideas have not enjoyed the longevity of the empirical demonstrations. Theory about the acquisition of stimulus control is important, and we shall have a lot more of it before we are done, but of greater importance was the fact that Thorndike, like Pavlov, pointed to the proper medium of behavioral causation and thereby obviated the logical difficulties that beset the work of earlier writers dealing with animal behavior. Pavlov and Thorndike left no doubt whatsoever about where to look for proximate causes of behavior; they left only the question of how the causal stimuli acquire their ability.

*Edward Chace Tolman's (1886-1959) Solution to the
Problems of Teleology and Retroactive Causation*

Understanding the nature and formation of associations was clearly recognized to be psychology's first task. Theories of learning popped up regularly between 1920 and 1960, this being a time of great optimism regarding the ability of learning theory to embrace virtually all of behavior, and a time of relatively little concern with boundary conditions or other factors that might set limits on the generalizability of behavioral laws. This was also a time when psychologists were least interested in ethology and instinctive behavior. Finally, this was a time when plurality developed regarding the kinds of mechanisms that were hypothesized to underly learning.[2] Guthrie (1935) and Watson (1925) had introduced the S-R connectionist tradition wherein learning was considered to result from the establishment of associations between responses and stimuli in whose presence they occurred. For Watson, all learning derived from classical conditioning. For Guthrie, the essential principle was simply contiguity between stimuli and responses and was therefore more general than classicial conditioning in that responses were presumed to be available for conditioning whenever or however they occurred (the US was nonessential provided that the UR could be produced in other ways). These frameworks, however much they may differ (Hill, 1971), view all behavior as responses to stimuli and view all learning as the establishment of associations between stimuli and responses. Other theorists

---

[2] Related to the plurality regarding mechanisms and processes considered to underly learning there is another plurality regarding "levels of analyses." It is difficult to understand the former without first understanding the latter. Some psychologists concentrate on the input aspects of learning situations, attempting to specify the necessary relationships that must exist between stimuli for learning to occur. Others focus on the relationships between responses (i.e., output aspects of the situation) or between stimuli and responses. Still others direct their attention toward organic processes that mediate between input and output. Accordingly, the language of various learning theories differs depending upon the extent to which the theories deal with inputs, outputs, or intraorganismic processes. Often the conceptual properties of a learning theory can be correlated in a fairly clear way with the level of analysis characterizing the theory. Input-oriented theories emphasize contiguity processes and consider them to be basic and essential. Output-oriented theories emphasize the role of reinforcement and related processes. Theories concerned with the mediation between stimuli and responses tend to emphasize various intervening processes, which may range from basic physiological mechanisms to hypothetical constructs to cognitive factors. It is not necessarily the case that theories are opposed to each other simply because they use different languages and concentrate on different aspects of learning situations. Indeed, plurality of explanatory concepts may derive from different levels of analysis corresponding to different aspects of the learning process.

(Thorndike, 1913; Skinner, 1938, 1950; debated the extent to which reinforcement and drive were necessary for the formation of associations, but did not dispute the more fundamental notion that behavior was essentially a matter of responding to stimuli. It is against this background that Tolman's (1932, 1938) contribution can best be appreciated.

He begins by arguing that human and animal behavior is not simply a collection of responses elicited by stimuli. Behavior has greater unity than this; it is a matter of striving for goals. Behavior sequences exhibit great complexity and flexibility, but always they are organized around some goal that specifies the adaptive purpose of the behavior and, thereby, gives a meaning to the behavior. Moreover, animals and humans often make compensatory adjustments when a favored response option is thwarted, thus enabling them to reach goals by avenues never before taken. Such considerations lead Tolman to the position that learning and performance included more than responses to stimuli; in his view, something that allowed far greater flexibility than S-R connectionist concepts was necessary. His solution was to include cognitions as intervening variables between stimuli and responses. Tolman accomplished three things by this maneuver: (a) he introduced the concept of intervening variables into psychology; (b) he used cognitive concepts in a way that avoided the problems of teleology; and (c) he provided a major alternative to connectionism, which inspired considerable research and which served as a model for subsequent generations of cognitive theorists.

One of Tolman's major ideas was that organisms form *field expectancies* (= sign-gestalt-expectations = cognitive maps), which are cognitions about the structure and organization of their worlds. Field expectancies form as a consequence of experience, but they do not require rewards and they can be limited by biological predispositions to attend to certain stimulus dimensions more than to others.[3] In this way, organisms build up expectations about which stimuli are associated with each other, about orderliness present in their worlds, and about how their behavior can interact with stimuli. Field expectations are different from teleological purposes in that Tolman provided statements dealing with the manner in which field expectations are formed and with how they influence behavior. In this way, Tolman was a behaviorist committed to objectivity and precision in theory construction. His desire always was to retain the exactitude of the connectionist tradition while adding to it the flexibility and the realism of cognitive concepts. Probably the most

---

[3] In this connection, it should be pointed out that Tolman was one of the first learning theorists to recognize the existence of biological constraints on learning. He spoke of *field cognition modes* and used this concept to refer to biases toward learning some things more readily than others. It would be 40 years before this idea would be generally accepted by psychologists (Hinde & Stevenson-Hinde, 1973; Seligman & Hager, 1972). See Chapter 5 by Marler et al. for examples of field cognition modes in honeybees.

important accomplishment of the theory was that it placed cognitions in a position to serve as causal agents vis-à-vis behavior. An animal placed into a familiar situation is presumed to possess a cognitive map containing representations of important features of the situation. Stimuli arising from the situation elicit or activate the cognitive map which, in turn contributes to the causation of behavior.

Tolman and his associates entered into a variety of scholarly debates with other workers wherein phenomena deducible from Tolman's theory were held up as counterinstances for competing theories. Among these phenomena were latent learning and place learning. Ensuing arguments were interesting and creative, but for present purposes it is not essential to go into them. What is important is that Tolman attempted to identify a kind of psychological process that can be activated by stimuli, and that occurs in the correct temporal position to serve as a cause for behavior. Although other theorists rejected field expectations, they did not reject the overall strategy of identifying psychological processes arising from experience and serving to influence subsequent activities in the same situation or in similar situations. Hull's (1943, 1952) intervening variables (D, K, sHr, etc.) and his fractional anticipatory goal reactions could have been used to illustrate this approach instead of Tolman's concepts (although the treatment would have been considerably longer because of the complexity of Hull's system). Similarly, ideas from Gestalt psychology, from mathematical psychology, and from various additional special areas could have been used equally well. In each case, the psychological process would be somewhat different, but all would interact with prevailing stimuli and would predispose the organism to respond in predictable ways. In all cases, the psychological process would avoid both the pitfalls of simplistic teleology and of retroactive causation.

Ethological Theory

Early Views

Prior to the 1930s, most observers of naturalistic behavior in animals were mainly concerned with demonstrating the innateness of at least some behavioral components or with specifying the interaction between innate and acquired components. Little systematic attention was given to an analysis of proximate causation of instinctive behavior until the work of Jacob von Uexküll (1921) and Konrad Lorenz (1937, 1950). Accordingly, we see in the early literature numerous accounts of adaptive behaviors that appear in individuals without practice or other experiential input that could involve conditioning. The following passage is prototypical of the early literature and illustrates the major concern of these writers.

How do the spider and the ant lion go about finding means of supporting themselves? Both can do no other than to live by catching flying and creeping insects, although they are slower in their own movement than is the prey which they seek out. But the former perceives within herself the ability and the drive to weave an artful net, before she had as much as seen or tasted a gnat, fly, or bee; and when one has been caught in her net she knows how to secure and devour it. . . . The ant lion, on the other hand, who can hardly move in the dry sand, mines a hollow funnel by burrowing backward, in expectation of ants and worms that tumble down, or it buries them with a rain of sand that it throws up in order to corner them and bring them into his reach. . . . Since these animals possess by nature such skills in their voluntary actions that serve the preservation of themselves and their kind, and that admit many variations, so they possess by nature certain innate skills. . . . A great number of their artistic drives are performed at birth without error and without external experience, education, or example and thus inborn naturally and inherited. . . . One part of these artistic drives is not expressed until a certain age and condition has been reached, or is even performed only once in a lifetime, but even then it is done by all in a similar manner and with complete regularity. For these reasons these skills are not acquired by practice. . . . But not everything is determined completely in the drives of the animals, and frequently they adjust, of their own volition, their actions to meet various circumstances in various and extraordinary ways [H. S. Reimarus, 1762, as cited in Eible-Eibesfeldt, 1975, pp. 4-5].

Instinct and Ontogeny

Observations such as these led to experiments where organisms were reared in environments deprived of various stimuli or opportunities for certain kinds of practice. The eventual appearance of specific adaptive patterns under such conditions was taken as evidence for their instinctive or inherited character (Carmichael, 1926, 1927, 1928; Grohmann, 1939; Spalding, 1873). Although there was so great an accumulation of descriptive and experimental facts by 1930 that no scientist could doubt the reality of instinctive behavior, there was nonetheless a controversy based partly on logical difficulties attending certain uses of the term *instinct*, and partly on the fact that some instinct theorists appeared to ignore important ontogenetic considerations (Lehrman, 1953). Deprivation experiments can shed light on the innateness of behavior, but not all instinct theorists bothered to execute such investigations before claiming specific behaviors to be instinctive. Without independent documentation, such claims often lead to circular arguments wherein a behavior is assumed to be instinctive and is then considered to be explained by this assumption. Actually, the unsupported claim that a behavior is instinctive amounts to little more than giving the behavior a name. Only when results of comparative and/or ontogenetic investigations are independently available does the claim contain any explanatory significance.

A deprivation experiment can tell us whether or not *certain particular stimuli* are essential for the appearance of the behavior under consideration (but see Bekoff, 1976, for important qualifications). For example, if an individual is reared in an environment containing no conspecifics, and if that individual nonetheless exhibits the target behavior at the appropriate time and under the appropriate circumstances, we can conclude that stimulation arising from conspecifics was irrelevant for the ontogeny of the target behavior. However, other conditions of stimulation that *were* present in the rearing environment may have contributed to the development of the behavior under consideration, and this point must always be kept in mind. Three issues derive from this point. First, experiential effects may occur in extremely subtle ways; important species-specific perceptual and/or behavioral processes can be influenced by experience in ways that do not depend upon the presence of conspecifics. Gottlieb's (1965, 1968, 1971) research on the development of parental-call recognition by ducklings is a good example. Duck embryos hear their own vocalizations and this turns out to be critical because embyos whose vocal cords were incapacitated were unable to recognize the maternal call after hatching. Gottlieb considers that the embryonic vocalizations serve to attune the duckling to particular components of vocalizations that are also present in the maternal call. Second, not all experiential effects involve learning. Gottlieb hesitates to conclude that his ducklings learned anything specific about maternal calls on the basis of listening to their own embryonic vocalizations. Rather, it is more likely that some form of sensitization occurred as a consequence of this experience, and that properly sensitized ducklings are subsequently better able to attend to auditory stimulation arising from their mothers. If we restrict the term learning to clear instances of classical and/or instrumental conditioning, then Gottlieb's findings as well as numerous other developmental findings do not qualify as instances of learning. Nonetheless we have found out something quite significant about the development of a particular adaptive pattern in the identification of a contingency of stimulation that must occur during a particular period for proper responsiveness to the maternal call to become manifest at the biologically appropriate time. The essential idea here is that behavioral development depends upon many kinds of experiential effects besides those that conform to the operational specifications we usually associate with learning. It would be shortsighted of us to single out only the latter and ignore all others. This leads to the third matter, namely, that all behavior has an ontogenetic history, even instinctive behavior. Saying that a behavior does not require learning is not the same as saying that all experience is irrelevant. Accordingly, we can conclude that development is an important dimension of behavior to study even if we know that learning is not involved. Dividing behavior into learned versus instinctive categories does not absolve us of the necessity to analyze development within either category, it merely indicates that certain particular kinds of contingencies of stimulation are restricted to one of the categories (Lehrman, 1953).

Now that we have a better idea of what it means to dichotomize behavior into learned versus instinctive categories, we can recognize that no real headway has yet been made concerning the issue of immediate causation of instinctive behavior. Given that an example of instinctive behavior is observed, we may be in a position to say some meaningful things about the development and the adaptive functions of that behavior, but what are we to say about proximate causation? We know that holding the adaptive functions to be the immediate causes would be inappropriate because we would be making statements involving teleology or retroactive causation. Also, knowing that certain kinds of experiential inputs were necessary at early points during ontogeny tells us little about the reasons why the behavior occurs at particular times or places in later life. To answer questions about immediate causation we need to explore the behavioral physiology of instinct in a manner analogous to the way we explored the behavioral physiology of learning. There we found that stimuli of various sorts act in concert with psychological processes of various sorts to elicit or otherwise influence the occurrence and/or intensity of learned responses. The same kind of inquiry has to be made vis-à-vis instinctive behavior.

Stimulus Control of Instinctive Behavior

The first person to appreciate the logical difficulties arising from attempts to explain immediate causation of instinctive behavior on the basis of the behavior's adaptive consequences was Jacob von Uexküll (1921). Although he, like Tolman, was sometimes accused of having vitalistic leanings, such claims were not only unfair, they tended to obscure the enormous metatheoretical contributions that von Uexküll actually made. He began by pointing out that animals do not respond to or perceive all possible stimulus properties in the environment surrounding them. Moreover, animals do not always respond to all the stimuli that they are, in fact, able to perceive. At this point, we recognize a similarity between von Uexküll and Tolman who also argued for the existence of selective attention in animals.

Then von Uexküll went on to point out that there exist aspects of the environment that bear special significance for organisms. These particular stimuli intereact with the "inner world" of the animal and thereby bring about behavior changes. The reason these stimuli have special significance has little to do with the stimuli themselves but rather depends upon the CNS of the organisms under consideration. The organism has evolved CNS processes that superimpose meaning upon stimuli and, by so doing, allows those stimuli to enter the inner world and influence behavior. Although von Uexküll's views are not usually described as comprising a cognitive theory, the similarities between his concepts and those of Tolman are unmistakable.

The most famous concrete application of von Uexküll's ideas was his analysis of the behavior of ticks. After mating, the female ascends vegetation, there to await the passage of a mammal. When a mammal walks beneath the tick, she looses her hold on the vegetation and drops on the fur of the passerby, eventually to take a blood meal for the purpose of nourishing her eggs. By careful experiments von Uexküll discovered that the tick can be induced to drop from the vegetation by presenting minute quantities of butyric acid, a substance associated with the integument of mammals. No other stimulus produces this behavior. Hence, he concluded that of all physical aspects of mammals, only butyric acid has any significance for the tick and that the significance of this substance is as much a part of the tick as any of her anatomical attributes.

Although the behavior of dropping from vegetation onto a mammal will have the greatest adaptive consequences for the tick, the thing that triggers or activates the behavior is butyric acid. Notice that butyric acid occupies a temporal position quite appropriate to serve as cause for the dropping behavior (unlike the eventual adaptive consequences of that behavior). Hence, by leading us to stimuli such as butyric acid, von Uexküll leads us away from temptation to make assertions involving retroactive causation. To my knowledge, von Uexküll was the first thinker to point out the stimulus control of instinctive behavior and to generate a conceptual framework that did for instinctive behavior what learning theory did for learned behavior. That is, he identified biologically relevant stimuli and psychological processes with which they interact to influence behavior.

Konrad Lorenz was the first modern zoologist to appreciate the metatheoretical accomplishments made by von Uexküll's approach to the analyses of behavior. He saw that it cleansed zoology, so to speak, of teleological and retroactive causational blunders in the same way that learning theory had done for psychology. As Lorenz continued his studies of animal behavior, he realized how general the fact of stimulus control of innate behavior actually was. Also, he realized how important such a general principle would be for guiding the activities of those scientists who wished to understand immediate causation of instinctive behavior. Accordingly, Lorenz abstracted two critical ideas that were embedded within von Uexküll's theory, and he served them up to the scientific community in an exceedingly palatable way. The ideas were: (a) fixed action pattern (FAP); and (b) sign stimulus or releaser. The first concept is essentially a formalization of what had been meant by "instinctive behavior" ever since the term crept into psychological parlance. An FAP is a coordinated sequence of responses cued by enviornmental stimulation but, once initiated, runs off without further dependence on any but taxic stimuli. As such, an FAP is an elemental unit of behavior; it cannot be subdivided into components that depend upon qualitatively different stimuli. Most FAPs are considered to be innate or to involve specialized kinds of learning, depending only on highly canalized contingencies through which all individuals of a species would normally pass. Finally, FAPs evolve through natural selection based on their adaptive consequences. An FAP is an instinct stripped of its teleological and mentalistic accompaniments; it retains only the objective and verifiable referents. In this regard, Lorenz was just as much an operationalist as any of the contemporary American psychologists.

A sign stimulus is that part of the environment that actually elicits the FAP. In the case of the female tick waiting for a blood meal, it was butyric acid arising from a mammal that releases the dropping response; visual, tactile, auditory, and other stimuli arising from the potential host are all irrelevant. In these two cases, as in all FAPs, the eventual adaptive functions of the behavior are irrelevant for their immediate causation. To understand causation, we must search for the sign stimulus from which both the FAP and its adaptive consequences follow. Furthermore, Lorenz (1950) and Tinbergen (1951) developed and popularized a methodology uniquely appropriate for the identification of sign stimuli. This methodology (i.e., the model experiment) has been profitably employed in hundreds of investigations and continues to be the main paradigm used by those concerned with the analysis of stimulus control of adaptive behavior.

At this point, Lorenz's instinct theory appears much like an S-R theory except that the "S" is a sign stimulus whose ability to control behavior derives from perceptual evolution of the responding organism rather than from an arbitrary choice by an experimenter; and the "R" is a FAP. From a purely methodological perspective, this was all Lorenz needed to do. But he went much farther by proposing several concepts to explain how sign stimuli articulated with behavior. The most important of these is the innate releasing mechanism (IRM) which refers to innate psychological processes that (a) recognize sign stimuli when they are present in the environment; and (b) respond by disinhibiting the appropriate FAPs. The IRM is in fact analogous to selective attention and associational processes hypothesized by certain learning theorists. It also has some features that are analogous to the sign-Gestalt-expectations of Tolman.

Comparison of Learning Theory and Instinct Theory

Current workers in animal learning and instinctive behavior have developed their respective theoretical systems considerably beyond where they were in the 1930s and 1940s. Several times, the two lines of inquiry have converged upon common problems (e.g., biological constraints on learning, behavioral embryology, and behavior genetics). Many early theoretical concepts in both fields have been modified or replaced. But, never has the underlying commitment to proper causal constructions been challenged in either field. Both are still very much involved in the analysis of stimulus control and in the specification of psychological processes that mediate it. So basic are these commitments that many workers appear to take them for granted and to behave as if things had always been this way. When we recognize that things have, in fact, only been this way since the first decades of the twentieth century, it is easier to see how very much has actually been accomplished. Moreover, it is also easy to see how much the two fields have in common at the levels of methodology and metatheory. Viewed in this light, the "instinct-learning controversy" was not only an overstatement of the differences between these two theoretical positions, it also obscured certain fundamental commonalities.

Developmental Plasticity in the Perspectives of Learning and Instinct Theory

The task of conceptualizing developmental plasticity is not as easy as it might at first appear to be. For example, one approach might be to imagine that developmental plasticity automatically implies learning or is synonomous with learning. The analysis of developmental plasticity would then reduce to the analysis of learning. However, there are serious problems with such a view.

We must readily acknowledge that some aspects of behavioral variability usually discussed under the heading of developmental plasticity surely represent the operation of ordinary learning mechanisms. Different individuals experiencing different contingencies of reinforcement end up behaving differently. But the existing literatures in ethology and in animal learning suggest that at least two other categories of developmental plasticity exist. First, the operation of learning mechanisms may be constrained by a wide variety of biological processes that restrict the time during which learning can occur (critical or sensitive periods), the stimuli available for conditioning (selective attention processes) or the responses available for conditioning. Such circumstances probably indicate that we are dealing with canalized systems and that evolution has emphasized the reduction of variability by placing constraints of various sorts on the learning process. If our goal is to understand developmental plasticity, we must be interested not only in systems that encourage variability by placing no constraints on learning processes, but we must also be concerned with systems that reduce variability by superimposing constraints. Understanding the contraints may give us insight into the range of plasticity that is possible even in cases where only a small part of that range is expressed under normal conditions.

Another kind of plasticity exists wherein individuals have two or more innate alternative behavior patterns (polyethisms) that satisfy a given ecological function in different ways. Some species may exhibit predatory behavior under certain conditions and carrion feeding under others. Neither pattern has to be learned. The decision about which aspect of the repertory will be engaged in at any particular time is made with reference to prevailing stimulus conditions. If we think of developmental plasticity as the set of processes responsible for major kinds of behavioral variability within and/or between individuals, then polyethisms must be included in our purview. Therefore, we distinguish three levels of developmental plasticity based on the degree to which variability in developmental outcome is constrained: (a) ordinary learning with no special constraint; (b) learning occurring within biologically determined boundary conditions; and (c) nonlearned polyethisms. The framework of contemporary learning theory provides adequate conceptual tools for dealing with the first and most straightforward level of developmental plasticity. However, these tools must be supplemented to accommodate the material on biological constraints. Thus, we must add various concepts from instinct theory to the learning framework to proceed to the second level of developmental plasticity. Finally, instinct theory and sociobiology are necessary if any headway at all is to be made on the third level. This will require a total merger of these disciplines with learning theory.

Learning theory and instinct theory could remain fairly distinct enterprises as long as the respective theorists concern themselves mainly with issues that are more or less traditional for their respective disciplines. Developmental plasticity is, however, an area demanding synthesis because its issues cut across discipline boundaries in fairly obvious ways. Accordingly, we can predict some degree of integration of learning and instinct concepts by developmentalists.

Evolution of Developmental Plasticity: Implications for the Analysis of Immediate Causation

Stenotopic versus Eurytopic Organisms

Some species appear to exhibit behavioral repertoires that contain sets of FAPs, each one nicely adapted to particular ecological conditions. Very little behavioral change occurs as a consequence of experience, and, because of this, the organisms are restricted to environments that (a) contain the appropriate releasing stimuli; and (b) contain the ecological factors to which the FAPs under consideration are adapted. Such animals are called *stenotopic* and usually represent a fairly high degree of species specialization. On the opposite side of the plasticity dimension are *eurytopic* species, which tend to be generalists and which exhibit great capacity for behavioral change throughout life. These are the species most readily approachable within the framework of traditional learning theory, but they constitute a small proportion of the animal kingdom. Most species probably occupy positions somewhere between the opposite extremes of the stenotopic-eurytopic continuum, and exhibit some degree of learning in at least some aspects of their repertoires at some times in their lives. Learning ability of some sort or degree has been demonstrated in virtually all mobile organisms that have so far been studied. Accordingly, the real implication of a species' position along the stenotopic-eurytopic continuum is that it tells us about the degree to which developmental outcomes are canalized and about the range of variability we can expect in behavioral traits.

That natural environments constrain variability and, hence, developmental plasticity, can be most readily seen by a comparison of individuals bred and raised in captivity with conspecifics bred and raised under natural conditions. The latter always show narrower ranges of variability in behavioral and morphological characteristics. When a species is removed from pressures arising from predation, and from intra- and interspecific competition, it is said to experience *ecological release*. This is essentially what happens to zoo animals, and this is what probably happened to *Homo sapiens* (Wilson, 1975). The rapid emergence of variability under such conditions tells us about the potential for learning and other forms of plasticity that had been latent in the species while it was under the influence of selective pressures. Accordingly, the fact of low behavioral variability and a stenotopic appearance of a species under natural conditions should not automatically be taken as evidence for low *potential* developmental plasticity. This point has been treated recently by Chauvin (1977) who showed that individuals exhibit only particular FAPs when they are studied under natural conditions, the conditions for which they are best adapted. But, when the same individuals were tested under unusual environmental circumstances, they exhibited novel forms of behavioral adjustment. Chauvin thinks of this as animal resourcefulness and warns that we may never know the extent of this potential behavioral variability unless we take pains to observe organisms under conditions that pose challenges for them.

These considerations allow us to speculate that under natural conditions, relatively high degrees of developmental plasticity will appear in permissive environments (those that remove certain selective pressures) and in environments that explicitly select multiple phenotypic outcomes (polymorphisms and polyethisms). The opposite kinds of environments will be associated with species showing reduced amounts of variability but not necessarily with species that are unable to exhibit plasticity under more permissive conditions. It must therefore be remembered that stenotopic and eurytopic are functional concepts referring to the degree to which species have been selected for adaptive modifiability of behavior; they are not genetic concepts describing the kinds or amounts of plasticity inherent in developmental systems that underlie behavior patterns.

Learning-Instinct Intercalation

Highly specialized species tend to show high degrees of intraspecific stereotypy in their behavior. Their FAPs tend to be canalized into specific ontogenetic outcomes, thus constraining variability. More generalized species tend to exhibit larger ranges of variability and less stringent canalizations. Often we see behavioral sequences containing some components that are genetically fixed and others that are experientially determined. This is referred to as learning-instinct intercalation, and it is a way for evolution to insure both the presence of some particular adaptive elements and the presence of some amount of variability. It is also a way for evolution to be relatively directive with respect to what is learned. Consider cases where learning is constrained to occur during particular sensitive times (e.g., imprinting). As individuals are usually found within predictable social conditions at such critical periods, the conditions of stimulation available for learning are quite dependable. In this way, evolution can select a developmental system that contains a gap to be filled by experience but yet insures some constancy with respect to the quality of experience likely to be available to the learner. Accordingly, canalization can still be achieved even though learning contributes to the adaptive pattern in question.

Learning does not imply arbitrariness with respect to stimuli and/or responses that are involved. Indeed, this is the major implication of the entire field of biological constraints on learning. Constraints usually act to canalize the learning process into adaptive directions.

In some textbooks on the psychology of learning it has been standard practice to distinguish between two questions: "What is learned?" and "How is it learned?" Learning theory as it was conceived especially in the 1930s and 1940s dealt with the second question and assumed that the same underlying process would be found to mediate all forms or instances of learning. The discovery of biological constraints has prompted a reevaluation of the independence of these two questions. Perhaps there is some correlation between what is learned and how it is learned. Carried to an extreme, this view amounts to a hypothesis that there are as many qualitatively different learning processes as there are instances of biological constraints. However, this view is not a necessary derivative of the demonstration of biological constraints. There is no reason to deny that a limited number of processes underly all instances of learning, provided that we are able to identify interacting factors that can explain the differential operation of these processes under constrained versus nonconstrained conditions.

This view makes the study of developmental plasticity easier than it otherwise would be. It suggests that our job does not necessarily involve the specification of qualitatively different forms of learning each time we discover an ontogenetically canalized system to which experience makes some important contribution. Instead, the task becomes one of identifying how familiar processes are being used in new and seemingly aberrant ways.

Polymorphisms

A species is said to be polymorphic with respect to an anatomical trait whenever individuals can be readily grouped into two or more categories based on their appearance vis-à-vis the trait in question. A frequency distribution describing the variance of a polymorphic trait will usually assume a clear multimodal shape, though the modes need not be of equal height. Modes may correspond to the sexes (in the case of sexually dimorphic species), different age groups, different castes, different status categories, or to morphological categories that are uncorrelated with obvious demographic factors but which continue to exist in some sort of selective equilibrium. In the latter case we are essentially admitting our ignorance concerning the adaptive significance of the polymorphism in question.

Polymorphisms can be of two sorts: obligate and facultative. In the first case, individuals differ from each other, but each individual is constrained to assume only one appearance. No switching is possible. In the case of facultative polymorphism, however, switching is possible. For example, animals may exhibit color and/or brightness changes depending upon alterations in background chromaticity, social circumstances, or age-related effects. Discovery of the functional cue(s) responsible for the switching and/or discovery of the physiological mechanisms mediating the switch often constitute fascinating and theoretically valuable ontogenetic analyses, especially when the trait under investigation is relatively complex. For example, at least nine orders of fish are known to contain species that undergo complete and functional sex reversals. Such sequential hermaphroditism (protogyny = female transforms into male; protandryny = male transforms into female; the former is most common) is a dramatic form of facultative polymorphism. In certain species protogyny appears to be socially controlled such that females will transform into males when formerly present dominant males are removed or killed (Robertson, 1972; Warner, Robertson, & Leigh, 1975). Hence, stimulation normally arising from a dominant male suppresses female transformation, whereas the removal of these cues will permit change within a period of several hours.

Discovery of the ecological circumstances favoring such abilities constitutes an important step in the direction of understanding the evolution of polymorphisms in general and of facultative polymorphisms in particular. In the case of piscine protogyny we often find a breeding system wherein large males enjoy great reproductive advantages relative to smaller males as well as to all females. In the cleaner wrasse (*Labroides dimidiatus*), breeding units usually consist of harems where one male has exclusive fertilization privileges with respect to six females. Under these conditions, the male has six times the reproductive success of each female, and females who are able to transform into males will have an enormous competitive advantage over other females. Although this argument does not explain why sequential hermaphroditism has failed to evolve in higher vertebrates where favorable socioecological conditions also exist, it does provide insight into the probable adaptive significance of this process for those fish which exhibit it.

The foregoing paragraphs indicate that analyses of facultative polymorphisms from ontogenetic and ecological perspectives can enhance our understanding of these complex adaptations. Both kinds of analyses are necessary as an understanding of proximal control (i.e., functional stimuli and/or physiological mediations) sheds little light on adaptive significance and evolution; and, conversely, an understanding of distal causation (selective pressures and the ecological circumstances in which they operate) sheds little light on the proximal mechanisms. Ontogenetic and phylogenetic insights are both necessary in our attempts to piece together a reasonably complete view of adaptation.

Obligate and Facultative Polymorphisms in Evolutionary Perspective

In general, natural selection can be expected to lead to facultative mediation of polymorphic traits, provided that requisite mutations occur and that they are nondeleterious. From the point of view of the individual, it will generally be preferable to be able to assume all morphs and, thereby, to be able to enjoy a wider range of resources. Facultative mediation of polymorphisms clearly allows an individual to exploit more aspects of its habitat than obligate mediation would allow. In times of local shortages or other temporary exigency, facultative morphs will have greater freedom of movement than obligate morphs. Accordingly, individuals capable of assuming several morphs will, in the long run, generally enjoy greater reproductive success than individuals with obligate morphology, and evolution should select facultatively polymorphic strategists over obligate forms (exceptions to this generalization have been reviewed by Hamilton, 1978).

Obligate and Facultative Polyethisms

Just as individuals of a species can be morphologically variable in discontinuous ways, they may be behaviorally variable in equally discontinuous ways. Hence, we can speak of polyethic species whenever we encounter individuals that exhibit categorically different behavioral adaptations. Furthermore, polyethisms may be obligate or facultative, and we can usually expect the latter to predominate whenever the requisite genetic mutations have arisen. Numerous aspects of behavior in various species can be tentatively conceptualized as facultative polymorphisms: dominance status, social roles, social structure, feeding modes in species showing more than one (e.g., predation versus carrion feeding), food choices, mating postures in species showing more than one, and so on. Behavioral ecologists and sociobiologists do very well at modeling such behaviors and describing the conditions that are necessary for the maintenance of polyethic strategies as well as the conditions that will introduce selection for one of the morphs over another. But, we also need to focus on the *conditions of stimulation* that trigger the appearance of ethomorphs. That is, we must have an analysis of the proximal causes as well as an analysis of the distal ones.

Behavioral physiology and ontogenetic research are likely to make major contributions to our understanding of proximate causation of facultative polyethisms, whereas ecological, genetic, and evolutionary analyses are likely to make inroads into distal causation. It is essential to see that both approaches are necessary; neither has the ability to solve the problems addressed by the other. Even though these aspects of behavior are strongly influenced by genetics and natural selection, we recognize that many questions remain regarding their immediate causation. In fact, we can conceptualize the evolution of a facultative polyethism as the selection of an ontogenetic pathway involving "OR gates" that respond to conditions of stimulation present during certain stages of development. The task of the researcher interested in these aspects of developmental plasticity is to identify both the physiological bases of the "OR gate" and the conditions of stimulation that interact with it.

The Confluence of Ideas: Emergence of Interest in Animal Perception

It is probably clear from the previous sections of this chapter that a major approach to the analysis of behavioral variability in general and developmental plasticity in particular depends upon the identification of stimuli that control or otherwise influence adaptive responses. This is necessary whether we are dealing with learned patterns, biologically constrained patterns, or polyethisms. In all cases, proper explanation of immediate causation requires the specification of these stimulus factors and the sensory-perceptual processes that subserve their transduction, their CNS representation, and their eventual impact on behavior. Accordingly, concepts such as selective attention from learning theory and innate releasing mechanisms from instinct theory are becoming increasingly important. Perhaps the time is right to attempt to bring these and other ideas into a common conceptual framework. One such framework is currently emerging, and the following discussion will present the general outlines of the argument.

Although classical conditioning is most frequently studied in laboratory situations with carefully controlled stimuli and temporal parameters, the operation of this phenomena under natural conditions would rarely have the same properties. For example, CSs would never be totally arbitrary and unrelated to USs. Whereas meat powder as the US could be paired with metronomes or buzzers as CSs in the laboratory, CSs in naturally occurring instances of conditioning could always be expected to bear some relationship to USs. Animals eat particular foods in particular conditions and consistent correlations between these define the kinds of stimuli that would ordinarily be made available for conditioning. As nestling birds are fed by their parents, stimulation arising from parent morphology and/or from behavior are consistently paired with the presentation of food. Hence, nestlings would be expected to become conditioned to make anticipatory feeding responses to the relevant parental attributes. This does not mean that the underlying associative mechanism cannot be made available to stimuli other than the ones naturally encountered by nestlings during feeding episodes. Rather, it means that real adaptive significance of associative processes are to be most clearly seen when they are observed under natural conditions. I do not mean to imply that laboratory analyses are unimportant. Indeed, if we want to focus upon the associative processes (e.g., their neurological, temporal, or other properties) rather than upon their specific adaptive significance, highly controlled laboratory situations are clearly the only way this can be done. It is only when we want to understand the contributions the processes make to behavioral adaptation that we must observe naturally occurring instances of conditioning such as has been done by Hailman (1967, 1969).

As we examine natural instances of classical conditioning, a particular feature of the stimulus situations emerges rather clearly. Natural CSs are always situational correlates of USs. Hence, we can conceive of classical conditioning as a process by which organisms are able to abstract such correlations and thereby draw reliably cooccurring stimuli into perceptual relationships with each other. Classical conditioning thus becomes a mechanism by which animals come to understand certain regularities of their worlds in a way that allows them to make predictions at least over modest temporal-spatial distances. This is, in short, a process of abstraction. We assume that having some ability to predict significant stimulus events on the basis of their correlates or causes is of the utmost adaptive value in at least some situations (Lorenz, 1965). Furthermore, we can acknowledge the possibility that associative processes may have evolved to subserve specific kinds of prediction in particular adaptive contexts. Subsequent evolution may have had the effect of making these associative processes available to other behavior systems in other adaptive contexts (Rozin, 1976). In this way, specific associative processes were transformed into general learning ability. This does not mean, however, that an organism cannot exhibit both specific and general associative mechanisms. Just as an instinctive response could be emancipated from its original behavioral-motivational context for use elsewhere and still continue to be used in the original context, a specific associative process could be emancipated from its original adaptive role for use elsewhere and yet continue to be used in its original way. An instinctive response emancipated from its primitive role may continue to bear some of its initial characteristics even in new contexts very different from the primitive one. This may also be true of conditioning and may help to explain some features of conditioning that are at first hard to understand. For example, blocking, masking, and related forms of selective attention might derive from primitive aspects of conditioning that have been transferred along with the associative mechanism from the original context to new ones. Such phenomena may have enhanced the original operation and adaptiveness of associative processes by focusing attention upon certain particular CSs to the exclusion of others. Whereas this may have been quite appropriate in the initial context, it is often inappropriate or even counterproductive in new contexts.

The essential part of the present conceptualization is that conditioning provides organisms with a way to represent stimulus-stimulus relationships and thereby to abstract at least certain aspects of the orderliness of their worlds. Indeed, we can think of conditioning as a basis for the perception of orderliness or lawfulness. In this way, the capacity for classical conditioning becomes a basis for the Kantian a priori category of causation; animals and humans perceive causation because they can do no other, as sensory impressions will automatically be organized in this way by classical conditioning (Lorenz, 1978). The only exceptions to this are those cases where selective attention draws some sensory impressions but not others into a common connection. The ones drawn together are thereby given the appearance of natural cooccurrence whether or not they are in fact connected by natural laws. The sensory impressions left out of the conditioning process by the operation of selective attention are not subsequently perceived as part of a causal or correlational relationship. We must assume that it was adaptive for some but not all potential CSs to be associated with USs, and some findings in the literature clearly suggest that this is so.

It is possible to view innate releasing mechanisms (IRM) in a manner quite analogous to the way we have just viewed classical conditioning. Just as the latter can be seen as a psychological process that constrains the organism to be sensitive to (and eventually to take advantage of) natural stimulus correlations, the IRM can be seen as a psychological process that constrains the organism to respond with adaptive behavior to a stimulus naturally associated with goal objects. That is, the relationship between sign stimuli and goal objects has the same properties as the relationship between CSs and USs. In both cases, the cuing stimuli are reliable predictors of some biologically significant event. Hence, the IRM can be understood as a perceptual process that makes available to organisms relationships between stimuli. These relationships are then put to good use in that antecedent stimuli are used to cue off adaptive behavior patterns that result in goal attainment. Innate releasing mechanisms can be thought of as perceptual adaptions to stimulus-stimulus correlations that have occurred regularly in the history of the species and that contain temporal and/or spatial properties such that responding to the antecedent event enables the organism to benefit by acquisition of the consequent event.

At the very least, this conceptualization allows us to see classical conditioning and IRMs as analogous mechanisms. They clearly serve the same sort of adaptive perceptual functions. This conceptualization also forces us to recognize that a major task facing animal behaviorists of all theoretical persuasions is that of providing an analysis of the perceptual mechanisms involved in both conditioning and innate recognition. We make a stab at the problem by understanding that the underlying adaptive function of both processes is to permit organisms to take advantage of naturally occurring stimulus-stimulus correlations. But, we will not be able to move much farther into the matter until we know more about how animal perception works. Some essential questions can be phrased as follows:

1. How are stimulus events represented not only in terms of sensory transduction but also in terms of higher levels of perception?
2. How are simultaneous stimulus representations brought into contact with each other? (i.e., do they merge into a single representation or do the individual elements retain some aspect of their initial integrity?)
3. How do stimulus representations interact with behavior?

These questions can be equally profitably asked about conditioning and innate recognition. Ethology and animal learning have long concerned themselves mainly with questions about stimuli, on the one hand, and responses, on the other. What is the effective stimulus? How must it be presented to the organism? What is the unit of response? How is it adaptive? Clearly, these are important questions, but they have neglected equally important issues regarding the perceptual processes that engage stimuli with responses.

We might also wonder about the extent to which innate recognition and learning are phylogenetically related. As they presumably have commonalities at the perceptual level, they may also have evolutionary commonalities. Learning and innate recognition may both have arisen from some common primitive representational capacity. Or, it is possible that one of these processes may have evolved from the other. My guess would be that innate recognition may have arisen from learning, rather than the other way around. Innate recognition might be a highly specialized form of representation that could have evolved from conventional forms of learning through the following steps. In the most primitive condition, the adaptive response had to be brought under the control of biologically appropriate stimuli through some form of experience. But, because primitive organisms have limited sensory capacities and detect only a fraction of the stimuli that are potentially available to them, there is not a great likelihood that inappropriate stimulus control would occur. As organisms evolved increased sensory-perceptual capacities, there would be an increased danger of forming inappropriate associations. Accordingly, some mechanisms would be needed to focus attention only upon those cues that are especially relevant to the adaptive response under consideration. Such biological and other constraints on learning are well known and do in fact predispose organisms to allow only certain stimuli to serve as CSs.[4] The next step would be for organisms to evolve an internalized representation of the appropriate releasing stimulus that would be available to guide adaptive behavior in advance of experience.

This does not mean that instinctive behaviors are higher-order adaptations than are learned behaviors, or that learning has not itself evolved considerably since its earliest appearance in primitive invertebrates. Rather, it means that instinctive behavior depends in part on specialized perceptual mechanisms that may have evolved initially in the context of specific

---

[4] I am using the term "biological constraints on learning" in a very broad manner. I mean to include all mechanisms that influence the availability of stimuli and/or responses for conditioning. In some cases, the constraint may be based on the existence of a special associative mechanism. In other cases, the constraint may be derived from attentional processes. In still other cases, the constraint may rest upon "mechanical factors" (e.g., if the US is electric shock, then certain responses and stimuli may be made unavailable for conditioning by virtue of preclusionary or mutually exclusive topographical features of the UR elicited by shock). No generally accepted taxonomy yet exists for biological constraints on learning. Accordingly, I prefer to use the term broadly instead of using it to refer to specialized associative mechanisms.

associative processes. In any case, the enormous variability that is so obvious when we consider learned behavior and the circumscribed variability characteristic of instinctive behavior should not necessarily be taken as prima facie evidence for a fundamental dichotomy. The real concern should be directed toward the underlying perceptual processes that define the plasticity of the respective systems and toward the variables (exteroceptive as well as interoceptive) that interact with and canalize the operation of these processes.

References

Bekoff, M. The social deprivation paradigm: Who's being deprived of what? *Developmental Psychobiology*, 1976, *9*, 499-500.
Carmichael, L. The development of behavior in vertebrates experimentally removed from the influences of external stimulation. *Psychological Review*, 1926, *33*, 51-58.
Carmichael, L. A further study of the development of the behavior of vertebrates experimentally removed from the influence of external stimulation. *Psychological Review*, 1927, *34*, 34-47.
Carmichael, L. A further experimental study of the development of behavior. *Psychological Review*, 1928, *35*, 253-260.
Chauvin, R. *Ethology: The biological study of animal behavior.* New York: International Universities Press, 1977.
Chiszar, D. They can do it in a hundred different ways. *Contemporary Psychology*, 1978, *23*, 654-655.
Eibl-Ebesfeldt, I. *Ethology: The biology of behavior.* New York: Holt, Rinehart & Winston, 1975.
Gottlieb, G. Prenatal auditory sensitivity in chickens and ducks. *Science*, 1965, *147*, 1596-1598.
Gottlieb, G. Prenatal behavior of birds. *Quarterly Review of Biology*, 1968, *43*, 148-174.
Gottlieb, G. *Development of species identification in birds: An inquiry into the prenatal determinants of perception.* Chicago: University of Chicago Press, 1971.
Grohmann, J. Modifikation odor funktionsreifung? *Z. Tierpsychol.*, 1939, *2*, 132-144.
Guthrie, E. R. *The psychology of learning.* New York: Harper & Row, 1935.
Hailman, J. P. The ontogeny of an instinct: The pecking response in chicks of the laughing gull (*Larus atricilla* L.). *Behaviour*, 1967, Supplement *15*, 1-159.
Hailman, J. P. How an instinct is learned. *Scientific American*, 1969, *221*, 98-106.
Hamilton, W. D. Genetic polyethism: Can it ever be best? Paper presented at Danz Symposium, Animal Behavior Society, Seattle, 1978.
Harlow, H. F., Blomquist, A. F., Thompson, C. I., Schiltz, K. A., & Harlow, M. K. Effects of induction age and size of frontal lobe lesions on learning in rhesus monkeys. In R. L. Isaacson (Ed.), *The neuropsychology of development: A symposium.* New York: Wiley, 1968.
Hill, W. F. *Learning: A survey of psychological interpretations.* Toronto: Chandler, 1971.
Hinde, R. A., & Stevenson-Hinde, J. (Eds.). *Constraints on learning: Limitations and predispositions.* New York: Academic Press, 1973.
Hull, C. L. *Principles of behavior.* Englewood Cliffs, N.J.: Prentice-Hall, 1943.
Hull, C. L. *A behavior system: An introduction to behavior theory concerning the individual organism.* New Haven, Conn.: Yale University Press, 1952.

Lehrman, D. S. A critique of Konrad Lorenz's theory of instinctive behavior. *Quarterly Review of Biology*, 1953, *28*, 337-363.

Lorenz, K. Z. Uber die Bildung des Instinktbegriffes. *Naturwissenschaften*, 1937, *25*, 289-300, 307-318, 324-331.

Lorenz, K. Z. The comparative method in studying innate behaviour patterns. *Symposium of the Society for Experimental Biology*, 1950, *4*, 221-268.

Lorenz, K. Z. *Evolution and modification of behavior.* Chicago: Chicago University Press, 1965.

Lorenz, K. Z. *Behind the mirror: A search for a natural history of human knowledge.* New York: Harcourt Brace Jovanovich, 1978.

Pavlov, I. P. *Conditioned reflexes: An investigation of the activity of the cerebral cortex.* London: Oxford University Press, 1927.

Reimarus, H. S. *Allgemeine Betrachtungen über triebe der tiere hauptsächlich über ihre Kunsttriebe.* Hamburg: 1762.

Robertson, D. R. Social control of sex reversal in a coral-reef fish. *Science*, 1972, *177*, 1007-1009.

Rozin, P. The evolution of intelligence and access to the cognitive unconscious. In E. Stellar & J. M. Sprague (Eds.), *Progress in psychobiology and physiological psychology* (Vol. 6). New York: Academic Press, 1976.

Seligman, M. E. P., & Hager, J. L. (Eds.). *Biological boundaries of learning.* New York: Appleton-Century-Crofts, 1972.

Skinner, B. F. *The behavior of organisms: An experimental analysis.* Englewood Cliffs, N.J.: Prentice-Hall, 1938.

Skinner, B. F. Are theories of learning necessary? *Psychological Review*, 1950, *57*, 193-216.

Spalding, D. A. Instinct, with original observations on young animals. *Macmillan's Magazine*, 1873, *27*, 283-293. (Reprinted in *British Journal of Animal Behaviour*, 1954, *2*, 2-11.)

Thorndike, E. L. Animal intelligence: An experimental study of the associative processes in animals. *Psychological Review, Monograph Suppliment*, 1898, *2*, No. 8.

Thorndike, E. L. *The psychology of learning.* New York: Teachers College, 1913.

Tinbergen, N. *The study of instinct.* Oxford: Oxford University Press, 1951.

Tolman, E. C. *Purposive behavior in animals and men.* New York: Naiburg, 1932.

Tolman, E. C. The determiners of behavior at a choice point. *Psychological Review*, 1938, *45*, 1-41.

von Uexküll, J. *Unwelt and Innevwelt der Tiere.* Berlin: 1921.

Warner, R. R., Robertson, D. R., & Leigh, E. G., Jr. Sex change and sexual selection. *Science*, 1975, *190*, 633-638.

Watson, J. B. *Behaviorism.* New York: Horton, 1925.

Wilson, E. O. *Sociobiology.* Cambridge, Mass.: Belknap Press, 1975.

4

The Infancy of Human Learning Processes[1]

LEWIS P. LIPSITT
JOHN S. WERNER

Introduction

From the beginning of scientific observation on the development of children's learning processes, it has been the first year of life that has held the greatest fascination. Most studies dealt initially with the question of the earliest ages at which certain learned responses could occur. As a consequence, even studies of *learned* behavior had a heavily maturationist emphasis (Lipsitt, 1963). For example, if an attempt failed to demonstrate in month-old babies a specific kind of learning, such as classical conditioning of salivation in the presence of a tone as the conditioning stimulus, the inference was drawn that cortical innervation was not sufficiently mature to allow acquisition of the response. Indeed, many of the Soviet studies (see, for example, Elkonin, 1957) were carried out with the intent of using the assessed learning capacities of different aged infants and young children as indirect indices of nervous system maturation.

Most studies of learning in human infants have been carried out in the past 20 years. Thus, there has been an upsurge of available information about the ways in which environmental stimulation or experiential factors alter the course of the baby's development and behavior. The effects of environmental parameters depend first, of course, upon the sensory and perceptual receptivity of the organism. Werner and Lipsitt, Chapter 2, this volume, have

---

[1] This chapter, completed while the first author was a Fellow at the Center for Advanced study in the Behavioral Sciences at Stanford, draws with permission on portions of an article by L. P. Lipsitt, The study of sensory and learning processes of the newborn, in J. Volpe (Ed.), *Clinics in Perinatology*, Vol. 4, No. 1, Philadelphia: W. B. Saunders, 1977.

reviewed the infant's sensory capabilities. These provide the initial constraints upon and opportunities for experiential influences.

In this chapter, we explore the behavioral changes that take place when the senses are stimulated in systematic ways during development. A major thesis of this presentation is that the pleasures and annoyances inherent in sensory stimulation help to determine whether there will be lasting influences of that experience (Young, 1936). With this acknowledgment of the important role of incentives in the modification and perpetuation of behavior, the hedonic basis of infant learning is emphasized.

Hedonic features of the infant's response repertoire have an honorable tradition. Darwin (1877) was so captivated by the intrinsic fascination of his son's face, and particularly the baby's hedonic expressions, that he wrote down his observations and made an evolutionary case for their adaptive significance. Similarly, Freud's (1927) attributions of psychosexuality in infancy, and the tremendous importance he placed upon the manifestation of sexuality in the first days and months of life, have been sufficiently celebrated to need no elaboration. Suffice it to say that Freud's "pleasure principle" as the impetus to behavior is central to his theory of personal development and socialization. Leave behind for now that the pleasure principle is also the instigator of social restraints that every society imposes and every child experiences; the fact is that the pleasure principle is a front-and-center proposition in Freudian theory. Moreover, if the first year of life is not traversed successfully, the inner psychic life and the overt behavior of the person may be deleteriously affected throughout the life span.

Freud's assumptions about the importance of childhood experience in determining later motivational and behavioral conditions were matched by those of Piaget (1932, 1952). Piaget was another architect of developmental theory with "a grand design" (Kessen, 1965). It is a basic tenet of Piaget's theory that genetically guided early sensory-motor experience is crucial in laying the groundwork for *all* later logical thought processes and intelligence. Moreover, if the child does not enter successfully into and emerge from the progressive stages of development, subsequent cognitive processes presumably will be impaired. Whereas Piaget's theory cannot be characterized easily as an incentive or reinforcement theory, the pleasures of successful experiences are certainly not overlooked. Piaget (1952) vividly described the surprise and gleeful reactions that children in the sensorimotor period manifest when they seem to grasp the way in which things work. Violations of accustomed events similarly result in the child then paying closer attention and being rewarded, presumably, by the novelty of the situation. In this latter regard, Kagan's discrepancy hypothesis (Lipsitt & Eimas, 1972, pp. 1-3) involves similar mechanisms. There is also considerable discussion by Piaget about the importance of the search for solutions and the mastery of problems.

The behavioral orientation to the understanding of child development has also emphasized early experiences as the building blocks from which later response capabilities and coping capacities would emerge. The hedonic basis of the behavioristic theory of development, as eventually enunciated by Watson (1928), was clearly foretold in the parsimonious characterization of Thorndike (1913), that those behaviors that are followed by a satisfying state of affairs will tend to be repeated in the future, whereas those followed by unsatisfying states would not.

For Watson, life began with three major response types or physiological reflexes (fear, rage, and love), which, through pairing with initially neutral stimuli, could come to be manifested on a conditioned or learned basis. Fear and rage were unconditioned aversive responses, and love was a congenital or unconditioned response of pleasure. This dichotomous characterization of responses, involving approach and avoidance behaviors, remains today as a basic classification scheme, although, to be sure, there have been numerous elaborations of complexities, and documentations of the simultaneous activation of diverse combinations of approach and avoidance tendencies in single situations (Lewin, 1935).

Let us celebrate the rapidity of human development in the first year of life. Growth rate is faster in the first 6 months than it will ever be again. The human baby doubles in weight during the first 3 postnatal months, and triples within the first year. From birth to 6 months, babies increase in weight about 2 gm every 24 hr; from 6 months to 3 years, the increase averages about .35 gm; from 3 to 6 years, the increase is .15 gm/24 hr. The concomitant behavioral changes are very important. The onset of smiling at about 2 months, the clear recognition of significant other persons by 6 months, the obvious attachment to some persons by 8 months, the taking of the first steps and the utterance of meaningful words by the end of the first year, are all exceptionally rewarding events to those caring for the infant. Achievements by the infant signal that "all is well." It is not surprising, therefore, that the earliest fascinations of child developmentalists were with milestone achievements. Anthropometrists charted the somatic growth of the infant from precise measurement with tapes and calipers. Child psychometrists recorded behavior in response to specific test items, to assure that the young child's mental age was keeping pace with chronological age. Individual differences were important; the needy, unusual child had to be detected to be accorded special treatment.

A natural progression exists from the study of developmental *status*, derived as a developmental quotient based upon the ratio of mental and chronological age, to developmental *progress*. The shift is from description to an understanding of process. There was an early emphasis in the field of child development on constitutional determinants of behavior. This emphasis, celebrated so successfully by Gesell and his colleagues (Gesell, 1948; Gesell & Ilg, 1946), has been only gradually supplanted by a more tempered view of the mutually dependent relationship between brain and behavior. The older view, generally, was that behavior waits upon the development of neural tissue. Little attention was given to the radical view, accepted more easily today, that nervous system tissue change is sometimes *attributable* to environmental stimulation. Recent findings show that this is so. Rats reared in enriched environments, for example, have greater depth of cortex and more dendritic spines than deprived control rats (Rosenzweig, 1971; Rosenzweig & Bennett, 1976). Similarly, the dendritic branches and spines of the neurons in fish have been shown to be influenced appreciably in their development by social stimulation, specifically, visual-tactile contact with conspecifics (Coss & Globus, 1978). Although analogous experiments from human behavior cannot be conducted, evidence exists that some features of sensory-neural mechanisms do require experience for their full development, and that brain growth sometimes *depends upon* behavioral development (Gottlieb, 1971; Purpura, 1974, 1975). It is thus of great eventual interest to know how the "developmental schedules" for neuronal, gross anatomical, and behavioral maturation communicate with one another. An important step toward that goal involves documentation of the ways in which experiential factors affect behavioral outputs, learning processes, and indeed, individual differences in maturation itself.

The Pleasures of Sensation as Incentives for Infant Learning

The general principle was enunciated by earlier philosophers, but it was Thorndike (1913) who formalized the law of effect for psychologists: Those behaviors that are followed by a satisfying state of affairs will tend to be repeated in the future, and those that are followed by no satisfaction, or by punishment, will not (Kling & Riggs, 1971). This satisfaction or "pleasure criterion" underlying the acquisition and extinction of behavior has been honored in learning theory through its emphasis upon the concept of reinforcement and incentive-motivation variables (Kimble, 1961). The "pleasures of sensation," to borrow Pfaffmann's (1960) phrase, are the building blocks of Thorndike's satisfying states of affairs. They provide the incentives that help to perpetuate instrumental behavior, and often determine the style of that behavior once it has begun. We will shortly see that this is so even for

4. THE INFANCY OF HUMAN LEARNING PROCESSES

newborns. Recent research on the effects of the sweet taste on the sucking behavior and autonomic nervous system functioning of newborns provides insights into the mechanisms by which these "pleasures of sensation" can come to serve as reinforcing events for learning.

Neonatal Sucking Behavior Is Largely a Matter of Taste

Recent studies have generated data about the approach and avoidance style of the newborn, the individual differences among newborns in such styles, and the reactions of babies to stimuli that adults would regard as pleasant and unpleasant (Steiner, 1979; Lipsitt, 1976). Numerous studies now demand the inexorable conclusion that the baby is keenly sensitive in the first few hours of life to subtle changes in gustatory stimulation, and that pronounced preferences exist for sweeter fluids (Crook, 1979; Lipsitt, 1977). The baby acts on its discrimination of these taste changes either to perpetuate the taste or to remove it, depending largely upon the amount and sweetness of the fluid delivered contingent on sucks.

The newborn is a hedonic creature, responding to the incentive-motivational properties of reinforcers with motor changes in behavior, such as in sucking and swallowing, and autonomic changes. The autonomic accompaniments of pleasant and unpleasant stimulation, such as heart-rate changes, are, of course, the rudiments of affect. There is no mistaking the most avid manifestations of such affect, as when the infant goes quickly quiet when offered a sweet fluid.

Many neonatal studies have been carried out at Women and Infants Hospital of Rhode Island (Lipsitt, 1977). The babies are placed in a special crib, housed in a white, sound-attenuated chamber with temperature about 80° F. Breathing is monitored by a Phipps and Bird infant pneumobelt around the abdomen, and respiration and body activity are recorded continuously on a polygraph. Electrodes are placed on the chest and one leg, permitting polygraphic monitoring of the primary heart rate, which is then integrated by a cardiotachometer and recorded on another channel.

Sucking is recorded on one of the polygraph channels using a "suckometer," which consists of a stainless steel housing with a pressure transducer, over which a commercial nipple is pulled. A polyethylene tube runs into the nipple from a pump source and delivers fluid under the experimenter's control and, in most of our studies, on demand of the subject. When delivering, the pump ejects into the nipple end a tiny drop of fluid, usually .02 ml, contingent upon the execution of a sucking response of preset amplitude. When the effect of the magnitude of the drop is under study, the drop amount may be varied from .01 to .04 ml. (See Figures 4.1 and 4.2 for the response-recording arrangements.)

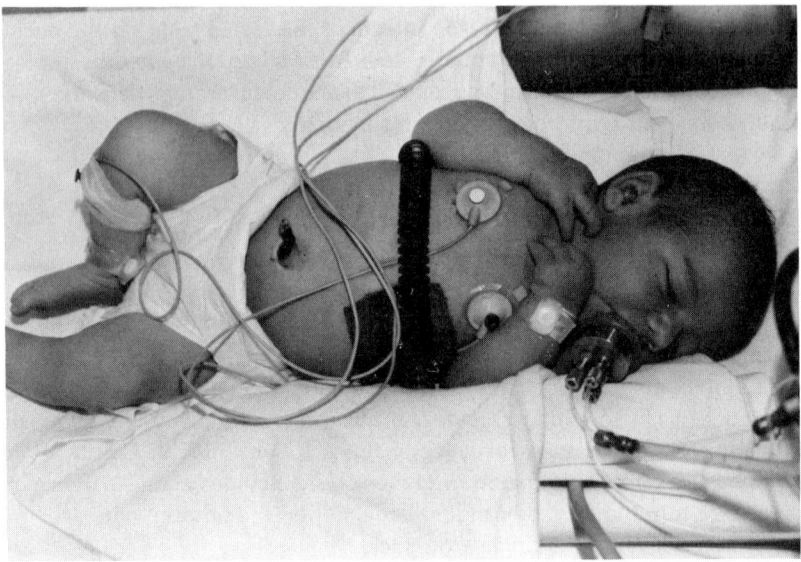

Figure 4.1. *Newborn infant prepared for simultaneous recording of respiration, heart rate, and sucking.*

The situation is arranged so that the infant may receive no fluid for sucking, or might receive a fluid such as sucrose or dextrose in any desired concentration. Contingent upon sucking, one drop of fluid is ejected into the baby's mouth for each criterion suck. A polygraph event marker records fluid deliveries, or the occurrence of a criterion suck during no-fluid conditions. A 74 dB background white noise assures a fairly constant acoustical environment in the infant chamber.

Newborns characteristically suck in bursts of responses separated by rests. Burst length and rest length both constitute individual difference variables under no-fluid conditions (i.e., some newborns engage in reliably longer bursts and pauses than others). Both of these parameters, as well as the sucking rate within bursts, however, are significantly influenced by the conditions that are prearranged to occur contingent upon the infant's behavior. With a change from a no-fluid condition to a fluid-sucking condition, or from sucking for a less-sweet solution to a sweeter solution, several behavioral consequences characteristically occur. There is a tendency for the sucking bursts to become longer, for the interburst intervals to become shorter, and for the intersuck intervals to expand. Because sucking rate within bursts becomes slower with increasing sweetness of the fluid, simultaneously with the infant taking fewer and shorter rest periods, more responses are typically emitted per minute for sweeter (hedonically more positive) fluids. It may be

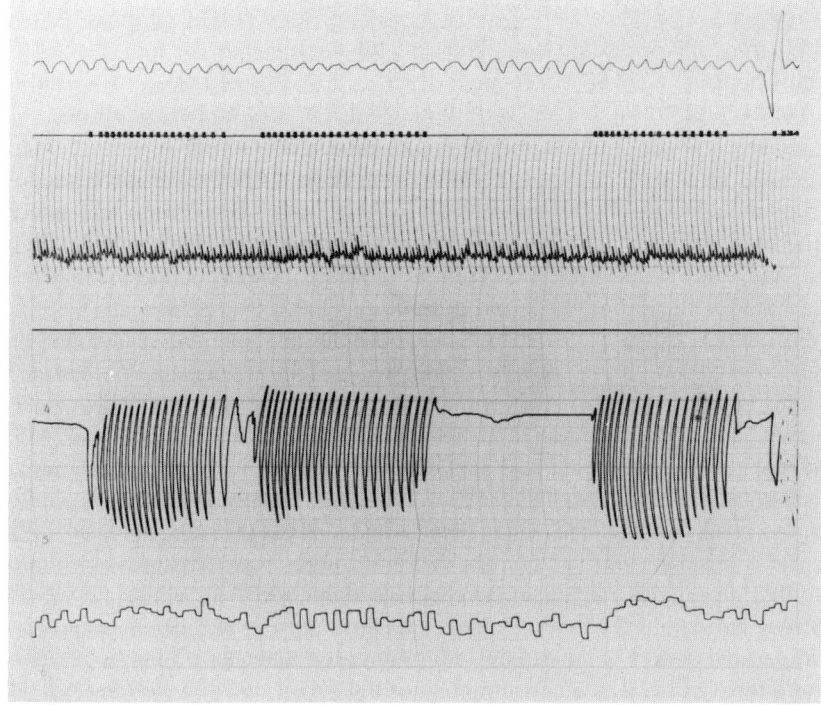

Figure 4.2. Polygraphic record from a newborn infant. From top to bottom, the successive records present: (1) respiration; (2) digital sucking response; (3) electrocardiograph; (4) basic sucking (analogic); and (5) cardiotachometer.

added that most of these effects are compromised in high-risk infants; the greater the severity of neonatal problems requiring intensive care, the less will be the effect of incentive shift as from 5-15% sucrose (Cowett, Lipsitt, & Vohr, 1978).

CONTRAST EFFECTS

The aforementioned regularities of response in relation to the prevailing incentive-reinforcement conditions during testing make it possible to explore the effects of previous experiences upon the infant's response during a subsequent taste experience. Similarly, we can investigate the interrelation among these various sucking-response parameters and, in turn, their relations to certain other response measures, such as heart rate.

Kobre and Lipsitt (1972) tested infants for 2 min on the nipple without any fluid delivery whatever. Only "good suckers" were retained in this study (i.e., no subject had a mean sucking rate lower than 30 per min during the 2-min test period). The 25 remaining subjects were divided into five groups. A total of 20 min of responding was recorded for each subject in four successive 5-min periods. Between each period, the nipple was removed for 1 min to allow the tube to be flushed with water and the child to be picked up.

These 25 infants, all about 3 days old, received one of five reinforcement regimens for the 20 min period. One group received only sucrose, a second group received water throughout, and a third received sucrose and water, alternated twice, in 5-min units. A fourth group received no fluid throughout the four 5-min periods and was compared with a group that received sucrose alternated with no fluid in 5-min periods.

Frequency polygons of the interresponse times, or intersuck intervals, were printed out on the console of an on-line PDP-8 computer. Comparison of the first three groups revealed that sucking rate within bursts slows down for a fluid-sucking condition relative to a no-fluid-sucking one, and that sucking rate within bursts becomes slower still for sweet fluid-sucking relative to sucking for plain distilled water. Thus, there is an orderly progression from no fluid to plain water to 15% sucrose sucking, with sucking response becoming slower and slower as the incentive value of the reinforcement delivered consequent upon the response increases (see Figure 4.3). Also, under the sucrose condition, the infants invested a larger number of responses during a comparable period of time than under either the water or no-fluid condition. The effect, which was a consequence of the infant taking fewer rest periods under the higher incentive condition, also occurred in the comparison of responses emitted for water compared with no fluid.

The infants sucking for sucrose throughout the 20-min testing period emitted significantly more responses per min than the groups receiving water throughout. Moreover, both groups showed stable response rates for their respective fluids through the four 5 min blocks. The most interesting finding in this experiment, however, concerned subjects who were alternated from one 5-min period to another between sucrose and water, or between sucrose and no fluid. These groups showed marked effects attributable to the alternating experience. For example, when sucking for sucrose, the sucrose and water group was essentially comparable to the group sucking for sucrose throughout. When switched to water, however, response rate during each of those 5-min periods was significantly lower than in the counterpart controls in the water-throughout group. Thus, when newborns have experience in sucking for sucrose, an immediately subsequent experience with water "turns them off." They display their apparent "aversion" for the water by a marked reduction in instrumental behavior that would put that fluid in their mouths. When the consequence of the response is changed, as from water to sucrose, response rate goes back to a normal level. The infant thus optimizes taste incentive experiences by modulating oral behaviors pertinent to their

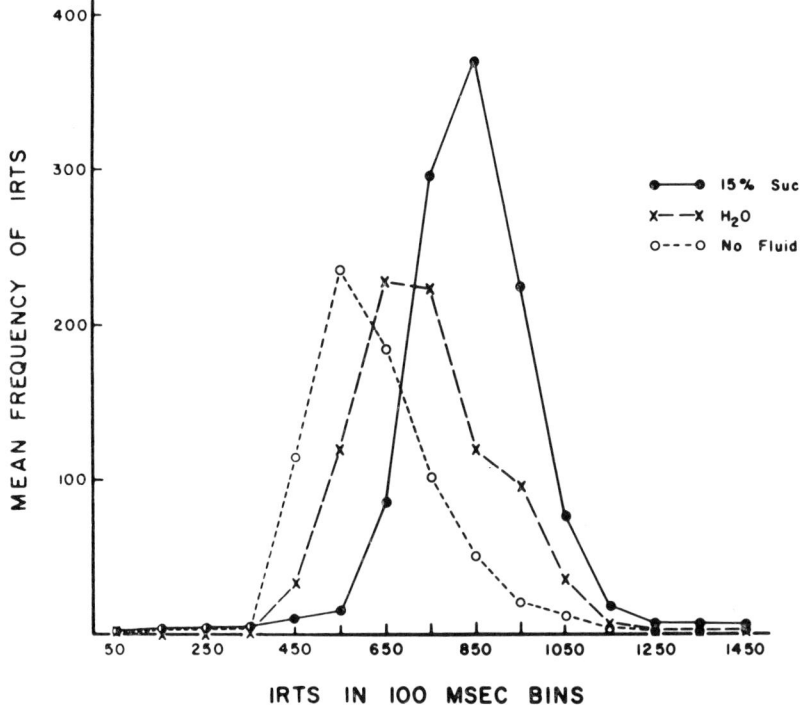

Figure 4.3. *Frequency polygon of interresponse times (IRTs) in 100 msec bins for three sucking reinforcement conditions: no-fluid, water, and 15% sucrose. Each IRT distribution is based on data from five infants, each tested for one 20-min session (From L. P. Lipsitt, The synchrony of respiration, heart rate, and sucking behavior in the newborn, Biologic and Clinical Aspects of Brain Development from Mead Johnson Symposium on Perinatal and Developmental Medicine, 1975, No. 6, pp. 67-72. Reprinted by permission.)*

occurrence. The same type of effect occurred in the sucrose and no-fluid group, which showed lower response rates when sucking for no fluid after experience in sucking for sucrose.

These negative contrast effects were reliable, and there is no reason to suppose that the phenomenon is not widespread throughout the range of incentive conditions to which neonates would be normally subjected. We would expect such effects to occur whenever the infant is called upon to "compare and contrast" two levels of incentive, such as formula versus plain water, or breast milk versus a sweeter formula.

Newborns, then, seem strikingly affected in their subsequent behavior by experiences within the immediately previous 5 min. The negative contrast effect demonstrated here is one of the most rudimentary types of behavioral alteration due to experiential circumstances. As with neonatal habituation to olfactory stimulation (Engen & Lipsitt, 1965), the suggestion is that memorial processes are already working in the newborn, such that there is a lasting impression made, admittedly of unknown duration, of the experience endured. These are the beginnings of learning processes.

RELATIONS BETWEEN SUCKING BEHAVIOR AND HEART RATE

We now know that at least some aspects of the motor behavior of the newborn change according to the incentive conditions to which the baby is exposed. Several studies have reinforced this impression, and further provide data on the rudiments of affect in the neonate in the form of autonomic changes dependent on incentive conditions.

A study of 44 normal full-term newborns, 24 males and 20 females, was conducted on 2 consecutive days using the polygraphic techniques described earlier (Lipsitt, Reilly, Butcher, & Greenwood, 1976). On the first day of testing, the mean age of the infants was 54 hours and, on the second day, 78 hours. Eleven of the infants were breast fed and the remainder bottle fed.

Immediately following calibration of the apparatus, a period of 10 min of sucking was recorded for each infant in five successive periods, each of 2-min duration. Three of these periods were spent sucking for no fluid, followed by two periods of 15% sucrose sucking. About 35 sec intervened between periods, during which time a computer printed out the interresponse time data (IRT) for the preceding period. The nipple was not removed between periods, and the infant continued sucking under the same condition as in the preceding period. The beginning of a period, following the 35-sec printout, was initiated after the infant stopped sucking for at least 2 sec and after the end of a burst. Following the second sucrose period, the nipple was removed. A 2-min period of polygraph recording then ensued during a "resting" state, defined as quiescent and with regular respiration, in which the infant neither sucked nor was stimulated in any way.

The data further substantiate our supposition that a hedonic or "savoring" mechanism is operative in the earliest days or even minutes of life. The findings with respect to sucking behavior in this study essentially replicated those of the previously reported study in showing that under no-fluid sucking, significantly more rest periods (defined as IRTs greater than 2 sec) occurred than for 15% sucrose sucking. Fewer responses per burst occurred for no-fluid sucking, moreover, and both the modal and mean IRT for no-fluid sucking were reliably shorter than for sucrose sucking. In addition, more responses per minute were emitted for sucrose than under the no-fluid condition. However, a very curious interplay occurred between sucking behavior and heart rate. It was observed that with increases in sweetness of the fluid, and even as sucking rate within bursts diminished as is typical for sweeter fluids, the heart rate increased. This was so even in the first burst after the switch to the sweeter fluid, and it took place within only a few sucks.

The seemingly paradoxical increase in heart rates during sucrose-sucking conditions, where the sucking rate within bursts was slower, is of considerable interest. This effect, like all of the sucking parameter effects, occurred on both the first and second days of the study. During basal recording, the heart-rate mean was approximately 116 beats per minute. When sucking for no fluid, the mean rate rose to 124, and when sucking for sucrose, the mean rate rose further to 147. Thus, although sucking rate within bursts was reduced when the infant sucked for sucrose, heart rate nevertheless increased reliably. Moreover, correlations from the first to second day of testing indicated that heart rate is a stable individual difference variable under all three conditions. Interestingly, the correlation coefficient rose from .29 to .46 to .71 in going from basal to no-fluid to sucrose sucking, respectively. An incidental suggestion from this finding, then, is that, as an individual difference parameter, heart rate might have greater utility when measured under a high-incentive sucking condition.

One interpretation of the increased heart rate during the sucrose period relative to the water-sucking period is that there is greater energy expenditure during the high-incentive condition, and the higher heart rates are secondary to this energizing phenomenon. This would seem compatible with the observation that while sucking under the higher incentive condition, more sucks per minute are emitted and fewer rest periods occur even though sucking rate is slower within bursts. Inspection of the polygraph records from the Lipsitt et al. (1976) study indicate quite clearly, however, that the enhanced heart rate during sucrose-sucking bursts, relative to no-fluid bursts, cannot be attributed to a generalized increase in heart rate over the entire period of sucrose sucking. The enhanced heart rate is almost always seen within a few sucks of the switch from no fluid to sucrose or from water to sucrose. It takes only a few seconds or a few sucks for the effect of the sweet taste on the tongue to be reflected in the higher heart rate. Moreover, the possibility that the heart rate effect is an artifact of differential sucking amplitudes under the different incentive conditions has also been ruled out (Ashmead, Reilly, & Lipsitt, 1980).

These observations were substantiated in a subsequent study by Crook and Lipsitt (1976). They showed that the enhanced heart-rate effect under sweet-sucking conditions occurs even when length of the sucking burst is controlled and when heart rates are measured only during actual sucking and not during interburst intervals. A detailed analysis in the Crook and Lipsitt study of heart rate and sucking was made possible by tape recording each interbeat interval and each intersuck interval for subsequent computer processing. Because heart rate accelerates to a stable level at the start of a sucking burst, this period of acceleration was excluded from analysis. Only bursts of 12 or more sucks were considered, and, for any such burst, the heart rate within it was taken as the mean rate between the eighth and final suck. This method of analysis thus concentrates upon the asymptotic heart-rate level under differential incentive conditions.

Half of the 22 full-term newborns in the Crook and Lipsitt study sucked for 9 min in three blocks of 3 min, first receiving a .02 ml drop of 5% sucrose for each criterion suck, then no fluid contingent upon such sucks, and finally a .02 ml drop of 15% sucrose for each suck. The other half received these conditions in reverse order. Regardless of the order in which the two nutrient conditions were administered, intersuck intervals were longer under the sweeter condition, but heart rate was also higher.

Crook (1976) also documented the effects of quantity of the response-contingent fluid upon sucking rhythm and heart rate to complement the extensive data now available on sweetness. The temporal organization of neonatal nutritive sucking and heart rate was studied in two consecutive 4-min periods to analyze the effects of two quantities of response-contingent fluid.

In this study of 53 full-term infants with uncomplicated delivery (23 males and 30 females between 48 and 72 hours old), one group experienced only the larger amount (.03 ml per suck), a second experienced the smaller (.01 ml per suck), and two other groups experienced both in counterbalanced order. Crook found that cumulative pausing time and intersuck intervals were both affected by the amount of fluid delivered at each response, just as with variations in sweetness. At the start of sucking bursts, heart rate accelerated to a stable level, and within-burst heart rates were higher with increased quantity of contingently delivered fluid.

It is apparent that sweetness and amount of fluid operate upon the baby in comparable ways. Sweetness and amount of fluid are essentially collapsible incentive-motivational variables. It seems reasonable to infer that in these situations it is the "pleasures of sensation" (Pfaffmann, 1960) that are controlling certain features of the newborn's motor behavior and autonomic nervous system processes.

Imitative Behavior of Infants

It is not unlikely that some pleasures of sensation are involved even in imitation behavior observed in babies. It has been observed for a long time, certainly before there were disciplines of psychology, pediatrics, or child development, that infants in the first year of life tend to "ape" the behavior of others. Adults interacting with infants often include games involving imitation. The disruption of a series of imitative acts to which the infant has become entrained often elicits surprise. Examples are the games of peek-a-boo and patty-cake, to which infants customarily respond with smiles, excited vocalizations, and other manifestations of glee.

Although such imitative behaviors are expected toward the end of the first year of life, and are even presumed by theorists as unlikely to occur before that time (Piaget, 1952), several interesting studies have demonstrated recently that very young infants can engage in rudimentary imitation, at least in response to a visual model (Maratos, 1973; Meltzoff & Moore, 1977). Infants even as young as 2-3 weeks of age have responded with enhanced tongue movements when another person, in reasonable visual range and after gaining the infant's attention, sticks his or her tongue out. Although there has been disagreement as to the requisite stimulation for the elicitation of such responsive behavior with some investigators (Jacobson & Kagan, 1979) believing that it is not "true imitation," there is little question that in pictures taken of such interchanges between the adult and the infant, the infant seems to be deriving pleasure from the experience.

In one of the pioneer experiments, in which controls were introduced to help assure that the "imitative behavior" was not merely the result of enhanced arousal, Maratos (1973) simultaneously recorded a number of different responses in infants 7-8 weeks of age, while serving as a model performing only one of these responses at a time (e.g., thrusting of the tongue, movement of the fingers, vocalization). When the model engaged in tongue thrusting, the infant's tongue movement was increased, and when the model engaged in finger waving, the infant's finger movement was enhanced.

On the basis of recent replications of the Maratos study, under different conditions and with even younger infants, it has been asserted that such imitative behavior is present by 3 weeks of age (Meltzoff & Moore, 1977) or even by 10 days of age (Bower, 1977). Meltzoff (1977) has reported that babies as young as 21 days of age can hold in memory the imitated model for at least 2.5 min. He first demonstrated tongue thrusting to the infants while they were sucking on a pacifier and presumably were thus inhibited from engaging in the imitative tongue thrusting. After an interval of 150 sec following the demonstration, the pacifier was then removed from the infant's mouth, whereupon a significant amount of tongue thrusting was seen to occur.

It is apparent from these studies that infants do engage in reciprocating interactions with other persons, and with objects, from a very young age. Our best information is that very young children capitalize in their behavior on basic pleasures of sensation, as when they savor sweet substances. Moreover, they seek to perpetuate tasks that yield satisfaction as indexed by signs of glee or surprise.

The Infant as Learner

Some Historical Considerations

In the past 2 decades, a burgeoning interest has arisen in the learning capabilities of the human infant. This is perhaps due to an increasing awareness that the peripheral sensory pathways are already functional at birth, and that the cerebral cortex is instrumental, even at birth, in mediating between stimulus inputs and response consequences (Peiper, 1963). Thus it became possible, in principle, for experiences of one moment to affect behavior in subsequent moments. Still, a pervasive pessimism about *infant* learning seemed to hang on from an earlier period, even in the midst of a rising enthusiasm for conditioning techniques in general. After all, one of the pioneers in the use of conditioning procedures with young children, Krasnogorsky, had said in 1913:

> In the normal newborn infant, the cortical innervations are developed to such an insignificant extent that conditioned connections cannot yet be found. In the second half of the first year, the formation of conditioned reflexes . . . is possible, but takes place more slowly than at a later age [from Elkonin, 1957, p. 50].

Fuller treatment of historical factors in the infant conditioning field is found in Lipsitt (1963), and in Horowitz (1969), and in Hulsebus (1973). It is probably fair to say that the older view that cerebral innervations are developed to an insufficient extent to permit complex behavioral changes with cumulative experience is obsolete, but it has to be acknowledged that there is some debate about the specific mechanisms involved, and about the appropriate terminology for those processes (Sameroff, 1972).

One can speculate that, to an extent, the earlier views about the incompetency of infants were fueled by the institutional ritualization of childbirth over the past 100 years. In the late nineteenth century, childbirth was removed from the home bedroom, attended mostly by friends and relatives, into the hospital, where medical personnel and a rapidly developing technology took over (Wertz & Wertz, 1979). The attendants at the birth were often persons who had only a remote, occupational connection to the mother and infant. They were cloaked in protective clothing and masks. The father was removed from the delivery room, and often from the mother and baby for many hours after the delivery. The mother was emotionally separated by anesthesia and analgesia, and she was not permitted to handle her baby immediately after the trials of birth. Both mother and baby were regarded as patients, and hence as sick. The professionalization of the birth process and the associated depersonalization of the early moments of the infant's life were reinforced by the lethargy that high dosages of maternal drugs can produce in babies. This contributed to an essential dehumanization of the neonate; denial of humanoid characteristics meant the exclusion, of course, of the ability to learn and to remember. These were the very characteristics that have been, until recent years, overlooked in the newborn.

It may be mentioned in passing that with the new awareness that normal neonates are capable creatures with all sensory channels functioning (although certainly not at a mature level), increasing attention is currently paid to the *experience* of birth and the role of the infant in promoting mutual attachments on the part of the mother and child (see, e.g., Klaus & Kennell, 1976). This new respect for the sensory and learning competencies of the newborn has helped to argue for a return, even in hospitals, to an approach to childbirth that honors the neonate as capable of affecting others and being affected within the first few days of life by the prevailing conditions.

Learning refers to change in behavior that accrues over time as a result of experience. Learning implies memory, and requires that, on subsequent occasions, the organism will behave differently because of previous experience with a given episode of stimulation. Learning by infants has been studied in the context principally of three paradigms, classical conditioning, operant learning, and habituation.

Classical Conditioning

In the process known as classical conditioning, recruitment of response to a previously neutral stimulus takes place. Some response to a given stimulus, which was not present at the outset, is evoked after training. Denisova and Figurin (1929) showed, for example, that infants begin to show anticipatory sucking movements, when placed in the feeding position, after they have been subjected to several trials in which the opportunity to suck has followed such "position placement," this being the conditioning stimulus.

It is not unlikely that some of the earliest manifestations of infants' behavior changes, even those occurring in the first days of life, are instances of or are complex elaborations of the simple classical conditioning paradigm. However, many of the early naturalistic observations of presumed classical conditioning, even such as that of Denisova and Figurin, have been subject to error of interpretation. The substantiation of a classical conditioning effect must always rule out the possibilities that (a) the "learned" response obtained was actually to the unconditioned rather than to the conditioning stimulus: and (b) the observed enhancement of response was not due to a maturational change in the response repertoire merely through increased age and associated changes in neurosensory transmission or changes in unconditioned response thresholds. Documentation of a classically conditioned response necessitates that the effect obtained can be strictly ascribed to pairings of an initially ineffective with an initially effective stimulus, with the result that the response previously called forth by the unconditioned stimulus is now elicited, at least in some measure, by the conditioning stimulus.

Sucking behavior and the eye-blink response have been successfully conditioned to tactile stimuli. Lipsitt, Kaye, and Bosack (1966) used a paradigm that differed somewhat from strict classical conditioning. Newborns were presented with an intraoral stimulus, actually a short piece of rubber tubing .625 cm in diameter, which mimicked a nipple but was known from a previous study to be a stimulus far less than optimal for eliciting sucking (Lipsitt & Kaye, 1965). The aim was to determine whether the sucking response to the tube could be enhanced by the presentation through the tube of a known effective elicitor of sucking. Thus the infant was presented with a 5% dextrose solution in association with presentation of the tube. In contrast to control infants who did not receive pairing of the conditioning and unconditioned stimuli, but instead received unpaired presentations of the same stimuli, the experimental or "trained" infants came to respond anticipatorily to the presentation of the tube alone with enhanced sucking responses.

The Babkin response (Babkin, 1960) has been successfully conditioned by Kaye (1965). This reflex involves several components, including gaping, sucking, turning the head to midline, raising the head, and, in some instances, eyelid closure and forearm flexion. The congenital stimulus for this behavior is pressure on the baby's palms. Kaye utilized tactile-kinesthetic stimulation, involving movement of the arms from the infant's sides up to the head, just before application of the pressure. This procedure resulted in classical conditioning of the Babkin response, a finding subsequently replicated by Connolly and Stratton (1969), who successfully demonstrated the acquisition of this response to both auditory and kinesthetic conditioning stimuli. Although the question has been raised as to whether real conditioning took place (Sostek, Sameroff, & Sostek, 1972), and much work remains to be done to refine our understanding of the necessary and sufficient conditions for the elaboration of different types of conditioning, these studies support the contention that cortically mediated alterations in behavior do take place in the newborn.

Temporal conditioning involves the passage of time as a conditioning stimulus. Early studies of temporal conditioning were conducted in connection with feeding schedules. If the infant is put on a time schedule for feeding, such as a 3- or 4-hour interval, some of the behavioral disruptions that occur subsequent to elimination of a feeding might be attributable to a temporally conditioned effect. Marquis (1941) did a study in which one group of newborns was changed from a fixed 3-hour schedule to a 4-hour schedule. Their behavior was compared with that of a control group on a 4-hour schedule throughout. Infants whose schedules were changed evidenced more bodily activity and general behavioral disruption than did the control group during the fourth hour, when the feeding ordinarily would have been given. Of course, infants on a 3-hour schedule may not eat as much as those on a 4-hour schedule, and the effect might be due in part to greater hunger in the switched group (Lipsitt, 1963). However, results of further studies, although not entirely free from artifact, have supported the general finding. Changes of feeding schedules have been shown to induce anticipatory increases not only in responses in bodily activity (Bystroletova, 1954), but in blood leukocyte count as well (Krachkovskaia, 1959).

Studies are scarce using visual conditioning stimuli with very young infants. This is probably due to the fact that, whereas tactual and auditory stimulation is essentially inescapable, visual stimulation is received, at least to some extent, at the "whim" of the observer. Nonetheless, in a study by Kasatkin and Levikova (1935), six infants ranging in age from 14 to 48 days developed stable conditioned visual responses. In this study, a colored light was presented in association with the feeding bottle. After a number of trials, the babies responded to the green light with many of the response components, such as turning the head and making mouth movements, characteristic of the feeding situation itself. Following conditioning to a specific color, generalized responses did occur to other colors close to the conditioning stimulus, but after a number of trials in which the conditioning stimulus and the alternate color were successively presented, response differentiation took place. Very stable classically conditioned differentiation was present between the ages of 88 and 116 days for all infants in the study.

Auditory stimuli have been used successfully in classical conditioning with newborns. Marquis (1941) used the sound of a buzzer preceding the offering of a bottle nipple during the first 10 days of life. Within 5 days of training, the 10 infants studied were making sucking responses to the sound of the buzzer. Kasatkin and Levikova (1935) were able to demonstrate conditioned differentiation of tones by the time infants in their experiment were 2-3 months of age.

Lipsitt and Kaye (1964) presented a low frequency loud tone (93 dB) in association with the insertion of a nipple in the mouths of infants 3 and 4 days old. Sensitization effects were controlled through the use of a control group to whom the tone and nipple were presented noncontiguously. On every fifth trial, the tone was presented alone as a test for conditioning. After training was completed, all babies received a series of extinction trials with the test tone alone. Evidence was found for classical appetitive conditioning, although the effects of training did not manifest themselves until the extinction condition. Other investigators have also been able to obtain such conditioning (Abrahamson, Brackbill, Carpenter, & Fitzgerald, 1970).

A practical application of the classical conditioning technique was introduced by Aldrich (1928) who sounded a small bell while stimulating the sole of the infant with a pin prick. After 12 to 15 such pairings, Aldrich found that the sound of the bell alone was effective in producing the response. He suggested that such behavior could provide a definitive test of deafness. A number of other studies demonstrate classical aversive conditioning not unlike that represented by the Aldrich study (Lipsitt, 1963). Failures to obtain aversive conditioning, which have also been reported, may perhaps be attributed to the possibility that rather intense noxious stimulation must be used to obtain the effect, and these conditions are seldom implemented in the laboratory.

It is a matter of clinical import for the neonatologist, as well as for the research psychologist, that the newborn is capable of retaining memories of experienced pain. That we do not recall, 20 or even 5 years later, painful experiences in the first few days of life does not mean that those painful experiences had no effects. It is possible for an infant to experience a painful episode in, for example, the third day of life that might have an effect upon perception of subsequent events later in the same day or in the next. For example, a 3-day-old might experience pain that causes irritation and crying for 5-10 hours, a sufficient length of time for him to be held many times by his mother. Mother as a stimulus would thus be "presented" naturalistically many times in association with pain. Such conditioning as might thereby occur *could* then influence the effects of still later experiences. By the time several days (or years) have passed, none of the experiences may be remembered specifically, but the painful episode and its sequelae may nonetheless have altered the organism in critical and even permanent ways.

This is not to say that unremembered painful experiences are necessarily deleterious. Indeed, it is possible that early stress and the recruitment of physiological and psychological resources for coping with that stress may have a beneficial effect upon the child's later ability to handle stress (Murphy, 1976). Very little is known about this. Pain perception and its significance have not been studied much in the very young infant; pediatric investigators and child psychologists would be loath to administer unnecessary pain stimulation. Nonetheless, in the natural course of caring for newborns, unusual levels of aversive stimulation sometimes *are* administered, as in circumcision, venipuncture, or excision of a supernumerary digit. Researchers would do well to capitalize more upon these rather routine hospital procedures to explore aspects of pain perception in the newborn and the possible conditioning effects that such procedures may have on the infant's reciprocal relationships with caretakers in the immediately subsequent hours and days. Some beginning attempts to do this have been made (see, for example, Emde, Harmon, Metcalf, Koenig, & Wagonfeld, 1971; Marshall, Stratton, Moore, & Boxerman, 1980).

Operant Conditioning

Whereas classical conditioning involves the learning of a new association between a neutral and an initially effective stimulus, operant learning results from the efficacy of reward and punishment, delivered contingently upon execution of some specific response. Although much remains to be explored concerning the underlying processes and principles of operant learning in infants, many studies with infants have been done and much has been discovered over the past 2 decades.

Sameroff (1972) has shown that the subsystems of the general sucking pattern of infants can be altered by specific environmental conditions. He demonstrated that when the delivery of fluid is made directly contingent upon the negative pressure component of the sucking response, this results in the infant using more negative pressure than when this contingency is not in effect. Such a demonstration is possible because there are two major components of sucking behavior in the newborn, one of these consisting of negative pressure in the buccal cavity, and the other simply the positive pressure created by gum action. The Sameroff study showed that these two components may be reinforced differentially to enhance one of them.

Brown (1972) showed that sucking in newborns can be brought under operant control even without using feeding. Using a blank nipple as a nonnutritive reinforcer, she based her study on the hypothesis put forth by Premack (1965) that a response that is higher in the habit hierarchy, or greater in strength, may be used to reinforce a weaker response. Lipsitt and Kaye (1965) previously had shown that nipples of certain shapes serve to optimize sucking behavior. Brown therefore had infants of 2 and 3 days of age suck on one nipple, either a regular commercial nipple or a blunt variation thereof, to gain access to another nipple, either regular or blunt. On the second day of conditioning, the rate of sucking on the regular nipple in the "nipple followed by blunt" group was reliably lower than the rate of sucking on the regular nipple in the "nipple followed by nipple" group. Sucking rates for the blunt oral stimulus in the "blunt followed by nipple" group, moreover, were significantly higher than such sucking in the "blunt followed by blunt" group.

A similar effect was found by Kobre and Lipsitt (1972) in a study of negative contrast. Newborns sucked less to receive water after having experienced the taste of sugar than they did after they had been sucking merely for water during a comparable period of time. Although this was a short-term study, as was Brown's, the implication is that a basic memorial process is already available to the infant, and that learning to behave on the basis of differential incentive conditions is possible.

Some studies of infant learning have involved elaborations or modifications of traditional operant conditioning procedures, with good effect. One such study (Siqueland & Lipsitt, 1966) adapted the conditioned head-turning techniques first reported by Papousek (1959; 1960). The Siqueland and Lipsitt (1966) elaboration of the procedure studied the effect of contingent reinforcement on ipsilateral head movements, but, instead of awaiting the natural or spontaneous occurrence of headturning, the head turns were elicited by a touch at the side of the infant's mouth. Such stimulation typically produces head turning as a part of the so-called rooting reflex, on about 30% of the preconditioning trials. In three methodological variations of the procedure, nutritive reinforcement in the form of a 5% dextrose solution was administered contingent upon the elicited head-turning movements. The result was that head-turning responses were enhanced considerably by the reinforcement condition.

In one of the Siqueland and Lipsitt studies, each infant served as its own control to obviate the possibility that the apparent conditioning effect obtained could be attributed to changes in state rather than to changes in cognitive or memorial factors. The study involved the use of two different auditory stimuli or tones. On alternate trials, the stimulus for head turning was a different tone-touch combination. On positive trials, the neonate received one of the tone-touch combinations to the left cheek, whereas on the other half of the alternated trials, a different tone-touch combination to the same cheek was used as the negative stimulus. On the positive trials, appropriate head turns were reinforced by opportunity to suck for dextrose, whereas on negative trials, no such opportunity was provided. The results of the study indicated that the positive stimulus combination facilitated the rise of responding from 30 to 83%, whereas, on negative trials, the response probability remained essentially the same as at the beginning. The total training effort required 30 min. The effect, shown in Figure 4.4, was reliable. Following this period of conditioning, the positive and negative stimulus conditions were reversed, such that the stimulus combination which was previously reinforced no longer was, and the previously negative stimulus combination was made positive. A gradual behavioral shift took place to accord with the changed stimulus conditions, producing a reliable effect of the contingent reinforcement once again. The reversal of behavior was significant. This and other studies demonstrate clearly that the contingent application of incentive events may alter the probabilities of elicitation of reflexive behaviors that are typical of infants within the first few days of life.

Visual reinforcement of high-amplitude sucking responses has been used to study operant learning in infants from the preterm period to 4 months of age (Milewski & Siqueland, 1975; Siqueland & DeLucia, 1969; Werner & Siqueland, 1978). In such studies, a screen is used for showing slides of varying illuminations, with the illumination being controlled by an operant

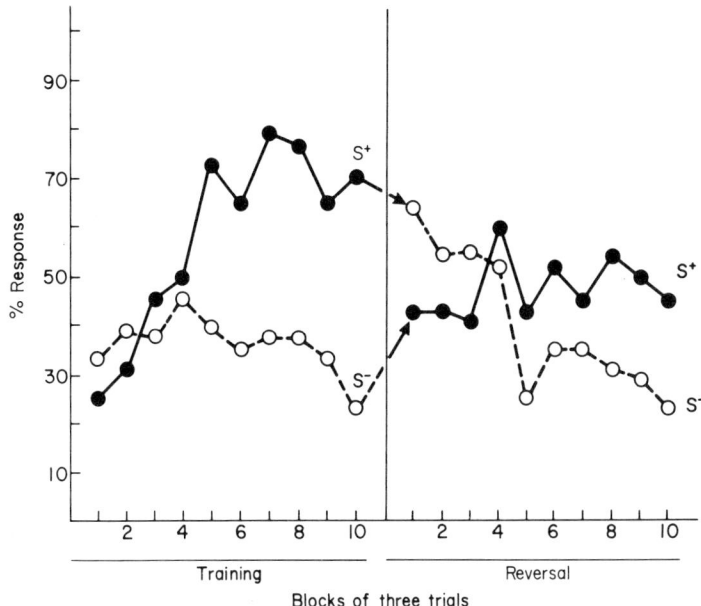

Figure 4.4. *Response percentages to positive (S^+) and negative (S^-) stimuli during the original training and reversal trials. (From E. R. Siqueland & L. P. Lipsitt, Conditioned head-turning behavior in newborns.* Journal of Experimental Child Psychology, *1966, 3, 356-376. Reprinted by permission.)*

behavior of the infant, specifically sucking responses. Under these conditions of so-called "conjugate reinforcement," infants show typical acquisition and extinction effects, in which they increase their sucking rates to receive visual incentives, and decrease their sucking either when the screen no longer provides contingently presented visual stimulation or after becoming habituated to the stimuli. The habituation aspects of the typical responses obtained enabled these researchers to capitalize upon the familiarization phenomenon to study the discriminability of very similar stimuli. A procedure implemented by Milewski and Siqueland (1975) has shown that infants as young as 1 month of age can discriminate between familiar and novel stimuli on the basis of both color and pattern, and that there is an increase in the reinforcing effectiveness of visual stimuli concomitant with an increase in their novelty.

The contingent reinforcement paradigm and the sucking response have been used to study the effect of auditory stimuli on very young infants as well. In one such study (Eimas, Siqueland, Jusczyk, & Vigorito, 1971) it was demonstrated that young infants will suck to hear auditory stimuli up to a certain point, whereupon interest lags and sucking wanes or even ceases. At that point a slight change in the auditory stimulus, such as from *ba* to *ga*, can be made and will cause a significant recovery of the sucking response relative to control infants for whom no such change in auditory stimulus is introduced. In such studies as these, the focus of attention is not so much on the operant learning process itself, but rather on the other perceptual or cognitive phenomena that operant techniques may elucidate.

Another response that has been shown to increase as a result of novel or interesting visual reinforcers, such as pictures of checkerboards, color patches, triangles, bull's eyes, and so on, has been headturning (Caron, 1967; Levison & Levison, 1967). Reinforcers combining several different modalities, such as visual and tactual, have also yielded positive results (Bower, 1965; McKenzie & Day, 1971).

Habituation

Classical and operant conditioning are essentially noncontroversial as involving experientially induced behavior changes that are persistent and perhaps essentially permanent (pending the superimposition of counterconditioning or extinction). The condition of the organism has been presumed to have been altered by the experience, and the residual of the experience is generally thought to have been mediated by brain structures. Habituation, on the other hand, does not clearly involve the development of new associations as does classical and operant learning—an association between a neutral and effective stimulus in the case of classical, and association between response-produced stimulation and reward stimulation, in the case of operant. Nonetheless, habituation does entail gradual changes in behavior with repetitive stimulation, and does thereby produce behavioral alterations which, like learning processes, last for a while.

Upon repetitive stimulation, and provided the stimulus is mild or only moderately intense, the elicited responses of human infants will gradually reduce in intensity. In some instances, response strength will actually reduce to zero, even though initial presentations of the stimulus produced marked startle responses, accelerated heart rates, and so on. After the waning of the behavior, the response will be seen again if a different stimulus is presented or merely after some time has passed in which the stimulus has been absent. The data are especially clear, and certainly more plentiful in the visual and auditory modalities, but in studies of olfactory habituation, progressive response decrement with repetitive stimulus presentation has been documented for infants (Clifton & Nelson, 1976; Engen & Lipsitt, 1965).

The literature on habituation in early infancy is extensive, and only exemplary studies may be touched upon here. Reviewers of studies on infantile habituation (Kessen, Haith, & Salapatek, 1970) have noted that the phenomenon of habituation is a valuable tool for the study of memory processes.

Thus habituation, as well as classical and operant conditioning, might illuminate the extent to which cortical functioning is present in the young human and possibly can reveal behavioral aberrations indicative of central nervous system (CNS) deficits.

Fantz (1961; 1964) has done extensive work on visual habituation of infants. He showed that the infant's gaze duration can be quantified, and that this measure may be used over repetitive trials to assess the infant's diminishing interest in specific visual stimuli. Although Fantz's early studies suggested that the newborn showed little visual habituation, whereas the infant beyond 10 weeks of age showed much of it, later studies have demonstrated visual habituation in neonates. Friedman, Nagy, and Carpenter (1970) presented 40 newborns with two checkerboard targets, one with 4 squares, the other with 144 squares. Both males and females showed response decrement over trials, but rather inexplicably, males demonstrated the phenomenon to the less dense visual stimulus, and females to the denser (144 square) stimulus.

Inasmuch as habituation relates to memory function, which is mediated by the brain, it should be possible to refine techniques for improved assessment of individual differences and of CNS impairments, by including indices of habituation (Brazelton, 1973). In fact, Lewis (1967) has shown that impaired infantile habituation is related to low Apgar scores and other measures of perinatal distress and brain damage. Hydrocephalic and anencephalic infants may show no habituation at all (Brackbill, 1971; Wolff, 1969), although one study of a single baby (Graham, Leavitt & Strock, 1978) is an exception.

A habituation technique involving the sucking response has been used to document the theory that infants born under conditions of excessive cranial pressure or with the cord around the neck have a deficiency in habituation (Bronshtein, Antonova, Kamenetskaya, Luppova, & Sytova, 1958). While sucking, the infants were presented a stimulus such as a sound. Normal infants interrupted their sucking momentarily when presented with such a stimulus, and then on repetitive presentations showed a reduction in (or habituation of) such sucking suppression. In this study it was shown that, with infants born at risk, it took many more stimulus presentations to habituate the sucking-suppression response.

A study of averaged auditory evoked responses in infants with Down's syndrome showed that, relative to normal infants matched in age (from 8 days to 13 months), the infants with Down's syndrome showed no significant response decrement (Barnet, Olrich, & Shanks, 1974). It was also demonstrated that none of the babies under 1 month of age, normal or otherwise, showed evoked response habituation.

Finally, effects of obstetrical anesthesia on the infant have been studied using habituation (Bowes, Brackbill, Conway, & Steinschneider, 1970). Infants were examined at 2 and at 5 days of age. Those whose mothers received high dosages of anesthesia required as many as four times more trials to habituate than did those whose mothers received little medication; this difference still existed when the infants were retested 1 month later. More studies are required that would capitalize upon the use of habituation and dishabituation procedures to document individual differences in behavioral functioning. Clearly, the habituation paradigm taps the organism's capacity for processing information for discriminating among modest differences in stimulus inputs, and for appropriating central nervous system resources more generally.

The Functions of Early Human Learning

Whereas it is unquestioned that much of early human behavior is endogenously generated, it is also apparent that behavior is determined and modified by the stimulation that is provided for the infant, either deliberately or inadvertently. Moreover, despite the fact that we tend, as discussed, to speak largely about the infant as a solitary organism in a world of experiential forces that impinge on him or her in well-organized regimens of stimulation, the fact is that the baby and its caretakers constitute for each other a control system in which each responds reciprocally to the other and each serves as a stimulant for the other. Although the essence of babyhood is perhaps dependency upon another, making it appear often as if the infant is merely the passive recipient of environmental inputs, individual differences among infants in their appearances, their psychophysiological tempos, and their behavioral characteristics all serve as determinants as well of other persons' reactions to them (see Thoman, 1979). It is in this context of reciprocating transactions that the infant acquires coping skills and gestures, maneuvers, manipulations, and "social graces."

By the same token, the early months of life are a proving ground for the adequacy of the infant's inherited and congenital response repertoire, the embellishment of which will be the by-product of early practice. In this view, learning disabilities and other developmental problems have their earliest origins in the interplay between the constitutional "givens" (the behaviors with which the infant comes equipped at birth as a gift of the species and of the fetal environment) and the experiential inputs that are accorded to or imposed upon the baby from the earliest moments of life.

Procedures have evolved in infant behavior laboratories that have had practical import. With greater investments in them and in the elaboration of the most promising techniques, further advances can be made in the care, treatment, and even education of the young child. Most such advances, of

4. THE INFANCY OF HUMAN LEARNING PROCESSES 125

course, relate either to the treatment or prevention of developmental anomalies.

Recent studies strongly suggest that some of the first manifestations of developmental disability may be found in the earliest days of life. It is now rather well accepted that there are certain events surrounding birth, such as maternal anemia, oxygen deficit at birth, or jaundice, that can have life-lasting influences on the infant who endures them (Field, 1979). It is well established in actuarial studies that, on average, those infants who are born prematurely, who are small and require prolonged administration of oxygen at birth, whose muscles seem limp or floppy (or conversely, overly tense and tight) do fare less well during the first year of life and in later years, physically and behaviorally, than infants whose "signs" or "indicators" were not so severe (Drillien, 1964; Sameroff, & Chandler, 1975). Many of the prenatal and birth risk factors have their most debilitating effects upon the sensory systems and CNS functioning. Thus behavioral deficits in general, and learning aberrations in particular, are not unexpected.

In summarizing a vast amount of data on the relationship between early competencies of the infant and later development, Appleton, Clifton, and Goldberg (1975) found reason to believe that the earliest manifestations of sensory competence do give rise in a causative way to later sensorimotor and intellectual proficiencies:

> The competence of the older child in manipulating and comprehending the environment is probably based on interactions with the environment which are dependent upon processing sensory information. Through development of perceptual and sensorimotor abilities the infant can learn about his or her surroundings and can develop a practical understanding of reality. Early learning of skills and expectations appears to contribute to further learning and development [p. 104].

Weak habituation manifested during the neonatal period may be suggestive of later developmental lag, and absence of habituation has been seen as a concomitant of serious brain deficiency (Brackbill, 1971; Bronshtein et al., 1958; Lewis, Goldberg, & Campbell, 1969).

Respiratory Occlusion and
Learned Anger in the Newborn

It is not uncommon for aversive or avoidance behavior to be noted in the first few days of life. It is reasonable to suppose, moreover, that defensive and even "angry" responses of the baby have some adaptive significance, and thus the total evasion or suppression of such behaviors would not seem to be well advised.

The English pediatrician, Mavis Gunther, pointed out a number of years ago (Gunther, 1961) that the "feeding couple" (as she called the mother and child pair) can affect one another in subtle ways that must be clearly understood to help both of them surmount the tensions of their earliest moments together. It is quite natural, she noted, for the newborn to become occasionally smothered for brief moments during the course of feeding. The nostrils come very close to the mother's breast and are sometimes stopped up while the baby has a tight latch on the nipple. This can cause momentary difficulty in breathing, which the infant in turn will object to, often strenuously. When this happens, the baby turns its head to and fro and pulls back from the nipple; the arms flail, the face often reddens, and, rather as a last resort if the blockage continues, there is a burst of crying, which throws the nipple from the baby's mouth with a gusto that can offend the mother. Mothers may become quite exasperated after a few such occasions, feeling themselves to be failures in coping with the baby's needs and frustrations (Newton, 1955).

As Gunther indicated on the basis of extensive observations of neonates with their mothers, a few simple pointers given to the mother will usually help her to adjust herself posturally, or to insert the nipple in a more felicitous way to prevent the occlusion from occurring. The seemingly angry response of the baby was, after all, a matter of awkward stimulation (Lipsitt, 1976). Even a bottle-fed infant can show the type of behavior described, as it is quite possible for the shield of the nipple to come against the baby's nostrils while feeding, especially with nipples that are short and stubby. Moreover, the baby's lips are fatty, and the nostrils are close to the upper lip. All of this tends to assure some respiratory blockage while feeding. Under circumstances of respiratory occlusion, newborns will usually go through the described pattern of behavior, culminating in releasing itself from the nipple, taking in air, and regaining its physical and behavioral stability (a high-incentive consequence). The response of uncoupling from the breast is thus reinforced, which means that the next time the baby is put to the breast the position will be found less desirable than previously. The aversive or negative consequences of being at the breast may override the positive features, which at the outset were very strong. To the extent that newborns can learn in such situations, the baby will adaptively decline the next opportunity to get smothered!

Simple instructions to the mother concerning the necessity of helping the infant cope with keeping its air passages clear can often produce rapid changes—first, in the behavior of the mother and, second, in the objections of the infant (Gunther, 1961).

There have been too few studies conducted to provide information about the crucial early moments of interaction between mother and infant, and about the lingering effects of these critical experiences. The few studies that do exist (e.g., Korner, 1974) suggest that mother behavior *and* baby behavior are powerfully important in setting the stage for later interactions and perhaps for the discovery of conditions related to developmental aberrations.

Crib Death: A Psychological Perspective

The sudden infant death syndrome (SIDS) takes as many as 8000 babies in America each year in the first year of life, excluding the especially hazardous first few days of life. The fallout in grief and despair of the thousands of close survivors compounds the tragedy immeasurably. Such deaths are especially difficult to endure for parents because of the absence of definitive answers regarding the basic mechanisms underlying SIDS. The historical neglect of research attention to the nature of the disorder, and the rather frequent popular confusion of SIDS with child abuse, have only recently abated.

We need to know more about the precursors of crib death and its psychological implications. Some progress is being made. The first line of defense against it is to define those conditions that are often present, prior to the death, in those infants that have succumbed. Even descriptions of such conditions, in the absence of satisfactory explanations, would be useful at this stage of our understanding. The casual and necessarily anecdotal report of psychophysiological attributes of infants who have been "near misses" would—or might—provide us with a lead or two into the biological processes that may be operating in infants who die (Steinschneider, 1972). Psychophysiological factors might well be involved in the final pathway to the condition that ultimately causes the death of the infant, usually between 2 and 4 months of age. The possibility must be considered that experience, and the effects that experiences have on the young child, may be implicated.

One study (Lipsitt, Sturner, & Burke, 1979) began with the extensive perinatal and pediatric records of 15 crib death cases, then composed two control groups, one of these consisting of the very next births of the same sex, and the other, the very next births of the same sex and race. It became apparent that the deceased group varied from the controls in several ways, all of them in a direction connoting greater perinatal stress and biological hazard in that group. There were reliable differences in Apgar scoring in the first few minutes of life, the Apgar test being a quickly administered scale for assessing vital signs, such as adequacy of respiration, heart rate, pallor, and muscle tone; the deceased group did more poorly than either control. Infants in the deceased group were identified significantly more often as having respiratory abnormalities. More of the deceased group had mothers with anemia. More of the deceased infants had required intensive care than the controls. The deceased infants were hospitalized longer. The infants that ultimately died required more resuscitative measures.

The infants who succumbed between 2 and 5 months of age were already showing, in general and on average, that they were beginning life with some fragility. We need to know more about that fragility, and about the developmental processes that apparently transform seemingly minor insufficiencies into morbid crises a few months later.

One line of research relating to the causes of crib death has to do with the possibility that a basic learning disability is implicated (Lipsitt, 1976; 1979b). This suggestion is based upon the observation that crib deaths have a peak period of occurrence in the range between 2 and 4 months of age. This is a critical time during development when many of the basic reflexes with which the baby is born are in transition. Whereas these responses, like the grasp reflex of the newborn, are very strong at birth, they begin to weaken soon after, and some of them are gone by 5 months. Many of these reflexes, such as turning the head to a touch near the mouth, or the grasp and swimming reflexes, go through an observable change. They are executed more slowly and seemingly on a voluntary rather than an obligatory basis (McGraw, 1943). In her careful documentations of the ontogeny of reflexes in the first year of life, McGraw's data show that the transitional period between the reflexive and voluntary phases of different reflexes (she called the transitional phases "disorganized behavior" or "struggling activity") usually occur around the ages of 100-150 days, just the age period in which infants are most at risk for crib death.

If a baby does not go through the transition smoothly from being an essentially reflexive creature to becoming a more intentional and reflective organism, a hazardous condition might develop in which the infant will not be able to adequately defend himself when, for example, threats to respiration present themselves. Although little is known about the origins of an adequate repertoire of defensive behaviors, it is not unlikely that, after the first weeks of life, much of this class of behavior is importantly dependent upon learning.

Summary

Various mechanisms of human infant learning have been reviewed, with particular attention to learning processes in the neonatal period. It was emphasized that the experiencing of stimulation, even by the young infant, takes place against a hedonic backdrop, partly dependent upon state and physical condition, in which some experiences are more pleasant than others. This capacity of infants for experiencing the pleasures and annoyances of sensation sets the conditions for learning, especially instrumental learning, by providing incentives for behavior. It was shown that incentive factors inherent in the environment, including those specially implemented by caretaking persons, produce variations in style and persistence of infants' behavior. Finally, it was suggested that some developmental aberrations, perhaps including crib death, may be the result of an unfortunate confluence of constitutional deficits resulting from perinatal risk factors, on the one hand, and experiential insufficiencies, on the other.

References

Abrahamson, D., Brackbill, Y., Carpenter, Y., & Fitzgerald, H. E. Interaction of stimulus and response in infant conditioning. *Psychosomatic Medicine*, 1970, *32*, 319-325.

Aldrich, C. A. A new test for hearing in the newborn: The conditioned reflex. *American Journal of Diseases of Children*, 1928, *35*, 36-37.

Appleton, T., Clifton, R., & Goldberg, S. The development of behavioral competence in infancy. In F. D. Horowitz (Ed.), *Review of child development research* (Vol. 4). Chicago: University of Chicago Press, 1975. Pp. 101-186.

Ashmead, D. H., Reilly, B. M., & Lipsitt, L. P. Neonates' heart rate, sucking rhythm, and sucking amplitude as a function of the sweet taste. *Journal of Experimental Child Psychology*, 1980, *29*, 264-281.

Babkin, P. S. The establishment of reflex activity in early postnatal life. *The central nervous system and behavior*, translated from the Russian by the U.S. Department of HEW, USPHS, U.S. Government Printing Office, Washington, D.C., 1960.

Barnet, A. B., Olrich, E. S., & Shanks, B. L. EEG evoked responses to repetitive stimulation in normal and Down's syndrome infants. *Developmental Medicine and Child Neurology*, 1974, *5*, 612-619.

Bower, T. G. R. Stimulus variables determining space perception in infants. *Science*, 1965, *149*, 88-89.

Bower, T. G. R. *A primer of infant development*. San Francisco: Freeman, 1977.

Bowes, W., Brackbill, Y., Conway, E., & Steinschneider, A. The effects of obstetrical medication on fetus and infant. *Monographs of the Society for Research in Child Development*, 1970, *35*, 3-25.

Brackbill, Y. The role of the cortex in orienting: Orienting reflex in an anencephalic human infant. *Developmental Psychology*, 1971, *5*, 195-201.

Brackbill, Y. Obstetrical medication and infant behavior. In J. Osofsky (Ed.), *Handbook of infant development*. New York: Wiley, 1979. Pp. 76-125.

Brazelton, T. B. *Neonatal behavioral assessment scale*. Philadelphia: William Heinemann Medical Books, 1973.

Bronshtein, A. T., Antonova, T. G., Kamenetskaya, N. H., Luppova, V. A., & Sytova, V. A. On the development of the functions of analyzers in infants and some animals at the early stage of ontogenesis. In: *Problems of evolution of physiological functions*. U.S.S.R.: Academy of Science, 1958. (U.S. Department of HEW, Translation Service, 1960.)

Brown, J. Instrumental control of the sucking response in human newborns. *Journal of Experimental Child Psychology*, 1972, *14*, 66-80.

Bystroletova, G. N. The formation in neonates of a conditioned reflex to time in connection with daily feeding rhythm. *Zhurnal Vysshei Nervnoi Veyatel'Nosti Imeni I. P. Pavlova*, 1954, *4*, 601-609.

Caron, R. F. Visual reinforcement of head-turning in young infants. *Journal of Experimental Child Psychology*, 1967, *5*, 489-511.

Clifton, R. K., & Nelson, M. N. Developmental study of habituation in infants: The importance of paradigm, response system, and state. In T. J. Tighe & R. N. Leaton (Eds.), *Habituation: Perspectives from child development, animal behavior, and neurophysiology*. Hillsdale, N.J.: Erlbaum, 1976.

Connolly, K., & Stratton, P. An exploration of some parameters affecting classical conditioning in the neonate. *Child Development*, 1969, *40*, 431-441.

Coss, R. G., & Globus, A. Spine stems on tectal interneurons in jewel fish are shortened by social stimulation. *Science*, 1978, *200*, 787-790.

Cowett, R. M., Lipsitt, L. P., Vohr, B., & Oh, W. Aberrations in sucking behavior of low-birth-weight infants. *Developmental Medicine and Child Neurology*, 1978, *20*, 701-709.

Crook, C. K. Neonatal sucking: Effects of quantity of the response-contingent fluid upon sucking rhythm and heart rate. *Journal of Experimental Child Psychology*, 1976, *21*, 539-548.

Crook, C. K. The organization and control of infant sucking. In H. W. Reese & L. P. Lipsitt (Eds.), *Advances in child development and behavior* (Vol. 14). New York: Academic Press, 1979. Pp. 209-246.

Crook, C. K., & Lipsitt, L. P. Neonatal nutritive sucking: Effects of taste stimulation upon sucking rhythm and heart rate. *Child Development*, 1976, *47*, 518-522.

Darwin, C. A biographical sketch of an infant. *Mind*, 1877, *2*, 285-294.

Denisova, M. P., & Figurin, N. L. K. Voprosu o pervykh sochetatlnykh pishchevykh refleksakh u grudnykh detei, in Russian (The problem of the first associated food reflexes in infants) *Vop. Genet. Refleks. Pedol.*, 1929, *1*, 1-88.

Drillien, C. M. *The growth and development of the prematurely born infant*. Baltimore: Williams & Wilkins, 1964.

Eimas, P. D., Siqueland, E. R., Jusczyk, P., & Vigorito, J. Speech perception in infants. *Science*, 1971, *171*, 303-306.

Elkonin, D. B. The physiology of higher nervous activity and child psychology. In B. Simon (Ed.), *Psychology in the Soviet Union*. London: Routledge and Kegan Paul, 1957. Pp. 47-68.

Emde, R. N., Harmon, R. J., Metcalf, D., Koenig, K. L., & Wagonfeld, S. Stress and neonatal sleep, *Psychosomatic Medicine*, 1971, *33*, 491-497.

Engen, T., & Lipsitt, L. P. Decrement and recovery of responses to olfactory stimuli in the human neonate, *Journal of Comparative and Physiological Psychology*, 1965, *59*, 312-316.

Fantz, R. L. The origin of form perception. *Scientific American*, 1961, *204*, 66-72.

Fantz, R. L. Visual experiences in infants: Decreased attention to familiar patterns relative to novel ones. *Science*, 1964, *146*, 668-670.

Field, T. (Ed.), *Infants born at risk*. New York: Spectrum, 1979.

Freud, S. *Beyond the pleasure principle*. New York: Boni & Liveright. 1927.

Friedman, S., Nagy, A. N., & Carpenter, G. C. Newborn attention: Differential response decrement to visual stimuli. *Journal of Experimental Child Psychology*, 1970, *10*, 44-51.

Gesell, A. *Studies in child development*. New York: Harper, 1948.

Gesell, A., & Ilg, F. L. *The child from five to ten*. New York: Harper, 1946.

Gottlieb, B. Ontogenesis of sensory function in birds and mammals. In E. Tobach, L. R. Aronson, & Shaw (Eds.), *The biopsychology of development*. New York: Academic Press, 1971.

Graham, F. K., Leavitt, L. A., & Strock, B. D. Precocious cardiac orienting in a human anencephalic infant. *Science*, 1978, *199*, 322-324.

Gunther, M. Infant behavior at the breast. In B. Foss (Ed.) *Determinants of infant behavior*. London: Methuen, 1961. Pp. 37-44.

4. THE INFANCY OF HUMAN LEARNING PROCESSES

Horowitz, F. D. Learning, developmental research, and individual differences. In L. P. Lipsitt & H. W. Reese (Eds.), *Advances in child development and behavior* (Vol. 4). New York: Academic Press, 1969. Pp. 84-122.

Hulsebus, R. C. Operant conditioning of infant behavior: A review. In H. W. Reese (Ed.), *Advances in child development and behavior*. (Vol. 8). New York: Academic Press, 1973.

Jacobson, S. W., & Kagan, J. Interpreting "imitative" responses in early infancy. *Science,* 1979, *205,* 215-217.

Kasatkin, N. I., & Levikova, A. M. On the development of early conditioned reflexes and differentiations of auditory stimuli in infants. *Journal of Experimental Psychology,* 1935, *18,* 1-19.

Kaye, H. The conditioned Babkin reflex in human newborns. *Psychonomic Science,* 1965, *2,* 287-288.

Kessen, W. *The child.* New York: Wiley, 1965.

Kessen, W., Haith, M. M., & Salapatek, P. H. Human infancy: A bibliography and guide. In P. H. Mussen (Ed.), *Carmichael's manual of child psychology* (Vol. 1). New York: Wiley, 1970.

Kimble, G. A. Hilgard and Marquis' conditioning and learning (2nd ed.). New York: Appleton-Century-Crofts, 1961.

Klaus, M. H., & Kennell, J. H. *Maternal-infant bonding.* St. Louis: C. V. Mosby, 1976.

Kling, J., & Riggs, L. A. *Woodworth and Schlosberg's experimental psychology.* New York: Holt, Rinehart & Winston, 1971.

Kobre, K. R., & Lipsitt, L. P. A negative contrast effect in newborns. *Journal of Experimental Child Psychology,* 1972, *14,* 81-91.

Korner, A. The effect of the infant's state, level of arousal, sex, and ontogenetic stage on the caregiver. In M. Lewis & L. A. Rosenblum (Eds.), *The effect of the infant on its caregiver.* New York: Wiley, 1974. Pp. 105-122.

Krachkovskaia, M. V. Reflex changes in the leukocyte count of newborn infants in relation to food intake. *Pavlov Journal of Higher Nervous Activity,* 1959, *9,* 193-199.

Levison, C. A., & Levison, P. K. Operant conditioning of head-turning for visual reinforcement in 3-month old infants. *Psychonomic Science,* 1967, *8,* 529-530.

Lewin, K. *A dynamic theory of personality.* New York: McGraw-Hill, 1935.

Lewis, M. The meaning of a response, or why researchers in infant behavior should be oriental metaphysicians. *Merrill-Palmer Quarterly,* 1967, *13,* 7-18.

Lewis, M., Goldberg, S., & Campbell, H. A developmental study of information processing within the first three years of life. Response decrement to a redundant signal. *Monographs of the Society for Research in Child Development,* 1969, *34,* (Whole No. 133).

Lipsitt, L. P. Learning in the first year of life. In L. P. Lipsitt & C. C. Spiker (Eds.), *Advances in child development and behavior* (Vol. 1). New York: Academic Press, 1963.

Lipsitt, L. P. Developmental psychobiology comes of age: A discussion. In L. P. Lipsitt (Ed.), *Developmental psychobiology: The significance of infancy.* Hillsdale, N. J.: Erlbaum, 1976. Pp. 109-127.

Lipsitt, L. P. The study of sensory and learning processes of the newborn. In J. Volpe (Ed.), *Clinics in perinatology* (Vol. 4, No. 1). Philadephia: W. B. Saunders, 1977.

Lipsitt, L. P. Perinatal indicators and psychophysiological precursors of crib death. In F. D. Horowitz (Ed.), *Early developmental hazards: Predictors and precautions.* Boulder: Westview Press, 1978.

Lipsitt, L. P. Infants at risk: Perinatal and neonatal factors. *International Journal of Behavioral Development,* 1979, *2,* 23-42. 1979a

Lipsitt, L. P. Critical conditions in infancy. *American Psychologist,* 1979, *34,* 973-980. 1979b

Lipsitt, L. P., & Eimas, P. D. Developmental psychology. *Annual Review of Psychology,* 1972, *23,* 1-49.

Lipsitt, L. P., & Kaye, H. Conditioned sucking in the human newborn. *Psychonomic Science,* 1964, *1,* 29-30.

Lipsitt, L. P., & Kaye, H. Change in neonatal response to optimizing and non-optimizing sucking stimulation. *Psychonomic Science,* 1965, *2,* 331-332.

Lipsitt, L. P., Kaye, H., & Bosack, T. N. Enhancement of neonatal sucking through reinforcement. *Journal of Experimental Child Psychology,* 1966, *4,* 163-168.

Lipsitt, L. P., Reilly, B. M., Butcher, M. J., & Greenwood, M. M. The stability and inter-relationships of newborn sucking and heart rate. *Developmental Psychobiology*, 1976, *9*, 305-310.

Lipsitt, L. P., Sturner, W. Q., Burke, P. Perinatal indicators and subsequent crib death. *Infant Behavior and Development*, 1979, *2*, 325-328.

Maratos, O. *The origin and development of imitation in the first 6 months of life.* Unpublished doctoral dissertation, University of Geneva, 1973.

Marquis, D. P. Learning in the neonate: The modification of behavior under three feeding schedules. *Journal of Experimental Psychology*, 1941, *29*, 263-282.

Marshall, R. E., Stratton, W. C., Moore, J. A., & Boxerman, S. B. Circumcision I: Effects upon newborn behavior. *Infant Behavior and Development*, 1980, *3*, 1-14.

McGraw, M. B. *The neuromuscular maturation of the human infant.* New York: Columbia University Press, 1943.

McKenzie, B., & Day, R. H. Operant learning of visual pattern discrimination in young infants. *Journal of Experimental Child Psychology*, 1971, *11*, 45-53.

Meltzoff, A. N. Gestural imitation in early infancy: Some implications for Piaget's theory of representation. Meetings of the Jean Piaget Society, Philadelphia. May, 1977.

Meltzoff, A. N., & Moore, M. K. Imitation of facial and manual gestures by 2-week-old infants. *Science*, 1977, *198*, 75-78.

Milewski, A. E., & Siqueland, E. R. Discrimination of color and pattern novelty in 1-month human infants. *Journal of Experimental Child Psychology*, 1975, *19*, 122-136.

Murphy, L. B., & Moriarty, A. E. *Vulnerability, coping, and growth: From infancy to adolescence.* New Haven: Yale University Press, 1976.

Newton, N. *Maternal emotions* (Part B). New York: Hoeber, 1955.

Papousek, H. A method of studying conditioned food reflexes in young children up to the age of 6 months. *Pavlov Journal of Higher Nervous Activity*, 1959, *9*, 136-140.

Papousek, H. Conditioned motor digestive reflexes in infants. II. A new experimental method for the investigation. *Cesk. Pediatrics*, 1960, *15*, 981-988.

Peiper, A. *Cerebral function in infancy and childhood.* New York: Consultants Bureau, 1963.

Pfaffmann, C. The pleasures of sensation. *Psychological Review*, 1960, *67*, 253-268.

Piaget, J. *The moral judgment of the child.* New York: Harcourt Brace, 1932.

Piaget, J. *The origins of intelligence in children.* (M. Cook, Trans.), New York: International Universities Press, 1952.

Premack, D. Reinforcement theory. In D. Levin (Ed.), *Nebraska Symposium on Motivation*. Lincoln: University of Nebraska Press, 1965.

Purpura, D. P. Dendritic spine 'dysgenesis' and mental retardation. *Science*, 1974, *186*, 1126-1128.

Purpura, D. P. Dendritic differentiation in human cerebral cortex; normal and aberrant developmental patterns. *Advances in neurology*, 1975, *12*, 91-116.

Rosenzweig, M. R. Effects of environment on development of brain and of behavior. In E. Tobach, L. R. Aronson, & E. Shaw (Eds.), *The biopsychology of development.* New York: Academic Press, 1971.

Rosenzweig, M. R., & Bennett, E. L. Enriched environments: Facts, factors and fantasies. In L. Petrinovitch & J. L. McGaugh (Eds.), *Knowing, thinking, and believing.* New York: Plenum, 1976. Pp. 179-213.

Sameroff, A. J. Learning and adaptation in infancy: A comparison of models. In H. W. Reese (Ed.), *Advances in child development and behavior* (Vol. 7). New York: Academic Press, 1972. Pp. 170-214.

Sameroff, A. J., & Chandler, M. J. Reproductive risk and the continuum of caretaking casualty. In F. D. Horowitz, M. Hetherington, S. Scarr-Salapatek, & G. Siegel (Eds.), *Review of child development research* (Vol. 4). University of Chicago Press, 1975, Pp. 187-244.

Siqueland, E. R., & DeLucia, C. A. Visual reinforcement of non-nutritive sucking in human infants. *Science*, 1969, *165*, 1144-1146.

Siqueland, E. R., & Lipsitt, L. P. Conditioned head-turning behavior in newborns. *Journal of Experimental Child Psychology*, 1966, *3*, 356-376.

Sostek, A. M., Sameroff, A. J., & Sostek, A. Evidence for the unconditionability of the Babkin reflex in newborns. *Child Development*, 1972, *43*, 509-519.

Steiner, J. E. Human facial expressions in response to taste and smell stimulation. In H. W. Reese & L. P. Lipsitt (Eds.), *Advances in child development and behavior* (Vol. 13). New York: Academic Press, 1979, Pp. 257-295.

Steinschneider, A. Prolonged apnea and the sudden infant death syndrome: Clinical and laboratory observations. *Pediatrics*, 1972, *50*, 646-654.

Thoman, E. (Ed.), *Origins of the infant's social responsiveness*. Hillsdale, N.J.: Erlbaum, 1979.

Thorndike, E. L. *Educational psychology*. New York: Columbia University Press, 1913.

Watson, J. B. *Psychological care of the infant and child*. New York: Norton, 1928.

Werner, J. S., & Siqueland, E. R. Visual recognition memory in the preterm infant. *Infant Behavior and Development*, 1978, *1*, 79-94.

Wertz, R. W., & Wertz, D. C. *Lying-in: A history of childbirth in America*. New York: Schocken Books, 1979.

Wolff, P. H. What we must and must not teach our young children from what we know about early cognitive development. Reprinted from *Planning for better learning*. London: Spastics International Medical Publications/William Heinemann Medical Books, 1969.

Young, P. T. *Motivation of behavior*. New York: Wiley, 1936.

5

Innate Programs for Perceptual Development: An Ethological View[1]

PETER MARLER
STEPHEN ZOLOTH
ROBERT DOOLING

Introduction

How does each kind of organism, with its unique spectrum of requirements for survival, growth, and reproduction, keep ongoing behavior attuned to environmental variation? For animals in nature, appropriate responsiveness to the environment is a matter of life or death. As Gibson (1969) indicates, "Without discriminative response, there would be no correlation with ongoing events, or with qualitative or quantitative differences in stimuli [p. 3]." To the extent that genetic mechanisms aid the organism in acquisition of the ability to respond appropriately, natural selection will play an inevitable role in strategies of perception and how they develop.

The embryologist C. H. Waddington speaks of patterns of development as "canalized." He used this term to characterize the innate tendency of developing organ systems to follow particular paths. The development of any organ reflects the influence of many environmental influences, and many competing and cooperating genes, which interact in a self-stabilizing and conservative fashion. It can be surprisingly difficult to divert such growth patterns from their developmental goals (Waddington, 1957). Is it possible that perceptual development is canalized? There is unquestionably interplay between genetically determined constraints and environmental influences, although the final product is not an organ but a perceptual system with a particular form of organization. The purpose of this chapter is to suggest that

[1] Research reviewed was supported by grants NSF BNS 77 16894, PHS MH 14651, MH 31386-01 and PHS MH 31165. Parts of this chapter appear in P. Marler, Comparative perspectives on ethology and behavioral development. In M. Bornstein (Ed.), *Comparative Methods in Psychology: Ethological, Developmental, and Cross-Cultural Perspectives*. Hillsdale, N.J.: Lawrence Erlbaum Associates: 1980.

young organisms of many species come armed with programs of perceptual development, which, in their own way, are as canalized as those for the development of a limb or a heart.

As we conceive of such programs, they focus the developing infant's attention on particular aspects of environmental stimulation that will form the fundamental building blocks of perceptual organization (Marler, 1977). Among the genetic specifications for such attention-focusing programs, some are broad, such as those that establish levels of receptor sensitivity. Others are highly specific, allowing selective responsiveness only to certain key features of the environment. Whereas some operate generally, others are state dependent. Some specifications are absolute, others are relative, and manifest only in choice situations. Such preferences may be durable, or they may be transient. They may be resistant to environmental influence, or highly subject to modification by experience.

We believe that the adaptive significance of many such innate specifications is intimately related to the nonrandom structure of natural environments. Preferences expressed in one stimulus domain, and at certain stages of life, will exert strong and pervasive influences on the likelihood of other particular stimuli being encountered, so imposing further probabilistic constraints on perceptual development. In this way, unique features of genetic constitution, social organization and ecology, each causally related to the other in a myriad of ways, all have an influence on the development of the adult perceptual "Umwelt," as distinct for each species as its morphology and way of life.

In past studies of the fundamental operations of the sensory systems, the emphasis in traditional psychology has been on the use of simple stimuli. The works of E. J. and J. J. Gibson 1969, 1966, illustrate the advances that ensue from appreciation of the ecological significance of stimuli employed in the analysis of perceptual development. Ethological studies of the stimulus control of behavior have a similar bias, as in use of the complex stimuli that emanate from inanimate models of members of the species in postures of display. In the auditory domain, recordings of natural and modified communicative sounds are often used, thus exploring the efficacy of "sign stimuli" in "releasing" natural patterns of behavior. Recent efforts to combine ethological and psychological approaches, especially in studies of responsiveness to biologically significant sounds, offer promise of new insights into the physiology and psychology of perception (Bullock, 1977). Here we seek to extend that approach to the study of general perceptual development.

One current theme of ethological studies on sign stimuli is the existence of hierarchies of responsiveness to particular features of complex stimuli. Certain stimuli are found to be optimal in achieving the release and orientation of particular behaviors. The specificity of such optimal responsiveness has been demonstrated in a wide variety of species (Tinbergen, 1951), but whereas the specificity of *optimal* stimuli for evoking certain responses is irrefutable, the class of *minimally* adequate stimuli is often large. As a result, deprivation from optimal stimuli need not by any means prohibit occurrence of the response. Instead, the organisms will respond to some less preferred stimulus. The importance of understanding the natural ecology of stimulus situations is immediately evident. In the laboratory, it is easy to overlook the varying probabilities of encountering stimulus situations in nature. Often it is only in the context of such natural, species-specific environments that genetic adaptations for perceptual development can be understood.

Perceptual Adaptations to Local Conditions: Honeybees

Honeybees can be trained to approach a great variety of scents in their search for nectar. It would be easy to conclude from laboratory experiments with odorous compounds drawn at random from the reagent shelf that the range of perceptible olfactory stimuli to which they can learn to respond is unlimited, and genetically unconstrained. The lack of potential limitation may be correct. The conclusion that there are no genetic constraints would be false (Lindauer, 1970).

Studies with natural stimuli demonstrate that, when given a choice, honeybees respond preferentially to odors that are common components in the perfume of nectar-bearing flowers. The rates with which bees of different races learn to associate odors with food vary. Each race learns to respond most rapidly to odors generally characteristic of its own native flora (Koltermann, 1973).

> Orange, fennel, jasmine and oil of rosewood are learned by *Apis mellifera carnica* and also by *A. m. ligustica* and *A. m. fasciata* much less readily than lavendar and rosemary: the Indian species *A. cerana*, however learns the scent from oil of rosewood most readily, followed by orange, fennel and jasmine. Thyme and anise are also learned better by *A. cerana* than by the honeybees of the West. These preferences correspond to the relevant native food plants, with the exception of the odor of orange. Oranges originated in southeast Asia and have been cultivated in Mediterranean countries for only 300 years. Evidently this time period has been sufficient to allow this learning disposition of western honeybees to become adapted to this addition to the available food plants [Lindauer, 1975, pp. 231-232].[2]

Although conditioning to odor cues is easier than to color cues, honeybees can be trained to approach many colors when they are associated with food reward. Once more there are preferences. They learn to respond more quickly to violet, blue, or yellow markings than to blue-green ones (Menzel,

---

[2] This quotation and all subsequent quotations cited to Lindauer, Evolutionary aspects of orientation and learning, 1975, are reprinted by permission of Oxford University Press.

1967). Again the preferences are ecologically appropriate, and different races show different learning predispositions.

> The races and species vary with regard to the strength of the reward required to achieve such alteration of preference through training. *Apis cerana* requires the greatest reward before any alteration of its natural preferences becomes measurable. It is followed by the Egyptian race *A. m fasciata*, and the Italian *A. m. ligustica*, with the mid-European *A. m. carnica* bees having the most easily modified preferences. Doubtless the various degrees of flexibility shown by the diverse populations are adaptations to the nectar supply of their native habitat. In the tropical regions of East Asia nectar is available in surplus throughout the year, while in more seasonal climates of mid-Europe it is advantageous to be able to adjust to seasonally varying food plant compositions and to accept food of low concentration [Lindauer, 1975, pp. 232-233].

More abstract learning predispositions are also manifest. Lauer and Lindauer (1973) have demonstrated a difference between the *carnica* and *ligustica* races of honeybees in their readiness to acquire responsiveness to a visual pattern at a feeding site when it is moved independently of other environmental features. *Carnica* bees do much better under these conditions, and Lindauer explains the difference as follows:

> *A. m. ligustica* originating in the Mediterranean area, are able to fly in sunny weather throughout the year. Sun-compass orientation can thus be relied upon for guidance directly to the goal. *A. m. ligustica's* learning capacities themselves are focussed on the recognition of patterns. *A. m. carnica* bees, natives of mid-Europe, have to orientate themselves more often under a partially or fully overcast sky and cannot rely on the sun-compass alone; thus it is adaptive for them to use landmarks which are associated with the food goal [Lindauer, 1975, p. 241].

Honeybees thus exhibit genetically controlled, ecologically appropriate stimulus preferences. Some are relatively fixed, whereas others are so malleable that experience can completely reorganize the relative ranking of different stimuli in a preference hierarchy. We believe that such innate stimulus preferences, manifest in a regular ontogenetic program, can have profound implications for species-specific perceptual development.

Sensitive Periods for Perceptual Change: Chick Pecking

The development of the pecking of young birds illustrates how color preferences may be translated into complex oriented behavior as a result of perceptual learning, but with modifiability varying greatly from one stage of development to another. As in the honeybee, adaptive differences in innate color preferences have been demonstrated. Several plant-eating ducks and pheasants have an innate tendency to peck preferentially at yellow and green (Kear, 1964). Grain-eating domestic chicks, however, are innately predisposed to peck at yellow or orangered, depending on the strain, with green as one of the least preferred colors (Hess, 1956; Hess & Gogel, 1954). They are also innately responsive to such features of form as grainlike shapes, three-dimensionality and illumination from above (Goodwin & Hess, 1969). In the first minutes of patterned visual experience, when chicks are presented with a graded series of solid forms seen through transparent plastic, they clearly favor round over angular forms. Round solid forms are also preferred to similar flat ones (Fantz, 1957, 1965).

Notwithstanding these preferences, young chicks will peck at a wide variety of objects differing in color, texture, and shape, if they fall within a certain size category. It is evident that they quickly come to discriminate between food and nonfood items. At least a partial explanation for the emergence of this discrimination is that the developing chicks begin to learn about two sorts of properties. In addition to the exteroceptive features we have already mentioned, at about the third day posthatching, they begin to learn to associate pecking with food ingestion. Hogan (1973) has reviewed in detail the complex interrelations of these two factors.

Among chicks' innate form preferences is a tendency to peck more at circles than at triangles (Hess, 1962). If naive chicks are given a choice between a white triangle on a green background and a white circle on a blue background, they clearly prefer the later. Hess set out to modify this preference by placing seeds behind the triangle on green. Chicks quickly discovered the seeds and shifted their pecking preferences accordingly as long as seed was present. After 2 hours of such rewarded pecking, the seed was withdrawn and they were tested on extinction with the same stimuli. Persistence of the new preference was strongly dependent on the age at which this 2-hour reinforcement was given, peaking sharply when this experience was given at 3-4 days (Hess, 1962). It is perhaps no coincidence that this is the age at which the yolk sac is finally exhausted. These data led Hess to postulate a "critical period" for the greatest modifiability of early food-object preferences.

Hogan subsequently demonstrated that the chick's ability to recognize food is more complex than a simple "critical period" hypothesis would suggest. He considered three types of experience, each capable of influencing pecking; the behavior of the mother hen; responsiveness to tactile features of food objects; and long-term effects of ingestion. Several experiments suggest that the last are most influential in permanently changing the pecking preferences of young chicks. Again the effects first appear around the third day of age. However a study of effects of reinforcement by forced feeding on discrimination between food and sand yielded the paradoxical result that reinforcement increased subsequent rates of pecking, but did not influence the discrimination of nutritive and nonnutritive objects (Hogan, 1973). Although more work is needed, it may be that the Hess result is more concerned with learning *where* to peck for food than with *what* to peck at. Also the relative timing of perceptual experience and reinforcement is another issue inviting further attention that we will not pursue here.

Although many questions remain, chicks obviously can acquire responsiveness to the new food items. One could not hope to understand the natural acquisition process, however, without an appreciation of innate predispositions brought to the learning task. Chicks have innate form and color preferences (Baerends & Kruijt, 1973). The speed of learning to discriminate between colors is significantly greater if innate preferences are reinforced than otherwise, as though unlearned preferences bear simple additive relationships to effects of experience, a conclusion suggested by studies on readiness of chicks to approach colors (Kovach, 1971; Kovach & Hickox, 1971). There are sensitive periods for maximal modifiability by food reinforcement. In nature, these innate proclivities will have inevitable consequences for the full emergence of feeding behavior, favoring responsiveness to natural food items, while permitting survival in circumstances in which only unpreferred stimulus situations offer themselves. This flexibility is an essential feature of the operation of innate constraints on perceptual learning in higher organisms.

Imprinting and Perceptual Development: Birds

It is ironic that in their major contribution to the understanding of animal learning, namely the discovery and analysis of imprinting, ethologists originally underestimated the role of innate factors. At least two distinct phenomena were incorporated under the rubric of imprinting as originally defined by Lorenz. One is concerned with parent-young attachment—filial imprinting—the other with sexual preferences in adulthood—sexual imprinting (Hess, 1973; Immelmann, 1972; Lorenz, 1935, reprinted 1970). Immelmann (1972) has reviewed numerous cases in which special rearing conditions affect mate choice either intraspecifically or across species. Domesticated strains of ducks, chickens, pigeons, and finches, foster reared by a different strain than their own, may exhibit a sexual preference for that strain in adulthood. In nature, different color phases of the snow goose exhibit selective mate choice, probably as a consequence of imprinting on the parental type (Cooke, 1978). By similar means, imprinted sexual attachments to members of other species have been demonstrated in many birds (review in

Immelmann, 1972). The sensitive period for sexual imprinting is different from that for filial imprinting, occurring later in life, with longer exposures required.

As Lorenz emphasized, the influence of imprinting on adult sexual behavior is preferential rather than absolute. Although the imprinted object may be preferred, sexual interaction with other partners is by no means excluded. There are many cases of ambivalent behavior in animals that breed successfully with a member of their own species while exhibiting an abnormal imprinted preference for another. Clearly, there is flexibility in the range of objects that are acceptable for sexual imprinting. Nevertheless, there are species-specific preferences for particular objects.

Immelmann concludes that birds generally imprint most easily on their *own* species. If no conspecifics are available, they imprint on species similar to their own more easily than on dissimilar ones. The lines of evidence for this conclusion are derived from a variety of experiments. In a choice situation for sexual imprinting, a conspecific is often preferred. Sexual preferences for the bird's own species are always more rigid than those for another, and the time constraints on reversal are different. With zebra finches Immelmann (1972) found that reversal of a sexual preference can be achieved up to about 40 days if the subject was foster reared by another species, but only up to about 20 days if it was reared by its own species.

Schutz (1965) has provided striking evidence of a capacity for innate recognition of sexual partners in birds renowned for their imprintability, namely ducks. Working with sexually dimorphic species in which the male has distinctive species-specific plumage patterns, he found that females were not imprintable. They became sexually responsive to conspecific males at maturity more or less irrespective of earlier experience. Males, however, were imprintable on other species, perhaps as a correlate of the greater visual similarity of female ducks of different species, as compared with males. Relevance of this correlation is implied by Schutz's further finding that, in a sexually monomorphic duck, the Chilean Teal, females are as easily imprinted as are males.

A contrasting result was obtained with the sexually dimorphic zebra finch, females being strongly imprintable on another foster species, the Bengalese finch (Sonnemann & Sjölander, 1977). Nevertheless the imprinted females still showed a considerable interest in their own species, whereas males showed a clear-cut preference according to imprinting. Yet even imprinted males will court and breed with females of their own species if the imprinted species is not available (Immelmann, 1972).

Given the evidence that at least females of most imprintable species have an innate capacity to recognize a conspecific sexual partner, and, as it seems likely that males have at least potential access to the same information, it is legitimate to ask why learning should intrude at all in the process of selection of a reproductive mate. One answer is that there is far more to mate selection than just a preference for members of the same species. As sociobiologists have reminded us, questions of subspecies, population, age, social status, and kin relationship all bear on the appropriateness of a partner as an ideal mate, to say nothing of a multitude of individual differences that also affect a mate's fitness. Responsiveness to many of these features could only be acquired through learning.

Bateson and Jaeckel (1976) have recently indicated one probable component of considerable subtlety. It is derived from their finding that chicks respond optimally to a parent surrogate that, although very similar to the original imprinted object, nevertheless exhibits a degree of novelty, superimposed on the preferred familiarity. The authors suggest that imprinting serves to aid young in recognizing close kin, and that the preference for novelty is designed to strike an optimal balance between inbreeding and outbreeding (Bateson, 1978).

A different interpretation for the functional significance of filial imprinting has been offered, although it too is concerned with development of the ability to recognize close kin. Bateson (1978) points out that many parent birds discriminate between their own offspring and others, and may attack and kill alien young. Thus it behooves young birds well to learn to discriminate parents from other adults early in life.

In filial imprinting, there is again strong evidence of interplay between perceptual learning, on the one hand, and innate constraints, on the other. Although the sensitive period for the first stage of the filial imprinting process is such that the learning process is often compressed in time, we can again view the organism as proceeding from an innate selective responsiveness, through a series of developmental stages. The simple, rather generalized initiating stimuli are gradually supplemented and superceded by a particular complex stimulus constellation. The end result is a recognition system uniquely matched to the demands of the social environment of the young organism.

If we consider the preferred stimuli for the early initiation of approach and following in, for example, ducks and geese, Gottlieb (1971, 1974) has shown that ducklings are innately capable of highly specific responsiveness to calls of a conspecific parent. In the absence of this optimal stimulation, they will also respond to a variety of other calls, as Lorenz showed with his proverbial "kom, kom, kom." By contrast, the present picture in the visual domain is one of innate responsiveness to relatively unspecific stimuli. Certain size conditions must be met—if the moving object is too small it will evoke pecking, and if too large, fleeing (Fabricius & Boyd, 1954). There are particular color preferences, but, for the most part, these have not yet been found to relate in any obvious way to species-specific adult coloration. This is surprising in view of the clear evidence from sexual imprinting that this information is innately available at a later age. One wonders whether further experimentation with live parents might reveal influences that have been overlooked thus far (cf. Fabricius, 1951).

Notwithstanding the evidence for innateness in responsiveness to stimuli that evoke following, responsiveness is modified rapidly through learning. In addition to the visual aspect (e.g., Miller & Emlen, 1975), individual auditory recognition between parent and young arises early in many precocial species (Beer, 1970; Cowan, 1974; Evans, 1970, 1972, 1977; Evans & Mattson, 1972; Impekoven & Gold, 1973; Mattson & Evans, 1974).

By the time a chick or duckling reaches adulthood it has acquired a great deal of perceptual information about its social companions, the vast bulk of it unquestionably learned. It is equally obvious that the direction taken by that learning is profoundly influenced by innate instructions, establishing selective responsiveness to particular subsets of stimuli at particular times. The selectivity is preferential and not absolute, exploiting certain probabilities inherent in the social relationship, such as the likelihood that an adult encountered by a newly hatched bird will indeed be close kin. Yet the potential plasticity is considerable, allowing for departures from the typical pattern of species-specific stimulation should they arise.

Configurational Features and Sign Stimuli: The Herring Gull

Tinbergen (1948, 1951) pointed out that a mere knowledge of the potential capacities of the sense organs never enables us to point out, in any concrete case, the actual complex of stimuli responsible for the release of a reaction. Beginning in the early 1930s, ethologists developed a variety of ingenious methods for characterizing such effective stimulus complexes, found to "release" given innate responses. By disguising live animals, and through the use of experimentally modified models, many innate responses were found to be triggered by a limited subset of the total array of stimuli presented by a given situation, dubbed as "sign stimuli." Tinbergen found that the red belly of a reproductive male three-spined stickleback was a sign stimulus for triggering attack in territorial males, and that the red underside would elicit this response in the face of a variety of abnormal distortions of other features of a normal male's appearance—features unquestionably visible to the fish, but ineffective in this situation.

Sign stimuli were found to be effective in a sufficient number of studied cases that dependence on them for elicitation was originally viewed as a diagnostic criterion of innate behavior. As such, this dependence was thought to differentiate innate reactions from conditioned ones, viewed as "not usually dependent on a limited set of sign stimuli, but on much more complex stimulus situations [Tinbergen, 1951, p. 37]." Viewpoints on this have changed with further research, as on the issue of innate responsiveness to configurational stimuli (Lorenz, 1970). There is evidence of such a transition in studies of one of the most celebrated ethological subjects, namely the pecking responses of gull chicks.

Tinbergen studied the preferred visual stimuli for pecking of herring gull chicks, and interpreted the results in terms of an ethological "innate release mechanism." As such, this would constitute a special perceptual mechanism adapted to match responsiveness of the pecking chick to coevolved stimulus features of the parent's bill. In the original conception, innate release mechanisms were viewed as implying innate perceptual schemata of stimulus objects (Lorenz, 1970). It has gradually become evident that the term *schema*, with its image-like connotations, is probably inappropriate for these innate predispositions. Instead, one typically finds "a mosaic of extremely simple receptor correlates activated by specific key stimuli [Lorenz, 1970, n. 56, p. 375]." Yet, although these initial triggering stimuli, to which responsiveness is innate, may undergo rapid supplementation or modification through learning, they are strongly valent in early occurrences of the pecking response. It follows that they play a significant role in the chick's acquisition of learned responsiveness to selected aspects of this stimulus complex presented by the parent, and thus in the acquisition of a full, learned, parental schema.

The herring gull feeds its young by standing above them with food. A chick pecks at the parent's bill and eventually hits the food. Stimulus properties evoking this pecking response include a particular bill shape, a red patch at the tip, characterized by both hue and contrast. Bill orientation and patterns of movement also have an influence (Tinbergen, 1953, 1973). In his now classical studies, Tinbergen emphasized the "relational" or configurational property of some of the effective sign stimuli such as the position of a red spot on a headlike shape.

When Hailman (1967) reanalyzed this situation, however, taking care to control the earliest experience of his gull subjects he found that the sign stimuli for pecking in fully naive chicks, now of the laughing gull rather than the herring gull, could all be interpreted in nonconfigurational terms. For example, Tinbergen emphasized the greater valence of a red patch on the bill tip over one on the forehead of the gull model. Hailman pointed out that the greater effectiveness of a bill-tip spot was attributable to the difference in arc traveled during pivotal movement of the stimulus model. Correcting for this, a spot on the forehead was as effective as one on the bill tip in fully naive laughing gull chicks. Optimal bill width also proved definable in absolute terms.

After a week or so of normal feeding experience, however, Hailman detected changes in the stimulus control of pecking. Now the chicks had become responsive to three-dimensional properties of the parent head that were irrelevant to the pecking of naive chicks. Most interestingly, configurational properties became significant, as though the chick was developing a more complex perceptual schema of the parent head. Now a red forehead spot was indeed less effective than one on the bill tip, irrespective of its rate of movement. Clearly the visual aspects of a week-old chick's perceptions of its parent have undergone considerable changes from birth as a result of learning. In addition, it has probably learned a good deal about its parents' voices. As in other organisms, certain preferences of young gulls may be easily modified by learning, others less so. Both domestic chicks and laughing gull chicks have clear innate pecking color preferences. As we have seen, the former can be trained to shift preferences to a new color. This is possible with gull chicks (Weidmann & Weidmann, 1958), although only with considerable difficulty.

One may ask whether the concept of sign stimuli has application beyond innate responsiveness? When the components of a stimulus complex to which a learned response has been established are analyzed critically, we typically find that here too a particular subset is most salient in its control, although the effective subset may vary from subject to subject, or, in the same subject, from time to time (Mackintosh, 1974). As with innate behavior, some stimulus features can be varied without affecting the response, even though they are known to be perceptible to the subject. When octopus, goldfish, and rats are trained to respond to patterned visual stimuli, and then required to generalize to other patterns that shared some features with the original, but differed in other respects, all prove to be responding more strongly to certain elements in the stimulus complex than to others (Lashley, 1938; Sutherland, 1968, 1973). As Mackintosh cautiously expresses it, "where several stimuli are simultaneously relevant, an increase in control by one stimulus is accompanied by a decrease in control by the remaining stimuli [Mackintosh 1974, p. 618]." It seems probable that all responses of organisms, innate and learned, depend most heavily on an abstracted subset of complex stimulus arrays that we might think of as sign stimuli.

Obviously the mode of stimulus control over innate and learned responses is not identical. In the former case, all individuals of a species, given equivalent histories of experience, will tend to form the same abstractions, in a species-specific fashion. In the control of learned responsiveness, we may be prepared for more intraspecific variation. However, even in this case, suggestions have repeatedly been found of within-species and even across-species uniformities in the kinds of stimulus features upon which learned responsiveness becomes dependent (Sutherland, 1973). "The similarity of the findings for species as different as the octopus, the goldfish and the rat is very much more striking than are the differences. This strongly suggests that there is some optimal way to process visual information and that this method has been discovered in the course of evolution independently by cephalopods and vertebrates [Sutherland, 1973, p. 164]." One is inevitably drawn to the implication of innate bias in the processes that underlie perceptual learning (Herrnstein, Loveland, & Cable, 1976).

Innate Constraints on Vocal Imitation: Birdsong

Bird vocalizations are communicative in function, and obviously exist to be perceived. Male songs in particular are the most complex social stimuli known from the animal kingdom. Given their complexity, and the fact that all oscine bird songs are learned (e.g., Marler & Mundinger, 1971; Thorpe, 1961), there is promise of novel insights into mechanisms of perceptual development. Consider the perception of song by an adult, experienced bird. The evidence is increasing that the intricate variability found within the song of almost any oscine bird is not accidental, but comprises a set of controlled variations that have significance and meaning to the birds themselves, to do with species and population, neighbors and strangers, varying levels of motivation, personal identity, and so on (Falls, 1969, 1978; Kroodsma, 1976, 1979). An adult's experimentally demonstrable responsiveness to such varied, subtle features of the acoustic structure of song, unique to each individual's experience, could have developed only through learning. Some involve simple properties, others complex; some are absolute, whereas others are configurational in nature (Thorpe & Hall-Craggs, 1976). It follows that the learned, mental imagery that a bird has of its species' song must be rich and complex.

Can we then ignore *innate* influences in song learning? The answer is unquestionably negative. For one thing there are "universals" in song structure. Although there is enormous variability of the song patterns of most birds, careful study reveals that this variation is restricted to certain parameters of the song, with other features more stable. These universals are the features that birds and ornithologists rely on in identifying the species of birds, often done more quickly and accurately by voice than by plumage.

There is direct experimental evidence of innate influences. If we look back into a bird's early life and examine the process by which the song was originally learned, we find an interesting paradox. Although most bird songs are learned, the learning is typically selective. Everyone knows that some birds learn sounds not only from their own species but from others as well, as in the mocking bird, the European starling, the lyrebird of Australia and other famous mimics. But these are exceptions, and it is more general to find that even in those songbirds in which a Kaspar Hauser male sings a highly abnormal song, a wild male only very rarely learns the song of other species (e.g., Baptista, 1972).

How do such birds avoid learning the wrong song? In some species this is done through a social mechanism. Young birds learn from an adult of their species with whom they have established a particular social bond, typically the father. They learn whatever sounds he produces and, as he will normally be a member of their species, the learning process is canalized so as to preserve a certain set of species-specific characteristics. Having learned the main skeleton of the father's song, the young will then often improvise on certain aspects so that individuality comes to mingle with stereotypy (Immelmann, 1969; Nicolai, 1959).

In other species, selective guidance of the learning process appears to be achieved by a physiological mechanism in possession of the individual bird. An example is the white-crowned sparrow. Figure 5.1 diagrams the development of song in young male white-crowned sparrows both under natural conditions and in the face of experimental variation of the acoustic environment. It illustrates three key points.

First, if a male is presented with a choice of two songs to learn during the sensitive period, one a song of his own species, the other a song of a close relative living in the same environment, the young bird will learn the conspecific model and will ignore the alien one. The second key point is that the song of a Kaspar Hauser, although abnormal in many respects, has some natural features, including the presence of long sustained whistles. Also relevant is the fact that this feature also distinguishes white-crowned song from song sparrow song (Marler, 1970a; Marler & Tamura, 1964). The final point of interest is that these normal features are lost if a young male is deafened early in life. The song he develops without the ability to hear his own voice is a noisy tuneless buzz (Konishi, 1965). We infer that the capacity to produce the sustained whistles rests upon a perceptual ability. The very same ability would also provide a simple way of distinguishing between white-crowned song and song sparrow song. In the face of some biologically significant natural choices, this ability thus focuses the attention of the young male upon songs of his species.

How might this perceptual process develop? The bird behaves as though there is an innate focus of attention on sounds with a particular set of simple properties. In the process of attending to "acceptable" sounds, other associated acoustic features are committed to memory. These include not

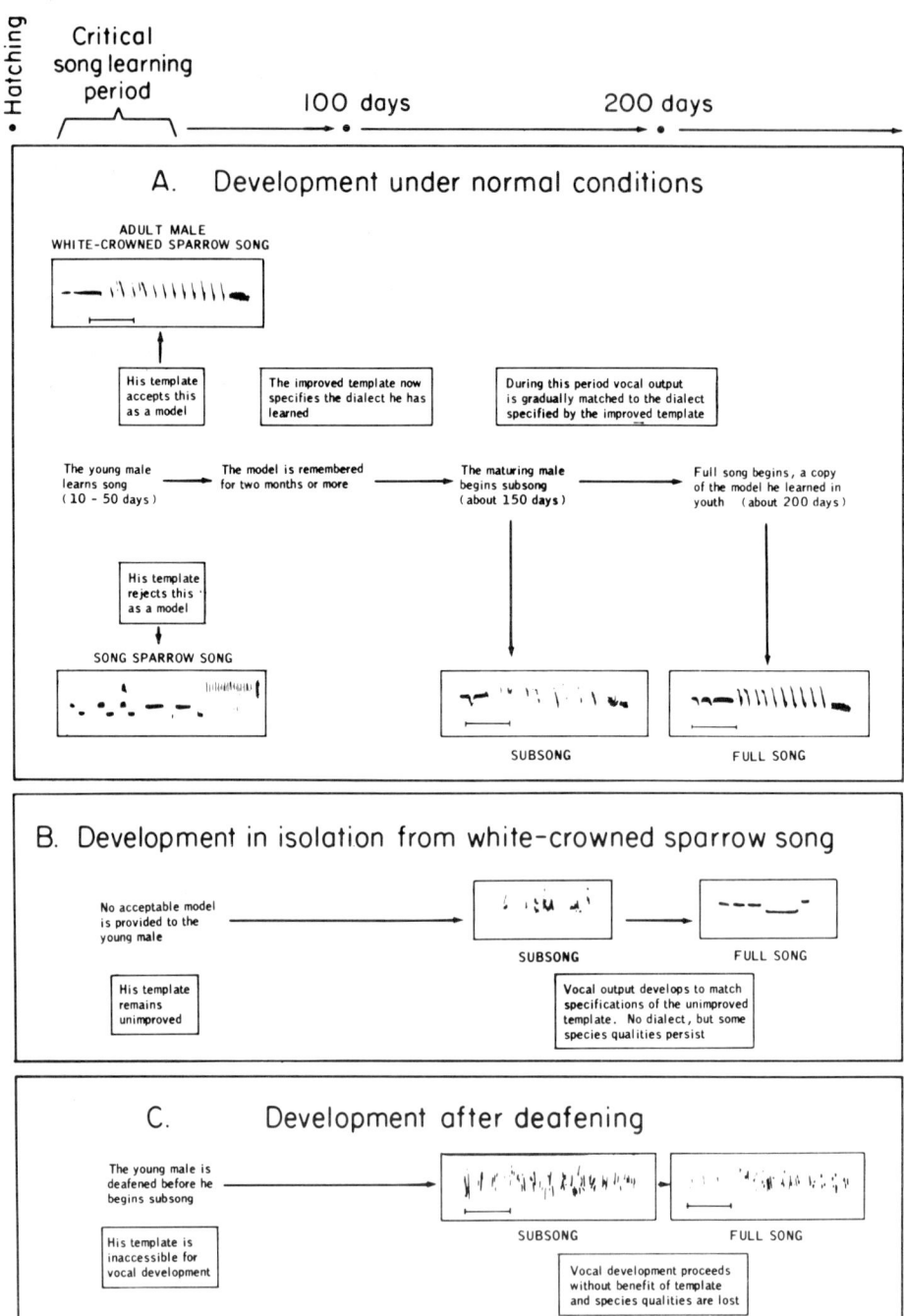

Figure 5.1. Song development of the male white-crowned sparrow: (A) under normal conditions; (B) reared in isolation from conspecific song from the early nestling stage; (C) after deafening early in life. (From P. Marler, On the origin of speech from animal sounds. In J. Kavanaugh & J. Cutting (Eds.), The role of speech in language. Cambridge, Mass.: M.I.T. Press, 1975. Reprinted by permission.)

only additional features of the species' song but also the particular dialect to which the young male has been exposed in his youth.

Like many song birds, the young male white-crown learns to sing from memory, only embarking on the gradual transition from subsong to full song after the sensitive period for the first, purely perceptual phase of the learning is completed. The male then behaves as though his next task is to match his vocal output to the memory of the particular pattern of sounds to which he was last exposed days, weeks, or even months earlier, during the sensitive period. In this sense then the learned auditory "engram" of song becomes used as a kind of template against which the young male matches his song. He goes through a series of increasingly accurate approximations to the pattern heard in youth until the representation is perfect, with improvisations introduced into those particular features of the song reserved for individual modifications and inventions (Marler & Mundinger, 1971). This interpretation is the essence of the modifiable auditory template hypothesis for selective vocal learning in birds (Marler, 1976a).

The male bird's behavior in the initial phase of the process in which he is first exposed to a medley of sounds from the environment is of special interest. He apparently rejects some sounds for the process of song learning, accepting only a limited set that have properties diagnostic of conspecific song. It is not inappropriate to invoke something equivalent to an innate release mechanism in the original Lorenzian sense. One might view the young male as possessing innate auditory imagery, albeit embodying only skeletal features of the song of the species. The young male would use this to select the appropriate model for learning, and then pad some of the detail through experience.

The nature and extent of the innate basis for perceptual learning in the white-crowned sparrow has yet to be studied in detail (but see Konishi, 1978). More is known about the swamp sparrow, which also learns its song in a selective fashion. It lives in close proximity with the song sparrow, a relative so close that it is placed in the same genus by bird taxonomists, yet whose song is quite different. In the area studied, almost every young male swamp sparrow has both swamp and song sparrows singing within earshot. How does it avoid learning the wrong song?

Marler and Peters used a series of semisynthetic songs to determine whether male swamp sparrows learn selectively when presented with a natural choice between swamp and song sparrow songs. Distinctively different sound clusters or syllables were edited from tape recordings of normal local songs of both species and then spliced together in a variety of simple temporal patterns, based on features of organization in which normal songs of the two species differ. Some patterns were "swamp sparrow-like". These included sequences of identical syllables at various steady rates. "Song sparrow-like" features included variable rates of delivery of syllable sequences (accelerating, decelerating) and a two-part structure (Marler & Peters, 1977).

After having been trained with the tapes between 20 and 50 days of age, male swamp sparrows learned to sing from memory, some months after termination of the training. They copied many synthetic songs and every one consisted of swamp sparrow syllables. No song sparrow models were copied. Thus the male swamp sparrows were extremely selective in their learning, accepting only conspecific syllables for imitation, and eschewing song sparrow syllables. Dooling & Searcy (1980) have since shown by monitoring heart-rate changes that 20-day-old swamp sparrows are much more responsive to conspecific than to alien song, long before they have started to sing themselves. In the song learning studies, the choice was clearly made at the level of components from which the song is constructed, and not the overall pattern of the song. Song sparrow syllables were rejected even though they were presented in similar temporal patterns. These results suggest that the first level of innate responsiveness seems to occur not to the overall pattern of species-specific song but rather to a more limited property of its component parts. This conclusion, reminiscent of Thorpe's (1958) inference from earlier studies of selective song learning in the European chaffinch, is also suggested by data on learning of synthetic songs by the white-crowned sparrow (Konishi, 1978).

Thinking again about the mechanisms that underlie the initial perceptual discrimination, it seems inappropriate to invoke a complete, innate, auditory image of the main features of the song. Rather what the bird seems to possess are some simple innate guidelines or signposts for a process of auditory learning. As with other innate release mechanisms manifest in infancy, the innate guidelines for avian vocal perception exist primarily to provide an ontogenetic basis for adult perceptual learning. By themselves, without supplementation by learning, they would be only minimally adequate as a basis for complex communicative behavior.

Species-Specific Processing of Vocal Stimuli: Macaques

Whereas innate constraints on perceptual learning are sometimes viewed as species-specific limitations, their phylogenetic distribution is often broader than this implies. If the issue is, for example, efficient form vision, it is reasonable to anticipate independent emergence of best solutions at more than one phyletic level (Sutherland, 1973). However, perceptual problems arise that are unique to a species as, for example, in social communication. In such cases, truly species-specific strategies will be required. Perceptual biases demonstrated by monkeys in auditory processing of their own vocal sounds illustrate such a case.

One of the most interesting characteristics of the sounds used in the communication of higher primates is the predominance of graded signals (Green, 1975; Marler, 1965, 1970b, 1975; Struhsaker, 1975). Graded vocal systems, as contrasted with the discretely organized systems found in some primates and many birds, are those where the amount of variability both within and between categories of signals is so great that they become connected by intermediate forms. In the extreme cases of grading, especially characteristic of such higher primates as the chimpanzee (Marler, 1976b; Marler & Tenaza, 1977) it becomes difficult for the observer even to assign vocalizations to a particular class. Often such judgments are based more on the observer's own comparisons with arbitrarily selected exemplars than on a detailed analysis of vocal morphology and usage. It is quite possible that, as human observers of the communication system of another species, we misinterpret the nature of the complexity of the sound systems used.

The functional significance of highly graded vocal systems could lie in their tremendous potential for conveying detailed information to others. Each variant could represent a unique and complex state of the vocalizing animal. By attending to such variation, a listener could accurately predict changing probabilities in the vocalizer's subsequent behavior (Zoloth & Green, 1979).

Before we can properly understand primate vocal repertoires and how they are used in communication, we must find some way to ask these primates to define for us the perceptual boundaries which they discern in a series of graded vocalizations. The task is reminiscent of the problem confronting investigators of speech perception by infants. A test is required that queries a nonlinguistic subject as to what it hears (cf. Snowdon, 1979; Snowdon & Pola, 1978).

An investigation of vocal perception in Japanese macaques (*Macaca fuscata*) sought to determine whether the acoustic parameters monkeys find relevant in auditory perception of calls in a laboratory situation are the same ones identified as useful classification guidelines from field studies of natural communication behavior (Beecher, Zoloth, Petersen, Moody, & Stebbins, 1976; Petersen, Beecher, Zoloth, Moody, & Stebbins, 1978; Zoloth, Petersen, Beecher, Green, Marler, Moody, & Stebbins, 1979). The latter were provided by Green's extensive field investigation.

Two call variants of the macaque *coo* vocalization were selected. Each variant of *M. fuscata's* graded repertoire is correlated with a unique, definable behavioral class (Green, 1975). One of the coo variants, the smooth early high (SEH), is given by animals isolated from companions and apparently desirous of making affinitive contact. The SEHs are given by all age and sex classes (excluding infants) and are often answered antiphonally. The second coo variant used, the smooth late high (SLH), is given by animals that are more highly aroused and actively seeking contact. They are given most typically by estrous females in the early stage of consort formation. However, all age and sex classes produce these calls, other than infants. As their names suggest, these calls are clear, tonal sounds with a single, unbroken frequency sweep. They can be typified by the temporal position of the frequency rise or "peak." In SEHs, the peak occurs within the first two-thirds of the call, in SLHs, in the final third. Thus it is possible to arrange SEHs and SLHs extracted from Green's recordings along the dimension of peak position and so form a graded series.

In the first experiment, using natural coos, Japanese macaques and monkeys of three other species, as controls, were trained to perform an operant response to playback of these calls (Beecher et al., 1979). Initially, it was relatively easy for all the monkeys to learn to distinguish between one SEH and one SLH in our test situation. As we introduced new examples of SEHs and SLHs, however, differences in rates of learning emerged. Generalization proceeded readily for the Japanese macaques. However, the other species, pigtail macaques (*M. nemestrina*), bonnet macaques (*M. radiata*), and vervet monkeys (*Cercopithecus aethiops*) seemed to find it difficult to acquire the ability to distinguish between calls. The calls were fully representative of the SEH and SLH types found in the field recordings of Japanese macaque vocalizations, and, as such, they varied considerably in acoustic dimensions other than the peak position, such as pitch, duration, and harmonic structure.

The task for each subject was to learn to identify vocalizations as belonging to the same class even though they differed along these other dimensions. The stimulus sets were so arranged as to be relatively stable only along the target dimension of peak position. We would argue, then, that as we introduced novel instances of SEHs and SLHs to our basic stimulus set, the animal's task changed from simple sensory discrimination of two different auditory patterns to the formation of a perceptual concept about auditory patterns.

After a great deal of "remedial" training, our control species monkeys eventually reached a level of accurate response similar to that of the Japanese macaques. Thus we could demonstrate that the task is not impossible for them, simply harder. The data are consistent with a hypothesis that these Japanese macaques were more capable than the other species of perceiving natural and inherent relationships existing among vocalizations taken from the *fuscata* repertoire.

In a second experiment, we rearranged the same coo vocalizations to present a discrimination task in which the target dimension is the onset frequency, with peak position as an irrelevant variation. Japanese macaques acquired this pitch discrimination much less easily than they solved the peak position task. The converse was true for the control monkeys tested (Figure 5.2). Other macaques and vervets learned the pitch discrimination more quickly than the peak relevant discrimination (Zoloth et al., 1979). As in the first experiment, these data suggest employment of differing strategies by these two species when discriminating between *Macaca fuscata* vocalizations.

One hypothesis generated by these listening tests is that the difference in the species' abilities to learn this discrimination has its basis in the existence of "natural categories." Rosch's (1973) data on cross-cultural patterns of color category learning, to be discussed next, suggest the existence of such perceptual divisions, which she defines as "nonarbitrary partitioning-domains whose stimuli are not discrete but composed of continuous physical variations."

A typical attribute of natural categories is that values or exemplars exist

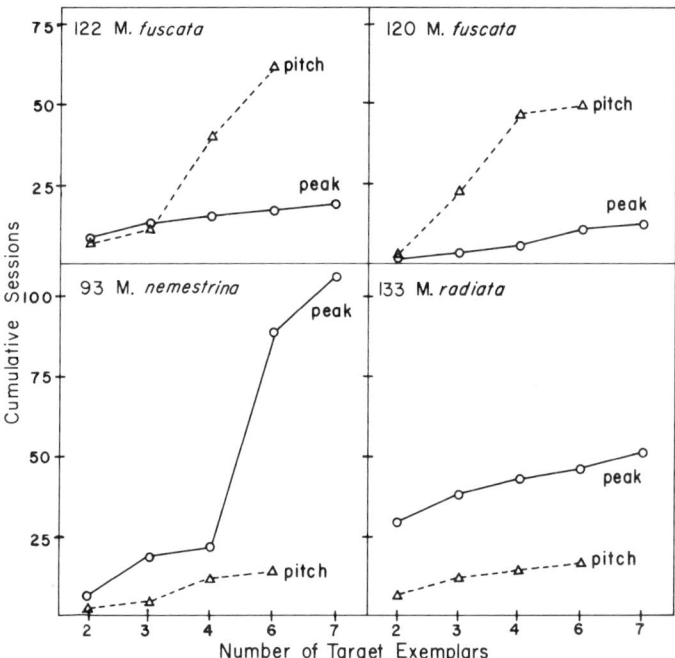

Figure 5.2. *Generalization rates of two Macaca species while being trained to discriminate between sets of natural calls that differ in frequency-peak-position. This feature is believed by Green to be employed in natural communication by Japanese macaques. With the "peak relevant" task the monkeys had to discriminate sounds on this basis, while ignoring variations in starting pitch. In the "pitch relevant" test, they had to sort them on the basis of starting frequency of the fundamental, while ignoring the peak position, something that the Japanese macaque obviously found more difficult. (After S. R. Zoloth, M. R. Petersen, M. D. Beecher, S. Green, P. Marler, D. B. Moody, & W. Stebbins, Species-specific perceptual processing of vocal sounds by monkeys. Science, 1979, 204, 870-873. Copyright 1979 by the American Association for the Advancement of Science.)*

which are more "perceptually salient" than others. These focal exemplars are said to more readily attract attention and to be more easily remembered than less salient stimuli (Rosch, 1973; Rosch-Heider, 1971, 1972). One might predict that between-category discriminations based on dimensions that diagnose natural categories would be easier to perform than those relying on other attributes of the same stimulus complexes. This would be concordant with the results of our experiments in which the two different dimensions of coo vocalizations were placed in competition.

There are echoes here of studies in speech perception, using a variety of tasks, in which competition is established between linguistic and nonlinguistic cues, such as phonetic identity and pitch (Kuhl, 1976; Springer, 1973; Wood, Goff, & Day, 1971). The results indicate that tasks requiring discrimination along linguistically relevant dimensions are often easier to perform, perhaps as a result of these dimensions being more attention getting than those along nonlinguistic dimensions. Although there is some disagreement over the phylogenetic status of the human case (Kuhl, 1976; Pisoni, 1978), in the monkey example it seems likely that truly species-specific perceptual predispositions are involved.

Color Vision, Naming, and Preferences

Behavioral universals are one source of information as to how perceptual worlds are organized. Some are derived from cross-cultural studies of the perceptual ability of infant and adult humans, posing such questions as—what features of the environment are attention-getting?; which are discriminable?; what is there in common among responses to different stimuli that reveals underlying tendencies for perceptual classification?

Human color vision and color naming illustrate how subtle, yet pervasive the effects on perception can be of what we may consider as predetermined constraints on sensation (Bornstein, 1978). Color naming was long thought to reflect the ultimate in environmentally determined influence on perceptual classification. The strongest statement was given by Whorf (1964), who argued that the environment has a direct influence on the labels given to various colors and that these names effectively determine perceptual organization. In opposition to Whorf's hypothesis was the viewpoint that mechanisms inherent to the process of visual perception preordain the basic color terminology. Resolution of this controversy came from three lines of evidence; seminal was an increased understanding of the physiological mechanisms underlying color vision (DeValois, 1973; Hurvich & Jameson, 1957; Ratliff, 1976) in humans and in other animals; another important line of evidence is derived from study of color naming across languages (Berlin &

Kay, 1969); of equal importance are studies of the processes of color classification in newborn human infants (Bornstein, 1975; Bornstein, Kessen, & Weiskopf, 1976a, b). Taken together, these studies demonstrate that constraints on the sensory apparatus have a strong influence on the perception of color and ultimately on the process of color naming.

When saturation and brightness are held constant and a series of color chips varying in hue are presented to English-speaking adults, they will label the entire set as if they were members of four basic categories: red, green, blue, and yellow, the primary colors. Labels can also be assigned experimentally for other species as diverse as bees, pigeons, and macaques, although the actual hue categories employed by these species are, of course, different from those used by people. Pigeons, for example, divide our visible spectrum into three primary categories with a substantially different placement of boundary wavelength values (Wright & Cummings, 1971). Whatever the differences, there seems to be a universal tendency to partition the spectrum into perceptual categories with regular boundaries between classes, a predisposition that is obviously independent of an ability to provide linguistic labels.

The effect of this categorization is also evident if a person is asked to discriminate between exemplars of different colors. Although we can discriminate between a great many different hues, it is possible to rank the relative ease of such discriminations. When this is done, discrimination is found to be more difficult with hue pairs near the center of the labeled color categories than with pairs across boundaries. As a result, perceptual categorization naturally imposes some restrictions on our perceptions of the visual wavelength continuum.

Research into the physiological mechanisms underlying color vision demonstrates how neural activity of the retinal cones is transformed by interactions within the retina and in higher visual centers into an opponent process color system. Much of the evidence is derived from research into visual processes in macaques (DeValois, 1973). Recording from single units within the lateral geniculate nucleus demonstrated the existence of four types of spectrally opponent neurons (DeValois, Abramov, & Jacobs, 1966). Each type is excited by one wavelength and inhibited by a second or opponent wavelength. The optimal wavelength values for these neurons divides them into groups that respond to two pairs of opponent hues, red versus green; yellow versus blue. Each group contains neurons that are excited by one member of the pair and inhibited by the other. From these data and from psychophysical evidence, it becomes possible to define an explicit model of color vision that accounts for the existence of the four primary hues (Abramov, 1977).

Further evidence of the primacy of the tetrachromatic interpretation comes from cross-cultural studies of color names. Berlin and Kay (1969) investigated color naming in a large number of language groups. By interrogating native speakers about color names and using standardized color chips they were able to determine which hues serve as best examples of the color categories, and where the boundaries between the categories fall. They demonstrate eleven color categories from which all color names are drawn. Most importantly, Berlin and Kay showed that, once names are given beyond black and white, the four color categories to appear first in any language are always the four primary colors. Thus the simplest color systems have names only for black and white; those languages with three color terms have words for black, white, and red. The order of linguistic development of color terms always progresses through the primary hues before extending to names for compound colors. The regularity of color naming systems and their parallel development across many languages suggests that this process reflects innately determined species-specific predispositions. Finally, using the Berlin and Kay cross-cultural data, Bornstein (1973) was able to show that the "best examples" they obtained for each hue are a good match with the DeValois data on the wavelength to which single units are most sensitive in the macaque.

In another cross-language study, Rosch demonstrated that the Dani of New Guinea, whose color-naming system is organized by brightness rather than hue were still able to recognize hue categories (Rosch, 1973; Rosch-Heider, 1972). Constructing categories around "focal colors" as defined by the Berlin and Kay study, Rosch demonstrated that these categories were learned faster and with fewer errors than were categories constructed around nonfocal colors. Thus, even when a color space is linguistically unlabeled, it does not remain undifferentiated.

These results demonstrate that the perception of color rests on a fundamental biological foundation, common to all members of the species rather than originating within a particular linguistic or social structure. The final piece of evidence on this issue is also the most compelling. Data gathered from prelinguistic infants strongly suggest that they perceive colors as belonging to the same categories as do adults. Bornstein et al. (1976a, b) habituated the fixation response of newborn infants to a colored stimulus by repeated presentations. When the subjects reached a certain habituation criterion, a probe stimulus was introduced at some other wavelength. If this were perceived by the infant as different from the first, it would produce dishabituation and fixation would be resumed. Stimulus pairs were chosen to straddle adult color boundaries or to fall within a single adult category. The latter resulted in less dishabituation than the former, demonstrating that infants partition the spectrum into the four basic hue categories. The implication from Bornstein's data is that the human infant brings into the world a set of innate perceptual proclivities that serve as guidelines for any subsequent linguistic organization of color space (Bornstein, 1978).

In summary, color categories in human and nonhuman species rest on the functional organization of the visual system. The tendency to divide the spectrum into perceptual categories, with focal regions and relatively fixed boundaries is ubiquitous in our species. As such, it provides a set of guidelines for the developing infant's organization of color space. The color naming systems of different cultures exhibit many idiosyncracies, such as the use by the Dani of intensity dimensions of light and dark as a basis for defining color space. We are reminded that, in the ontogeny of links between color and behavior, perceptual predispositions such as we have described are not immutable. They are by no means completely deterministic, and can be overridden by environmental and cultural effects. Yet we cannot begin to understand the ontogeny of color perception without taking them into account.

The coexistence of a universal set of predispositions together with the great variety of color-naming systems used in human societies illustrates the subtlety of visual constraints on the development of behavior associated with the division and labeling of color space. The "preferences" of humans for focal hues in a color-naming experiment resembles the type of animal experiment in which one aspect of a compound stimulus seems to have more salience and tends to take priority as a conditioning stimulus when choice is permitted. We have already suggested that many animal preference experiments can be reinterpreted as providing evidence for innate guidelines for the subsequent development of learned behaviors. As the growing organism encounters new problems in visual discrimination tasks, it will tend to persevere with stimulus dimensions already successfully employed in previous discriminations and to resist transfer of control to other dimensions (Mackintosh, 1974). Thus, if they are expressed at critical stages of behavioral development in environments that provide the appropriate choices, innate stimulus preferences will almost inevitably have consequences into adulthood.

The Ontogeny of Speech Perception

Having considered the ontogeny of human color vision and naming, we now extend our argument to the more complex behavior of speech perception (cf. Bornstein, 1978). In our view, research on the perception of speech by adults and human infants provides further support for the notion of canalized perceptual learning. According to this interpretation, development is guided first by innate responsiveness to certain stimuli, which subsequently becomes integrated into the process of formation of schemata for speech sounds, the latter so complex that we still cannot fully specify the acoustic carriers of the phonetic message.

Given the remarkable acoustic complexity of speech, it is surprising to a nonlinguist to learn that there are universals, recurring across all languages, in certain physical features of speech patterns that define boundaries between functionally distinct patterns of sound. We can best illustrate this by reference to the distinction in many unrelated languages between critical pairs of voiced and unvoiced and prevoiced consonants. The property known as "voice-onset-time" (VOT) has been a focus of special study, as it is a characteristic of speech that can be reliably measured from the frequency-time sound spectrograms on which so many bioacoustical studies are based.

As an example from English, the cross-cultural studies of Lisker and Abramson (1964) have shown that all languages studied employ VOT as one basis for differentiating stop consonants in speech. Furthermore, when there are VOT boundaries, they always fall in approximately the same place, at one of two locations (Figure 5.3). Universals have also been found in the patterns of formant onset that differentiate speech sounds produced at different points of articulation; labial, alveolar, and velar (e.g., [ba]—[da]—

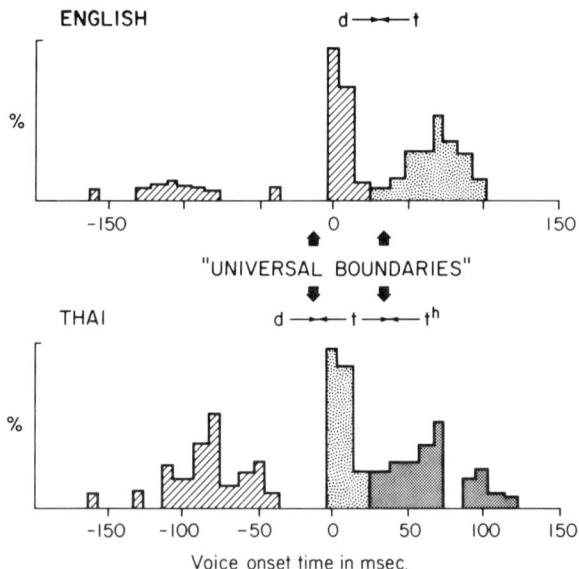

Figure 5.3. *Measurements of speech sounds: histograms of VOTs in stop consonants in English and Thai. The large arrows indicate the position of the perceived "universal" boundaries. Inserts show three examples of synthetic speech with VOTs of −150 msec, +10 msec, and +100 msec. (After Lisker & Abramson, 1964, and Cutting & Eimas 1975.)*

[*ga*]). There is a long list of other universals (Greenberg, 1969; Studdert-Kennedy, 1977), but these properties of consonants have the advantage that they are specific and lend themselves to precise analysis and experimental control. When such universals are discovered in ethograms of animal behavior, an ethologist is likely to entertain the possibility of genetic developmental controls.

A further relevant finding is the recurrence of "categorical perception." We have already described its characteristics in color vision, although it was first described in speech perception (Liberman, Cooper, Shankweiler, & Studdart-Kennedy, 1967; Liberman, Harris, Hoffman, & Griffith, 1957; Liberman, Harris, Kinney, & Lane, 1961). Such experiments employ stimulus series such as the VOT continuum, in 10 or 20 msec steps from [*pa*] to [*ba*] to [*mba*].

Asked to label sounds on such a continuum, an English-speaking subject divides them into two parts, labeling one side [*pa*], the other [*ba*], with a sharp boundary between that coincides with the trough in VOT productions. This boundary recurs in different languages, although with details that vary consistently from one to another. In some languages, such as Thai, there is a second boundary, around −20 msec VOT shared by the speech patterns of other cultures.

If adult subjects are tested for the discriminability of sound pairs differing by small increments on the VOT continuum, they display greater sensitivity to variations in the zone of the boundary than to within-category variations (Studdert-Kennedy et al., 1970), thus illustrating the other characteristic of categorical perception. Although within-category speech variations can be detected, especially with practice, (Strange & Jenkins, 1978), as adults, we nevertheless behave as though we are desensitized to them while being acutely sensitive to small changes at the boundary. This perceptual "quantization" of certain dimensions of complex stimuli contrasts with the more classical "continuous" perception of such properties as pitch or loudness. Perceptual discontinuities such as these, occurring along stimulus dimensions that are acoustically continuous, seem to be a consequence of change in a subset of components in a stimulus complex while remaining components are kept constant (Miller, Wier, Pastore, Kelly & Dooling, 1976; Pastore, 1976).

Categorical processing has the consequence of grouping stimuli in classes, imposing a particular kind of order on varying patterns of stimulation. Thus acoustically different sounds are treated by the listener as functionally equivalent. Although not unique to the perception of speech sounds (e.g., Cutting & Rosner, 1974; Miller et al., 1976; Pisoni, 1977) nor, as we have seen, restricted to the auditory modality (Pastore, 1976), it is especially well exemplified in responses of human subjects to complex acoustic continua from speech. In adults, categorical perception may be viewed as a component in the larger issue of the perceptual constancy of speech-sound categories. It also seems to make a critical, innate contribution to the development of speech perception.

A variety of measures of human infant responses to speech sounds, including habituation of a sucking response, heart-rate changes, and evoked brain potentials indicate responsiveness to similar boundary values between functionally distinct speech sounds to those observed by adults in subjects as young as 1 month of age or less (Eilers & Minifie, 1975; Eimas, 1974, 1975; Kuhl, 1976; Morse, 1972). The early age at which these responses are manifest gave rise to the speculation that responsiveness to some of these boundary properties may be innate.

Still firmer evidence for an innate component was obtained in studies of speech perception in 4-6-month-old infants living in a Spanish-speaking environment (Lasky, Syrdal-Lasky, & Klein, 1975). There are small but consistent differences in VOT boundaries in adult production and perception in English and Spanish. These led to the prediction that infants would demonstrate boundary limits different from those obtained by Eimas with children living in English-speaking environments, if these were acquired through infantile experience of speech patterns. The infants proved to be responsive to boundaries in both regions of the VOT continuum that are universals, the so-called "English" and the "Thai" boundaries, with no sign that experience of the distributions used in Spanish had affected their speech perception at this age.

In a well-conceived cross-cultural study in Africa, infant perception of boundaries along the VOT continuum was studied in children exposed to Kikuyu in infancy (Streeter, 1975). This language has the interesting feature that there is only one labial stop consonant, with a VOT of about -60 msec. Perhaps as a consequence of exposure to the pattern of usage in their culture, 2-month-old infants were responsive to one boundary along the VOT continuum somewhere between 0 and -30 msec. They seemed more responsive to this boundary than the subjects Eimas had studied in an English-speaking environment, although he reports signs of such a boundary as well. However the Kikuyu-exposed infants, although lacking experience of anything equivalent to a [p], also proved to be responsive to a boundary somewhere between $+10$ and $+40$ msec, thus resembling infants exposed to English and various other languages. Streeter concluded that, even at this early age, there is evidence of interaction between nature and nurture, and that some phonetic or acoustic discriminations are universal and innate.

Responsiveness that is apparently innate has also been demonstrated in infants between 1 and 6 months of age to the variations in second- and third-formant transitions that establish boundaries between the different points of

5. INNATE PROGRAMS FOR PERCEPTUAL DEVELOPMENT: AN ETHOLOGICAL VIEW **161**

articulation of labial, alveolar, and velar stop consonants (e.g., [ba], [da], and [ga]). Infants also seem responsive to differences between vowel sounds.

The potential lability of predispositions that human infants bring to segmentation of speech sound continua is also clear. Discriminations that are difficult at a younger age, such as those involving the fricatives [sa] and [za], are made with ease at 6 months (Eilers & Minifie, 1975; Eilers, Wilson, & Moore, 1977). The [ra]-[la] distinction that Japanese adults find so difficult, unemployed in Japanese, is probably easier for infants, though only American infants have been tested thus far (Eimas, 1975).

The properties of speech sounds on which learned responsiveness in adulthood is based are obviously more elaborate and abstract than those infants respond to, with more redundancy, perhaps involving configurational features rather than simple properties, and sometimes so changed that the effective stimulus sets no longer match the original predispositions of infancy. Nevertheless, the latter must surely play a significant ontogenetic role in setting the trajectory of the perceptual learning process.

One way in which such trajectories may be achieved is indicated in a study by Kuhl and Miller (1974). The formant patterns that distinguish different vowel sounds are complicated by variations in the fundamental frequency of different voices, as between men and women. These must be a serious distraction for an infant embarking on the linguistic analysis of speech. Given the importance of vowel coding in speech, we might perhaps expect a predisposition to focus more strongly on formant patterns than on pitch in early responses.

By independently varying the two features in sound presented to infants, while monitoring high amplitude sucking, Kuhl and Miller were able to show that variations in formant pattern are indeed more salient or arresting for human infants than variations in pitch. When formant patterns were varying randomly, the infants habituated more slowly and were more distracted from a pitch change than in the opposite condition, when they were required to attend to a vowel change in the face of random pitch variations. This is not to say that they are unresponsive to pitch variations—far from it. However the salience of pitch changes to infants of this age is lower than that of variations in vowel patterns, thus imposing some order in the process of learning to extract different features from the complex array of stimuli that speech sounds present, especially insofar as success in using this dimension in sorting speech sounds leads them to persevere with it, at least for a time, in further speech-sound discriminations (cf. Mackintosh, 1974, p. 615).

Further experiments using a conditioned head-turning response suggest that by 6 months of age, infants can perceive similarity between vowels produced by different size vocal tracts and between a fricative consonant when it occurs in different vowel environments and spoken by different talkers.

Clearly innate responsiveness to simple properties of consonants and vowel formants quickly becomes modified and enriched as a consequence of further experience with speech behavior.

Human infants thus bring well-defined perceptual predispositions to the task of developing responsiveness to the immensely complex pattern of sound stimuli that speech represents. Some predispositions are innately manifest in initial encounters, developing without prior experience of the stimuli involved. Although these innate contributions are clear, we are hardly tempted to view them as developmental instructions for designing infants as human automata. It is more appropriate to think of them as helping the infant to learn, by providing initial instructions that set the trajectory for development of learned responsiveness. Eventually elaborate arrays of abstracted features become embodied in mental images of schemata, as a basis for perceiving the meaningful phonological components of mature speech.

Conclusions on the Ethology of Perceptual Development

Of all contributions of classical ethology to behavioral biology, the emphasis on innate factors in behavioral development is perhaps most far reaching in its significance, particularly as manifest in the ontogeny of responsiveness to environmental stimuli. In theory, ethologists have also emphasized the intercalation of innate factors with learning in the development of behavior, especially in vertebrate animals. Yet the major emphasis has been on innateness. Except for special cases, such as imprinting, and with reservations about the dominance of innate contributions in the ontogeny of behavior of such higher vertebrates as nonhuman primates and man, there has been a tendency to view the major lineaments of the natural, species-specific behavior of animals as shaped primarily by innate developmental factors. A contrast has often been struck with the human situation, where innate influences are thought to play a minor, even vestigial role.

We argue here that the extremity of this contrast has been exaggerated. Especially in birds, there has been a tendency for ethologists to overstress the role of innate determinants of behavioral development. Psychologists have been guilty of a complementary bias in underestimating the importance of genetic constraints on human behavioral development. If we are right in imputing such errors of judgment to students of behavioral biology, this may be in part attributable to the curious reluctance of many comparative psychologists to accept the human species as a proper subject for study. This chapter stems in part from a growing preoccupation on our part, as students of animal behavior, with current revelations about early human behavioral development. A new synthesis of animal and human research may be coming into view.

As Tinbergen has indicated, early ethologists sought for explanations for the selectivity of responsiveness of different animals to environmental stimulation in part by characterizing species differences in the potential capacities of their sense organs (Tinbergen, 1951). Jacob von Uexkull stressed the uniqueness of the perceptual world of each species, consequent upon the design of its receptor systems (von Uexkull, 1921). Even the most extreme environmentalists have always appreciated the obvious fact that no organism is potentially capable of perceiving all possible changes in the external world. Limits are set by the structure of its sense organs. Because these develop through processes of growth, in which genetic controls play a major role, no one questions the importance of innate constraints at this level.

The sensory world of an insect is obviously different from that of a mammal. We all know of the honeybee's abilities to see the intricate patterns of ultraviolet coloration on many flowers that are invisible to us without the trick of photographing them through a quartz lens that allows the ultraviolet to pass (Daumer, 1956, 1958; von Frisch, 1967). It is impossible for us to imagine the subleties of ultrasonic echolocation at which bats are so incredibly adept (Griffin, 1974). There is growing evidence that birds hear infrasound, so low pitched that we cannot hear it. It may be important to birds because it attenuates so little with distance, providing possible long-range cues for orientation of migration (Yodlowski, Kreithen, & Keeton, 1977). Equally mysterious for us is the electrical sense that some fish use both for object location and communication (Hopkins, 1974; Lissmann, 1958).

The literature of ethology and comparative physiology is full of examples of specialized sensory systems so highly developed in certain species as to open up new sensory domains for them to which others, lacking such innate specializations are insensitive or even totally blind. It is self-evident that *some* innate features pervade all perception by imposing structural limitations on the organs of sensation. Less obvious are the contributions of innate processes to the development of more central stages of stimulus processing than mere sensation, phases in which impressions are organized and interpreted, reflected upon, and finally manifest in some appropriate action—in other words in the development of perception.

Being so far removed by innumerable developmental transformations from the chemistry of particular genes, and bearing in mind our ignorance as to which particular structures of an organism are responsible, for, say, the transmutations of mental imagery, it is exceptionally difficult to grasp how genes might influence the more subtle aspects of perceptual development. As we have tried to show here, a *comparative* approach sometimes helps in coping with this problem.

Experiments and theoretical discussions of Lorenz and Tinbergen made clear the need to invoke more subtle innate constraints on perception than those imposed by the structure of peripheral sense organs. As a recurrent finding in ethological studies, developing young animals manifest responsiveness to particular environmental stimuli. These especially salient stimuli are associated with events that are fraught with special biological significance for all species members, such as predator detection or sexual communication. The selectivity of such innate responsiveness is sometimes so narrow that it is hard to imagine peripheral structures that could explain the specificity of responsiveness, especially when mediated by receptor systems known to be responsive to broader ranges of stimuli. Thus the concept of innate release mechanisms as developed by Lorenz and Tinbergen (see Baerends & Kruijt, 1973; Schleidt, 1962) seeks to involve both peripheral and central influences in selective perception (Marler, 1961).

To the extent that environmental events with special significance in the life of an organism are predictable over transgenerational time, adaptive species-specific genetic control over stimulus responsiveness becomes feasible. The frequency responsiveness of the auditory systems of many frogs seems to be adapted to match species-specific features of the calling songs of males of the species, so that species recognition is achieved in part by a kind of complex frequency filter (Capranica, 1965; Capranica, Frishkopf, & Nevo, 1973). There are other examples, from invertebrates of even more specific selectivity of responsiveness imposed by adaptations of entire sensory systems (Marler, 1961; Marler & Hamilton, 1967), as with mosquitoes whose sound-sensitive antennae are tuned to resonate mechanically to the wing tone of conspecific females (Roth, 1948; Tischner & Schief, 1954).

In cases such as these, where much information about stimuli is discarded in the very process of sensory transduction, selectivity of responsiveness is obviously bought at considerable cost. This cost may be measured in terms of perceptual versatility. For an organism embarking on a stage of its life cycle dominated by a few special behavioral requirements, it will be efficient if irrelevant environmental changes impinging on its receptors are rejected or highly attenuated at an early stage. This will serve to focus immediate attention on the subset of stimuli that is biologically appropriate, as in long-range lepidopteran olfaction, specialized especially for detection and orientation to female sex pheromones (Schneider, 1970; Shorey, 1976). However, for organisms whose structure, behavior, and ecology allows them to benefit from responsiveness to many kinds of environmental information, such a price would be exorbitant. Instead, more versatile receptor systems will be favored, as exemplified by the olfactory system of the honeybee (von Frisch, 1967).

These conditions are met in many organisms. Active, nonspecialized predators such as octopus and dragonflies, and species with elaborate social behavior, such as the honeybee are cases in point. Above all, most higher vertebrates qualify, especially birds and mammals, which have been the main focus of this review, to the unfortunate neglect of a vast and important literature on the behavior in invertebrates and lower vertebrates.

The essential feature of "versatile" perceptual systems as we have characterized them is the dynamic quality of stimulus selectivity. Here the same receptor system can mediate selective responsiveness to many patterns of stimulation, the particular selection being adjusted according to changing needs of the organism. Sometimes these changes will reflect reversible, often cyclic changes in the hierarchically organized motivational states of an organism (Tinbergen, 1951). On other occasions, there will be progressive noncyclical changes in the selectivity of responsiveness, accruing from the cumulative experience of continuing interactions between the organism and its environment. We have been especially concerned here with the role of innate processes in these adjustments of selective responsiveness.

Although innate responsiveness plays a dominant role in the developing behavior of many animals, enrichment through learning is often extensive. In such cases, the challenge is to understand how genetic and environmental influences interact. We have pressed the viewpoint here that some "innate release mechanisms" as described by ethologists in experiments on the behavior of young animals should be viewed not so much as components for designing animals as efficient automata, but rather to provide developmental guidelines, modifiable through experience, for learned selective responsiveness. Thus we seek to unify concepts developed by ethologists for understanding innate behavior with those of psychologists arising from studies of animal learning.

According to this view, young birds and mammals have the potential to acquire responsiveness to most if not all perceptible features of a stimulus object. In many circumstances, they will nevertheless be prone to attend to certain features of natural situations in preference to others, as though these were endowed with an innate salience, thus serving to canalize perceptual development. In species-typical environments, the consequences for adult perceptual organization may be highly predictable in certain respects. Even though there is a potentiality for a range of types of perceptual organization, as manifest in individuals growing up in atypical environments or with unusual social histories, most species members will come to share a similar perceptual organization, as a result of the nature and timing of these innate perceptual constraints, operating in environments typical of the species.

We may expect comparative studies to reveal a spectrum from completely determinate, immutable stimulus control over certain behaviors, at one extreme, to open access to learned control, potentially by any perceptible stimulus, at the other. In given cases, the degree of openness favored will vary according to the ethological and ecological history of the species. Human behavior will be as subject to these influences as any other. Whatever the degree of openness, it is our expectation that perceptual learning will always be in some degree subject to the kind of innate but modifiable stimulus constraints we have described here.

As guidelines for perceptual learning, rather than prescriptions for automata, the effects of innate release mechanisms are often subtle. They are designed to operate in concert with stimulation from species-specific environments, physical and social, the importance of which cannot be overestimated. Out of this ecological interaction, the developing organization of perception of the external world of each species emerges in order and predictable procession, assured of the adaptiveness of its major lineaments, yet flexible enough to benefit from the vagaries of individual experience.

Acknowledgments

We are indebted to Judith Marler for editing and proofing and to Esther Arruza for typing the manuscript.

References

Abramov, I. Interactions among chromatic mechanisms. In H. Spekreijse & L. H. van der Tweel (Eds.), *Spatial contrast: report of a workshop*. New York: North Holland, 1977.

Baerends, G. P., & Kruijt, J. P. Stimulus selection. In R. A. Hinde & J. Stevenson-Hinde (Eds.), *Constraints on learning*. Cambridge: Cambridge University Press, 1973.

Baptista, J. Wild housefinch sings white-crowned sparrow song. *Zeitschrift für Tierpsychologie*, 1972, *30*, 266-270.

Bateson, P. P. G. Sexual imprinting and optimal outbreeding. *Nature*, 1978, *273*, 659-668.

Bateson, P.P.G., & Jaeckel, J. B. Chicks' preferences for familiar and novel conspicuous objects after different periods of exposure. *Animal Behaviour*, 1976, *24*, 386-390.

Beecher, M. D., Zoloth, S. R., Petersen, M., Moody, D., & Stebbins. W. Perception of conspecific communication signals by Japanese macaques (*Macaca fuscata*). *Journal of the Acoustical Society of America*, 1976, *36*(1), 26-58.

Beer, C. G. Individual recognition of voice in the social behavior of birds. In D. S. Lehrman, R. A. Hinde, & E. Shaw (Eds.), *Advances in the study of behavior* (Vol. 3). New York: Academic Press, 1970.

Berlin, B., & Kay, P. *Basic color terms, their universality and evolution*. Berkeley: University of California Press, 1969.

Bornstein, M. Color vision and color naming: a psychophysiological hypothesis of cultural difference. *Psychological Bulletin*, 1973, *80*, 257-285.

Bornstein, M. H. The influence of visual perception on culture. *American Anthropologist*, 1975, 77, 774-798.
Bornstein, M. H. Perceptual development: stability and change in feature perception. In M. H. Bornstein & W. Kessen (Eds.), *Psychological development from infancy*. Hillside, N.J.: Erlbaum, 1978.
Bornstein, M. H., Kessen, W., & Weiskopf, S. The categories of hue in infancy. *Science*, 1976, 191, 201-202. (a)
Bornstein, M. H., Kessen, W., & Weiskopf, S. Color vision and hue categorization in young human infants. *Journal of Experimental Psychology, Human Perception and Performance*, 1976, 2, 115-129. (b)
Bullock, T. H. (Ed.). *Recognition of complex acoustic signals*. Berlin: Dahlem Konferenzen, 1977.
Capranica, R. R. The evoked vocal response of the bullfrog. *Research Monographs*, 1965, 33, Cambridge, Mass.: M.I.T. Press.
Capranica, R. R., Frishkopf, L., & Nevo, E. Encoding of geographic dialects in the auditory system of the cricket frog. *Science*, 1973, 182, 1272-1275.
Cooke, F. Early learning and its effect on population structure. Studies of a wild population of snow geese. *Zeitschrift für Tierpsychologie*, 1978, 46, 344-358.
Cowan, P. J. Selective responses to the parental calls of different individual hens by young *Gallus gallus*: Auditory discrimination learning versus auditory imprinting. *Behavioral Biology*, 1974, 10, 541-545.
Cutting, J. E., & Eimas, P. D. Phonetic feature analyzers and the processing of speech in infants. In J. Kavanagh & J. E. Cutting (Eds.), *The role of speech in language*. Cambridge, Mass.: M.I.T. Press, 1975.
Cutting, J. E., & Rosner, B. Categories and boundaries in speech and music. *Perception and Psychophysics*, 1974, 16, 564-570.
Daumer, K. Reizmetrische Untersuchung des Farbenschens der Bienen. *Zeitschrift vergleichende Physiologie*, 1956, 38, 413-478.
Daumer, K. Blumenfarben, wie sie die Bienen sehen. *Zeitschrift vergleichende Physiologie*, 1958, 41, 49-110.
DeValois, R. Central mechanisms of color vision. In R. Jung (Ed.), *Handbook of sensory physiology* (Vol. 7). New York: Springer-Verlag, 1973.
DeValois, R., Abramov, I. L., & Jacobs, G. Analysis of response patterns of L.G.N. cells. Journal of the Optical Society of America, 1966, 56, 966-977.
Dooling, R. J., & Searcy, M. A. Early perceptual selectivity in the swamp sparrow. *Developmental Psychobiology* 1980, 13, 499-506.
Eilers, R. E., & Minifie, F. D. Fricative discrimination in early infancy. *Journal of Speech and Hearing Research*, 1975, 18, 158-167.
Eilers, R. E., Wilson, W. R., & Moore, J. M. Developmental changes in speech discrimination in infants. *Journal of Speech and Hearing Research*, 1977, 20, 766-780.
Eimas, P. D. Auditory and linguistic processing of the cues for place of articulation by infants. *Perception & Psychophysics*, 1974, 16, 513-521.
Eimas, P. D. Speech perception in early infancy. In L. B. Cohen & P. Salapatek (Eds.), *Infant perception: from sensation to cognition* (Vol. 2). New York: Academic Press, 1975.
Evans, R. M. Imprinting and mobility in young ring-billed gulls, *Larus delawarensis*. *Animal Behavior Monographs*, 1970, 3, 193-248.
Evans, R. M. Development of an auditory discrimination in domestic chicks (*Gallus gallus*). *Animal Behaviour*, 1972, 20, 77-87.
Evans, R. M. Auditory discrimination-learning in young ring-billed gulls (*Larus delawarensis*). *Animal Behaviour*, 1977, 25, 140-146.

Evans, R. M., & Mattson, M. E. Development of selective responses to individual maternal vocalizations in young *Gallus gallus*. *Canadian Journal of Zoology*, 1972, *50*, 777-780.

Fabricius, E. Zur Ethologie junger Anatiden. *Acta Zoologica Fennica*, 1951, *68*, 1-178.

Fabricius, E., & Boyd, H. Experiments on the following reaction of ducklings. *Report of the Wildfowl Trust*, 1954, *6*, 84-89.

Falls, J. B. Function of territorial song in the white-throated sparrow. In R. A. Hinde (Ed.), *Bird vocalizations*. Cambridge: Cambridge University Press, 1969.

Falls, J. B. Bird song and territorial behavior. In L. Krames, P. Pliner, & T. Alloway (Eds.), *Aggression, dominance and individual spacing. Advances in the study of communication and affect IV*. New York: Plenum, 1978.

Fantz, R. L. Form preferences in newly hatched chicks. *Journal of Comparative and Physiological Psychology*, 1957, *50*, 422-430.

Fantz, R. L. Ontogeny of perception. In A. M. Schrier, H. F. Harlow, & F. Stollnitz (Eds.), *Behavior of non-human primates II*. New York: Academic Press, 1965.

Frisch, K. von *The dance language and orientation of bees*. Cambridge, Mass.: Belknap Press of Harvard University Press, 1967.

Gibson, E. J. *Principles of perceptual learning and development*. Englewood Cliffs, N.J.: Prentice-Hall, 1969.

Gibson, J. J. *The senses considered as perceptual systems*. Boston: Houghton Mifflin, 1966.

Goodwin, E. B., & Hess, E. H. Innate visual form preferences in the pecking behavior of young chicks. *Behaviour*, 1969, *34*, 223-237.

Gottlieb, G. *Development of species identification in birds*. Chicago: University of Chicago Press, 1971.

Gottlieb, G. On the acoustic basis of species identification in wood ducklings (*Aix sponsa*). *Journal of Comparative and Physiological Psychology*, 1974, *87*, 1038-1048.

Green, S. Communication by a graded vocal system in Japanese monkeys. In L. A. Rosenblum (Ed.), *Primate Behavior* (Vol. 4). New York: Academic Press, 1975.

Greenberg, J. H. Language universals: a research frontier. *Science*, 1969, *166*, 473-478.

Griffin, D. R. *Listening in the dark*. New York: Dover, 1974.

Hailman, J. P. Ontogeny of an instinct. *Behaviour Supplement*, 1967, *15*, 1-159.

Herrnstein, R. J., Loveland, D. H., & Cable, C. Natural concepts in pigeons. *Journal of Experimental Psychology, Animal Behavior Processes*, 1976, *2*, 285-302.

Hess, E. H. Natural preferences of chicks and ducklings for objects of different colors. *Psychological Reports*, 1956, *2*, 477-487.

Hess, E. H. Imprinting and the "critical period" concept. In E. L. Bliss (Ed.), *Roots of behavior*. New York: Harper, 1962.

Hess, E. H. *Imprinting*. New York: Van Nostrand Reinhold, 1973.

Hess, E. H., & Gogel, W. C. Natural preferences of the chick for objects of different colors. *Journal of Psychology*, 1954, *38*, 483-493.

Hogan, J. A. How young chicks learn to recognize food. In R. A. Hinde & J. Stevenson-Hinde (Eds.), *Constraints on learning*. New York: Academic Press, 1973.

Hopkins, C. D. Electric communication in fish. *American Scientist*, 1974, *62*, 426-437.

Hurvich, C., & Jameson, D. An opponent-process theory of color vision. *Psychological Review*, 1957, *64*, 383-404.

Immelman, K. Song development in the zebra finch and other estrildid finches. In R. A. Hinde (Ed.), *Bird vocalizations*. Cambridge: Cambridge University Press, 1969.

Immelman, K. Sexual and other long-term aspects of imprinting in birds and other species. *Advances in the Study of Behavior*, 1972, *4*, 147-174.

Impekoven, M., & Gold, P. S. Prenatal origins of parent-young interactions in birds: a naturalistic approach. In G. Gottlieb (Ed.), *Behavioural Embryology*. New York: Academic Press, 1973.

Kear, J. Colour preference in young Anatidae. *Ibis*, 1964, *106*, 361-369.
Koltermann R. Rassen- und artspezifische Duftbewertung bei der Honigbiene und ökologische Adaptation. *Journal of Comparative Physiology*, 1973, *85*, 327-360.
Konishi, M. The role of auditory feedback in the control of vocalization in the white-crowned sparrow. *Zeitschrift für Tierpsychologie*, 1965, *22*, 770-783.
Konishi, M. Auditory environment and vocal development in birds. In R. D. Walk and H. L. Pick, Jr. (Eds.), *Perception and experience*. New York: Plenum, 1978.
Kovach, J. K. Interaction of innate and acquired: Color preferences and early exposure learning in chicks. *Journal of Comparative and Physiological Psychology*, 1971, *75*, 386-398.
Kovach, J. K., & Hickox, J. E. Color preferences and early perceptual discrimination learning in domestic chicks. *Developmental Psychobiology*, 1971, *4*, 255-267.
Kroodsma, D. The effect of large song repertoires on neighbor "recognition" in male song sparrows. *Condor*, 1976, *78*, 97-99.
Kroodsma, D. Aspects of learning in the ontogeny of bird song: where, from whom, when, how many, which and how accurately? In M. Burghardt and G. Burghardt (Eds.), *Development of Behavior*. New York: Garland, 1979.
Kuhl, P. K. Speech perception in early infancy: the acquisition of speech-sound categories. In S. K. Hirsh, D. H. Eldredge, I. J. Hirsh, & S. R. Silverman (Eds.), *Hearing and Davis: Essays honoring Hallowell Davis*. St. Louis: Washington University Press, 1976.
Kuhl, P. K., & Miller, J. D. Speech perception in early infancy: discrimination of speech-sound categories. *Journal of the Acoustical Society of America*, 1974, Supplement 1, *58*, 56.
Lashley, K. S. The mechanism of vision XV. Preliminary studies of the rat's capacity for detail vision. *Journal of General Psychology*, 1938, *18*, 123-193.
Lasky, R., Syrdal-Lasky, A., & Klein, R. VOT discrimination by four-to-six-month old infants from Spanish environments. *Journal of Experimental Child Psychology*, 1975, *20*, 215-225.
Lauer, J., & Lindauer, M. Die Beteiligung von Lernprozessen bei der Orientierung. *Forschritte der Zoologie*, 1973, *21*, 349-370.
Liberman, A. M., Harris, K. S., Hoffman, H. S., & Griffith, B. C. The discrimination of speech sounds within and across phoneme boundaries. *Journal of Experimental Psychology*, 1957, *54*, 358-368.
Liberman, A. M., Harris, K. S., Kinney, J. A., & Lane, H. The discrimination of relative-onset time of the components of certain speech and nonspeech patterns. *Journal of Experimental Psychology*, 1961, *61*, 379-388.
Liberman, A. M., Cooper, F. S., Shankweiler, D., & Studdert-Kennedy, M. Perception of the speech code. *Psychological Review*, 1967, *74*, 431-461.
Lindauer, M. Lernen und Gedachtnis-versuche an der Honigbiene. *Naturwissenschaften*, 1970, *57*, 463-467.
Lindauer, M. Evolutionary aspects of orientation and learning. In G. Baerends, C. Beer, & A. Manning (Eds.), *Function and Evolution in Behaviour*. Oxford: Clarendon Press, 1975.
Lisker, L., & Abramson, A. S. A cross-language study of voicing in initial stops: acoustical measurements. *Word*, 1964, *20*, 384-422.
Lissmann, H. W. On the function and evolution of electric organs in fish. *Journal of Experimental Biology*, 1958, *35*, 156-191.
Lorenz, K. *Studies in animal behavior* (Vol. 1). Cambridge, Mass.: Harvard University Press, 1970.
Mackintosh, N. J. *The psychology of animal learning*. New York: Academic Press, 1974.
Marler, P. The filtering of external stimuli during instinctive behavior. In W. H. Thorpe & O. L. Zangwill (Eds.), *Current problems in animal behaviour*. Cambridge: Cambridge University Press, 1961.
Marler, P. Communication in monkeys and apes. In I. DeVore (Ed.), *Primate behavior*. New York: Holt, Rinehart & Winston, 1965.

Marler, P. A comparative approach to vocal learning: song development in white-crowned sparrows. *Journal of Comparative and Physiological Psychology*, 1970, *71*, 1-25. (a)

Marler, P. Vocalizations of East African monkeys. I. Red colobus. *Folia Primatologica*, 1970, *13*, 81-91. (b)

Marler, P. On the origin of speech from animal sounds. In J. Kavanagh & J. Cutting (Eds.), *The role of speech in language*. Cambridge, Mass.: M.I.T. Press, 1975.

Marler, P. Sensory templates in species-specific behavior. In J. Fentress (Ed.), *Simpler networks: An approach to patterned behavior and its foundations*. New York: Sinauer, 1976. (a)

Marler, P. Social organization, communication and graded signals: The chimpanzee and the gorilla. In P.P.G. Bateson & R. A. Hinde (Eds.), *Growing points in ethology*. Cambridge: Cambridge University Press, 1976. (b)

Marler, P. Development and learning of recognition systems. In T. H. Bullock (Ed.), *Recognition of complex acoustic signals*. Berlin: Dahlem Konferenzen, 1977.

Marler, P. & Hamilton, W. J. III *Mechanisms of animal behavior*. New York: Wiley, 1967.

Marler, P., & Mundinger, P. Vocal learning in birds. In H. Moltz (Ed.), *The ontogeny of vertebrate behavior*. New York: Academic Press, 1971.

Marler, P., & Peters. S. Selective vocal learning in a sparrow. *Science*, 1977, *198*, 519-521.

Marler, P. & Tamura, M. Culturally transmitted patterns of vocal behavior in sparrows. *Science*, 1964, *146*, 1483-1486.

Marler, P., & Tenaza, R. Signaling behavior of apes with special reference to vocalization. In T. A. Sebeok (Ed.), *How animals communicate*. Bloomington and London: Indiana University Press, 1977.

Mattson, M. E., & Evans, R. M. Visual imprinting and auditory-discrimination learning in young of the canvasback and semiparasitic redhead (Anatidae). *Canadian Journal of Zoology*, 1974, *52*, 421-427.

Menzel, R. Untersuchungen zum Erlernen von Spektralfarben durch die Honigbiene. *Zeitschrift vergleichende Physiologie*, 1967, *56*, 22-62.

Miller, D. E., & Emlen, J. T. Individual chick recognition and family integrity in the ring-billed gull. *Behaviour*, 1975, *52*, 124-144.

Miller, J., Wier, C., Pastore, R., Kelly, W., & Dooling, R. Discrimination and labeling of noise-buzz sequences with varying noise-lead times: an example of categorical perception. *Journal of the Acoustical Society of America*, 1976, *60*, 410-417.

Morse, P. The discrimination of speech and nonspeech stimuli in early infancy. *Journal of Experimental Child Psychology*, 1972, *14*, 477-492.

Nicolai, J. Familientradition in der Gesangsentwicklung des Gimpels (*Pyrrhula pyrrhula* L.). *Journal für Ornithologie*, 1959, *100*, 39-46.

Pastore, R. E. Categorical perception: a critical re-evaluation. In S. K. Hirsh, D. H. Eldredge, I. J. Hirsh, & S. R. Silverman (Eds.), *Hearing and Davis: Essays honoring Hallowell Davis*. St. Louis: Washington University Press, 1976.

Petersen, M. R., Beecher, M. D., Zoloth, S. R., Moody, D. B., & Stebbins, W. C. Neural lateralization of species-specific vocalizations by Japanese macaques (*Macaca fuscata*). *Science*, 1978, *202*, 324-327.

Pisoni, D. B. Identification and discrimination of the relative onset time of two component tones: Implications for voicing perception in stops. *Journal of the Acoustical Society of America*, 1977, *61*, 1352-1361.

Pisoni, D. B. Speech perception. In W. K. Estes (Ed.), *Handbook of hearing and cognitive processes* (Vol. 6). Hillsdale, N.J.: Erlbaum, 1978.

Ratliff, F. On the psychophysiological bases of universal color terms. *Proceedings of the American Philosophical Society*, 1976, *120*, 311-330.

Rosch, E. On the internal structure of perceptual and semantic categories. In T. E. Moore (Ed.), *Cognitive development and the acquisition of language*. New York: Academic Press, 1973.

Rosch-Heider, E. "Focal" color areas and the development of color names. *Developmental Psychology*, 1971, *4*, 447-455.

Rosch-Heider, E. Universals in color naming and memory. *Journal of Experimental Psychology*, 1972, *93*, 10-20.

Roth, L. M. A study of mosquito behavior. *American Midland Naturalist*, 1948, *40*, 265-352.

Schleidt, W. Die historische Entwicklung der Begriffe "angeborenes auslösendes Schema" und "angeborener Auslösemechanismus" in der Ethologie. *Zeitschrift für Tierpsychologie*, 1962, *19*, 697-722.

Schneider, D. Olfactory receptors for the sexual attractant (bombykol) of the silk moth. In F. O. Schmitt (Ed.), *The neurosciences: Second study program*. New York: Rockefeller University Press, 1970.

Schutz, F. Sexuelle Pragung der Anatiden. *Zeitschrift für Tierpsychologie*, 1965, *22*, 50-103.

Shorey, H. H. *Animal communication by pheromones*. New York: Academic Press, 1976.

Snowdon, C. The response of non-human animals to speech and to species-specific sounds. *Brain, Behavior and Evolution*, 1979, *16*, 409-429.

Snowdon, C., & Pola, Y. V. Interspecific and intraspecific responses to synthesized pygmy marmoset vocalizations. *Animal Behaviour*. 1978, *26*, 192-206.

Sonnemann, P., & Sjölander, S. Effects of cross-fostering on the sexual imprinting of the female zebra finch Taeniopygia guttata. *Zeitschrift für Tierpsychologie*, 1977, *45*, 337-348.

Springer, S. Memory for linguistic and non-linguistic dimensions of the same acoustic stimulus. *Journal of Experimental Psychology*, 1973, 101(1), 159-163.

Strange, W., & Jenkins, J. J. Role of linguistic experience in the perception of speech. In R. D. Walk & H. L. Pick, Jr. (Eds.), *Perception and experience*. New York: Plenum, 1978.

Streeter, L. A. Language perception of 2-month old infants shows effects of both innate mechanisms and experience. *Nature*, 1975, *259*, 39-41.

Struhsaker, T. *Behavior and ecology of red colobus monkeys*. Chicago: University of Chicago Press, 1975.

Studdert-Kennedy, M. Universals in phonetic structure and their role in linguistic communication. In T. H. Bullock (Ed.), *Recognition of complex acoustic signals*. Berlin: Dahlem Konferenzen, 1977.

Studdert-Kennedy, M., Liberman, A. M., Harris, K. S., & Cooper, F. S. Motor theory of speech perception: a reply to Lane's critical review. *Psychological Review*, 1970, *77*, 234-249.

Sutherland, N. S. Outlines of a theory of visual pattern recognition in animals and man. *Proceedings of the Royal Society* B, 1968, *171*, 297-317.

Sutherland, N. S. Object recognition. In E. C. Carterette, & M. P. Friedman (Eds.), *Handbook of perception III*. New York: Academic Press, 1973.

Thorpe, W. H. The learning of song patterns by birds, with especial reference to the song of the chaffinch. Fringilla coelebs. *Ibis*, 1958. *100*, 535-570.

Thorpe, W. H. *The biology of vocal communication and expression in birds*. London and New York: Cambridge University Press, 1961.

Thorpe, W. H., & Hall-Craggs, J. Sound production and perception in birds as related to the general principles of pattern perception. In P.P.G. Bateson & R. A. Hinde (Eds.), *Growing points in ethology*. London and New York: Cambridge University Press, 1976.

Tinbergen, N. Social releasers and the experimental method required for their study. *Wilson Bulletin*, 1948, *60*, 6-52.

Tinbergen, N. *The study of instinct*. Oxford: Clarendon Press, 1951.

Tinbergen, N. *The herring gull's world*. London: Collins, 1953.

Tinbergen, N. *The animal in its world* (Vols. 1, 2). Cambridge, Mass.: Harvard University Press, 1973.

Tischner, H., & Schief, A. Fluggeräusch und Schallwarnehmung bei *Aedes aegypti* L. (Culicidae). *Verhandlungen Deutsche Zoologie Gesellschaft,* 1954, *51,* 453-460.

Uexkull, J. von *Umwelt und Innenwelt der Tiere.* Berlin: Springer-Verlag, 1921.

Waddington, C. H. *The strategy of the genes.* London: Allen and Unwin, 1957.

Weidmann, R., & Weidmann, U. An analysis of the stimulus situation releasing food-begging in the black-headed gull. *Animal Behaviour,* 1958, *6,* 114.

Whorf, B. *Language, thought and reality.* Cambridge, Mass.: M.I.T. Press, 1964.

Wood, C., Goff, W., & Day, R. Auditory evoked potentials during speech perception. *Science,* 1971, *173,* 1248-1251.

Wright, A., & Cummings, W. Color-naming functions for the pigeon. *Journal of Experimental Animal Behavior,* 1971, *15,* 7-17.

Yodlowski, M. L., Kreithen, M. L., & Keeton, W. T. Detection of atmospheric infrasound by homing pigeons. *Science,* 1977, *265,* 725-726.

Zoloth, S., & Green, S. Monkey vocalizations and human speech: Parallels in perception? *Brain, Behavior and Evolution,* 1979, *16,* 430-442.

Zoloth, S. R., Petersen, M. R., Beecher, M. D., Green, S., Marler, P., Moody, D. B., & Stebbins, W. Species-specific perceptual processing of vocal sounds by monkeys. *Science,* 1979, *204,* 870-873.

IV

ASYMMETRIES AND VARIATION

The relationship of cerebral asymmetries to behavior has become a major focus of speculation and research in recent years. Not only has the topic appealed to the scientific community but it has also roused a great deal of popular interest. Among professionals and laymen alike, speculation, sometimes bordering on the mystical, has been rife. In the following chapter, Jerre Levy, while not eschewing speculation, presents a carefully researched statement on cerebral asymmetry and its relationship to developmental plasticity. She traces the origins of asymmetry, describes the generality of the phenomenon, and provides the outline of a model of lateralization that may become very useful in explaining many features of individual differences in biological and behavioral development. She incorporates the evolutionary themes expressed elsewhere in this volume and most effectively demonstrates the absurdity of the hoary nature-nurture argument.

The hemispheres, even in infancy, are not equipotent, and in the course of ontogenesis there is a tendency for an increasing lateralization of function to develop. Thus, as individuals age there is increasing vulnerability to cognitive dysfunction from unilateral cerebral injury.

Levy also reviews the literature on sex differences in lateralization and concludes that the evidence overwhelmingly supports the notion that females manifest a smaller degree of lateralization than males. The chapter includes an evaluation of the possible sources of this sex difference as well as a consideration of its psychological significance.

Lastly, the evolutionary significance of lateralization is discussed and the advantages and disadvantages of asymmetry are considered. Suggestions for ideal organizations of brain function are presented in a comparative context.

6

Lateralization and Its Implications for Variation in Development

JERRE LEVY

The Origins of Asymmetry

Asymmetric traits are extremely prevalent in almost all animal phyla. This can be seen in the coiling direction of snails, in claw size in lobsters, and in a large number of other features. In people, the nose tends to deviate toward the left or right (Sutton, 1967), the two gonads are of unequal size (Mittwoch & Kirk, 1975), and the two sides of the scrotum are of unequal length (Morgan & Corballis, 1976). The question is: Where do these asymmetries come from; what causes their development?

When there is a species-specific asymmetry, the simplest assumption that one can make is that this is a fixed genetic trait having no variation within the species and that the information coding for the asymmetry resides in the DNA sequence. There is little argument among most researchers regarding this determination, although only a developmental study could specify how genes are translated to morphology. In the absence of genetic variability, there could obviously be no correlations among relatives for the trait, and population genetic studies would be incapable of revealing its genetic basis. We assume that species-invariant features, such as bipedalism, the number of fingers, or location of the heart, are specified by genes, but the assumption is more an act of faith than a conclusion forced by empirics.

Morgan (1977) made the claim that no asymmetric trait, regardless of what it is, can be encoded in DNA, that, in principle, this is impossible. His argument, I believe, is fallacious. Morgan reasons that for there to be a differential specification of left and right, there must be a reference point at some early time in development that cannot, itself, be encoded in the genes.

This argument would also hold for anterior-posterior development, mediolateral development, or for any coordinate mapping of the organism, but the gravitational field, the point of entry of the sperm, the place of attachment to the uterine wall, or any number of possible factors could conceivably serve as a reference for these coordinates. In contrast, there can be no communication of a left-right asymmetry unless there is an underlying asymmetry that serves as a relational specification (Corballis & Beale, 1976).

In fact, however, DNA itself is asymmetric, and when DNA is transcribed into RNA, the transcription always proceeds in a single direction on one strand of the DNA doublet. It proceeds from the 3' end toward the 5' end in any given DNA segment, and this directional transcription represents a further asymmetry. The order in which codons are copied is biologically important because the triplet sequence spells a different genetic "word" depending on the direction in which it is read. When messenger RNA is translated to protein, the same condition holds. Furthermore, the amino acids are all of a single isomeric form so that a given primary structure of protein yields a tertiary structure that is either levo- or dextrorotatory. Thus, the genes themselves are asymmetric molecules, as are their products, and because of this, the coding for lateral asymmetry is possible.

Nevertheless, Morgan claims that all instances of directional variations within a species of any laterally differentiated trait are due to cytoplasmic inheritance. By cytoplasmic inheritance is meant the transgenerational transmission of information from mother to offspring via elements contained in the cytoplasm of the ovum. Apparently, what Morgan fails to recognize is that all cytoplasmic inheritance is the result of gene products of the mother's genome. In most animals, when cytoplasmic inheritance occurs, it is mediated by messenger RNAs formed during oögenesis when the mother-to-be is still an embryo or fetus. These mRNAs, and their protein products, are critical for the early development of the zygote. They are typically identical in all members of the species but on rare occasions may vary and, when they do so, cytoplasmically determined polymorphisms may result. One beautiful example of this, in terms of control of an asymmetry, is the coiling direction of a certain species of snail. Some snails coil to the right and some to the left, and coiling direction is determined entirely by the maternal genome (see Srb, Owen, & Edgar, 1965). Dextral coiling is dominant and sinistral coiling is recessive. If the mother is a dominant homozygote or a heterozygote, all her offspring will be dextral coilers, regardless of the mother's phenotype, and if she is a recessive homozygote, all her offspring will be sinistral coilers. The paternal genome plays no role in the coiling direction of his offspring, but does play a role in the coiling direction of his daughters' progeny. Coiling direction is determined at the very first cleavage of the fertilized egg of the snail, and is governed by cytoplasmic factors in the ovum. Cleavage in a snail is not the same as cleavage in a vertebrate, and when the original cell cleaves for the first time, there is an asymmetrical allocation of cytoplasm to the two daughter cells. The asymmetrical cell division continues through subsequent cleavages, giving a spiral pattern to the developing zygote. This is clearly an instance in which the direction of an asymmetry is, in fact, encoded in DNA—not the DNA of the organism itself, but the DNA of its mother. It is quite irrelevant to the question of the source of information specifying the asymmetry that the morphologic realization of that information occurs in an organism's offspring rather than in the organism itself. This has to do with the timing of genetic expression and is a developmental issue. In contrast to Morgan's arguments, the snail example provides definitive proof of within-species genetic variation in the direction of an asymmetrical trait.

Direct genetic determination of asymmetry direction in the organism itself has been observed in the fruit fly, *D. melanogaster*. In the wild-type fly, the external abdomen is perfectly straight and untwisted. There are two loci, however, one on chromosome IV, the abdomen rotatum (*ar*) locus, and one on chromosome I, the twisted (*tw*) locus that can produce rotation of the entire abdomen if mutant alleles are present. The direction of rotation, clockwise or counterclockwise, is a function of which particular mutant allele is present (Lindsley & Grell, 1967). Thus, not only can variations in asymmetry direction be under genetic control, but the genes can act on the organism possessing them and not merely through cytoplasmic constituents on the progeny.

Although Morgan is wrong in his conclusions, he has highlighted the importance of understanding the relation between genetic information and environmental factors in the process of phenogenesis. Genes are analogous to a computer program and the environment to a computer. Depending on the consequences of the computer's analysis of a specified step in the program, subsequent processing may continue at any one of a number of choice points in the series of statements. No aspect of an organism is wholly a function of its genes or wholly a function of its environment. There is a constant interplay between the two and it is this that governs development. No better example of this type of transaction can be given than Hibbard's (1965) experiments with salamanders. Hibbard implanted an extra medulla in developing animals, just anterior to the normal one, but in reversed anterior-posterior orientation. In the salamander medulla are two giant cells, the Mauthner neurons, whose axons normally grow posteriorly down the whole length of the animal. Hibbard found that the axons of the Mauthner neurons in the extra medulla started to grow *anteriorly*, according to the gradient of the implanted tissue, but upon reaching the midbrain, they reversed direction and grew posteriorly in accordance with the overall bodily gradient. These observations demonstrate the potency of the tissue environment in reversing the normal direction of growth, and illustrate the plastic capacity of genetic expression to accommodate the changed environmental conditions. This plasticity should be kept in mind when human cerebral asymmetry itself is discussed. The nature-nurture argument will be seen to lose much of its force as a full appreciation of the intimacy of the gene-environment transaction is gained. Genetic determination of any trait depends on the controlling environment in which genes express themselves, and expression is determined at all moments, even prior to fertilization, on the environment in which those genes are embedded.

With respect to left-right asymmetry, we are faced with a peculiar problem. In innumerable embryological experiments, a great deal of evidence has been gathered for the existence of anterior-posterior, dorso-ventral, and medio-lateral gradients of determination, but almost none for the existence of a left-right gradient, and, indeed, certain findings seem to suggest that left and right are indistinguishable in terms of morphogenetic control. Sperry (1945) severed the optic fibers from the eyes of amphibians and blocked the region where these fibers normally cross to the opposite side. The optic axons regenerated, but because of the blockage, were forced to grow ipsilateral to their cell bodies. Growth proceeded normally and the fibers made appropriate synaptic contact with neurons in the ipsilateral optic tectum. The fact that left eye fibers synapsed on the left tectum, rather than on the right, and vice versa for fibers from the right eye, did not disrupt functional organization. The sole consequence of this abnormal growth was a complete left-right reversal of the perceptual world. Frogs would strike at a spot precisely 30° to the left when a fly was presented 30° to the right. The problem these results raise for the asymmetry issue is that axons arising from one or the other side of the body behaved as if they were totally unspecified biochemically with respect to left and right. Fully functional, although highly maladaptive, synaptic connections were formed and no evidence was seen that axons from a given eye had any difficulty in establishing synaptic contact with the wrong half of the tectum. Although this lack of left-right differentiation might reflect symmetry of the frog eyes and tecta, no evidence of a left-right gradient has been reported in other bodily regions that are clearly asymmetric.

However, at least one embryological study has revealed a difference in the left and right halves of the frog blastula that may represent the basis for asymmetric development. Spemann and Falkenberg (1919) tied a string around the midsagittal planes of amphibian blastulas. Although to the naked eye, the blastula appears to be a smooth ball of cells, closer examination reveals that various regions of the ball differ, and the anterior-posterior and dorso-ventral axes can be specified. This means, of course, that the blastula has a left and a right side. By constricting blastulas into an hourglass shape, Spemann and Falkenberg induced zygotes to split into pairs of monozygotic twins. Animals formed from the left half developed normally, displaying normal asymmetry relations, but animals formed from the right half were random with respect to the direction of internal asymmetries, half being normal, half being mirror reversed, a condition known as *situs inversus*. These results indicate that at the stage of blastular development, there was something different about the left and right halves of the organism, and furthermore, they probably also indicate that the left side of the animal was more differentiated than the right. It would appear that genes for asymmetry determination were being expressed on the left and that induction of differentiation was proceeding toward the right. Derepression of asymmetry-determining genes on the right may normally have occurred at a later stage, triggered by induction at the midsagittal border. Division of the zygote seems to have blocked such derepression, resulting in a random asymmetry development.

If the Spemann and Falkenberg (1919) observations have generality for other vertebrates, variations in the direction of asymmetries may, in part, reflect variations in the developmental rates of the two halves of the body, and these, in turn, may be a manifestation of genetic variation. The timing at which genes become expressed has profound effects on morphogenesis, and if, for any reason, this timing is changed, radical changes may be expected in development. It is obvious that, as an organism grows and changes, its future course becomes more narrowly defined. As critical stages are passed, genes may become irreversibly repressed, plasticity will be reduced, and even if certain genes become active, their effects will differ as a function of developmental stage. Although the amphibian blastula may not be a universal model for asymmetry development, it shows, nonetheless, a probably universal plasticity of response to environmental events. A naive investigator studying Spemann and Falkenberg's dextral-derived amphibians would incorrectly conclude that no genes existed in these animals that could affect lateral differentiation. He could not know that a rather extreme environmental manipulation had produced a permanent repression, and he would be likely to accept the view that Morgan has propounded. Another naive investigator studying Spemann and Falkenberg's sinistral-derived amphibians, would be likely to conclude that asymmetry development was completely specified by genetic information and that the environment could have no effect. He would be equally wrong. The single-cell zygote consists of a set of contingency statements, usually designated as its genome, and from the moment of its conception until the death of an aged organism, the environment selects what will be manifested. Natural selection, acting over eons, not only designed the genetic structure, but acting over the lifetime of the organism, also designs the phenotype.

Lateralization of the Brain

Factors Affecting Lateral Development

The central issue of this chapter is lateral differentiation of the human cerebral hemispheres. The foregoing discussion of the possible origins of asymmetric traits and the control mechanisms by which they are realized was felt to be a necessary introduction because of current arguments regarding the causes of human cerebral asymmetry and its various manifestations in different people and groups. If, as Morgan has claimed, no asymmetric trait, even in principle, can be affected by the nuclear DNA sequence, then the entire causal matrix of genetics become totally irrelevant for a discussion of the etiology of brain lateralization. The evident fact, as outlined previously, that both species-invariant and species-variable asymmetries can be and are subject to genetic control, means that both genetic and environmental factors must at least be considered in a reasonable analysis of mechanisms underlying the phenogenesis of the asymmetric human brain.

For a species-invariant trait (e.g., having two legs, two eyes, one heart), quantitative genetic studies provide no information with respect to the mechanisms of transmission, and only developmental studies can identify how particular sequences of DNA lead to particular phenotypic results. Though modern biological research has made remarkable advances in specifying the relationships between the genetic code and proteins, and between proteins and certain biological microstructures and metabolic activities, it is still very distant from being able to tell us how genes are translated into macrostructures having particular characteristics. At present, we are limited to noting covariations between a phenomenon of interest and various possible antecedent conditions, either genetic or environmental, and either naturally occurring or experimentally manipulated.

When we seek to identify the historical causes of some process or structure, by implication, we are assuming that there exist conditions under which the process or structure does not obtain. Thus, the question, "What causes (in the historical sense) a?", has meaning only if, in some circumstances, non-a occurs. Or, if we ask, "What causes a to have the magnitude it does?", then we are assuming that a can occur with differing magnitudes, depending on antecedent events. For these reasons, rare and atypical phenomena provide critical information for understanding the development of the usually observed phenomena. This is particularly true for understanding human brain organization and the lateral differentiation of the cerebral hemispheres: rare prenatal developmental anomalies, unusual genetic histories, early brain damage, deviant postnatal sociocultural environments, all provide rich sources of information for comprehending development in the typical individual.

For these reasons, there will be a strong emphasis in this chapter on the unusual and atypical, but with a view toward understanding normal development.

Environmental Determinants of Cerebral Lateralization

The environment begins to act on lateral determination in very early zygotic stages, when the potential human being is merely a few cells. Under normal conditions, embryogenesis proceeds so as to produce species-typical asymmetries (e.g., heart on the left), but under unusual and not clearly understood circumstances, *situs inversus totalis* occurs in which the heart, stomach, and all internal organs are mirror imaged. Complete reversal of all normal asymmetries appears in about 1 in 10,000 people (Torgersen, 1950), is associated with various anomalies of development that show familial associations (Pernkopf, 1937; Torgersen, 1946, 1949), and is not associated with twinning (Torgersen, 1950).

Because of the failure to find an excess of twins among people having *situs inversus*, some have argued that asymmetry reversal is not homologous with induced mirror imaging in animals, but, rather, reflects a peculiar genetic pathology. That genetic factors are implicated is suggested by the increase in developmental anomalies observed in families of individuals with *situs inversus*, but it is extremely rare, in fact, to find *situs inversus* itself in more than one member of a family. What may be under genetic control are disorders of coordinate mapping during embryogenesis, leading, in most cases, to structural pathologies, but no asymmetry reversal, but in a few cases, resulting in loss of asymmetry specification, which then comes under control of random environmental factors.

Certain strains of mice, homozygous for the recessive gene, *iv*, manifest a normal direction of asymmetry development in 50% and mirror-reversed asymmetries in 50% (Layton, 1976). The probability of *situs inversus* is unrelated to parental phenotype, showing that asymmetry determination in these strains is neither under genetic control (since the mice are homozygous at the critical locus) nor affected by parental phenotype. Evidently, the homozygous *iv/iv* genotype has a developmental course in which all genes for asymmetry determination are repressed, and the directional asymmetry that is eventually manifested is a consequence of intrauterine events that have a 50/50 probability of yielding the normal asymmetry or its mirror image. Such repression of genes for asymmetry determination, and the assumption by random environmental events of control of lateral development, may also occur in people and affect cerebral laterality.

Human *situs inversus* is probably associated with a more general loss of coordinate mapping, given the associated developmental anomalies, than simply failure of asymmetry determination. In the absence of expressed genetic information, morphogenesis would be expected to result in various pathologies.

Another early prenatal event that is related to lateral anomalies is monozygotic twinning. Boklage, Elston, and Potter (1979) obtained a variety of dental measurements on monozygotic and dizygotic twins and examined the relationships obtaining among teeth on the two sides of the mouth, as well as those for teeth on the same side. Monozygotic twins were found to have much lower correlations among tooth measurements, both between sides and within sides of the mouth, than dizygotic twins. There were significant differences in the left and right sides for monozygotic twins, but not for dizygotic twins. The identical twins manifested disharmonies among structures of the teeth not seen in fraternal twins, and multiple discriminant analysis could differentiate, with 82% accuracy, the zygosity of a twin from the pattern of dental relationships.

Boklage (1980) has argued persuasively that monozygotic twinning itself is a reflection of abnormal and poorly controlled coordinate mapping, and, particularly, of midline specification. He notes that in the normal zygote, there is a strong tendency for cell-to-cell adhesion, so that even physical separation of cells does not regularly lead to twinning, since the cells migrate back together to form a single organism. Further, the twinning event itself apparently precedes physical separation, as indicated dramatically in cases of Siamese twinning. Thus, physical separation of cells is insufficient to guarantee twinning, and adhesion among cells is insufficient to preclude twinning.

Some anomaly of coordinate specification evidently can lead both to twinning and to disharmony among ectodermally derived features within a twin. The body of evidence (Bulmer, 1970; Guttmacher, 1937) suggests that monozygotic twinning is not genetic, though Carter-Saltzman (1979) has found that when twins display mirror imaging of ectodermally derived traits, familial twinning is found in most families, in contrast to families in which a twin pair does not manifest mirror imaging. Further, there was an elevated incidence of left-handedness among the relatives of mirror twins, and the frequency of low birthweight in mirror twins was about twice that for other twins. These observations are intriguing in their suggestion that there are two types of twinning events, one being independent of genetic factors, associated with high birthweight and lack of mirror imaging, the other being partly genetic, associated with low birthweight and mirror imaging.

A reasonable interpretation is that nongenetic twinning occurs when unusual intrauterine factors serve to force twinning of a normal zygote at a very early stage of development, with each resultant twin then developing normally, and that genetically mediated twinning represents an abnormal genotype, lacking, in some degree, good control over the specification of coordinates. In consequence, the zygotic midline, left and right, and possibly other coordinates as well, are ill-defined, often lead to twinning, and whether or not twinning occurs, asymmetry determination, to a greater extent than is normally the case, comes under control of random environmental events. One would presume, also, that the twinning event occurs relatively late in mirror twins, given their low birthweight as a possible indicant of an insufficiency of cytoplasm in early zygotic development. Supporting this inference is evidence that mirror imaging of ectodermally derived tissue is associated with late twinning (Boklage, 1980).

Since the nervous system is derived from ectoderm, it would be expected that in some cases of monozygotic twinning, the typical directional lateralization will be reversed, in one or both twins. If genes for asymmetry determination are only partially repressed with respect to ectodermal tissue, and environmental events are such as to push toward a reversal of the typical brain asymmetry pattern, then the phenotype might express an unusual degree of functional bilateralization, differentiation between the two sides of the brain being especially weak. If, in contrast, genes important for normal ectodermal asymmetry determination are completely repressed, and the environment happens to produce either the normal or the reversed asymmetry, the degree of lateral differentiation of the hemispheres might be as strong as for the typical individual.

Because of evidence that schizophrenia is associated with a specific disorder of the left cerebral hemisphere (Flor-Henry, 1976; Gruzelier & Venables, 1973), Boklage (1976) examined concordance for schizophrenia in monozygotic twins as a function of whether pairs were concordant for right-handedness or not. If left-handedness in one or both twins is related to an early embryonic history in which abnormal asymmetry development occurs, then a typically lateralized disorder in such twins should deviate from that normally found in singletons. First, pair-wise concordance for schizophrenia was found to be 92% in twins concordant for right-handedness, but was only 25% in those who were not. Second, fully right-handed pairs had more severe illness and much higher frequency of nuclear schizophrenic diagnoses than pairs having at least one left-handed member. Among these latter, left-handers were more often schizophrenic (but rarely of the nuclear subtype) than right-handers.

In appears that in twins concordant for right-handedness, genes predisposing to the lateralized disorder of schizophrenia gained their typical expression, with nuclear schizophrenic illness being manifested in both twins. In contrast, the usual neural asymmetric substrate in which schizophrenia develops was not present in other types of twin pairs, resulting in a high rate of discordance for schizophrenia, nonnuclear and less severe schizophreniform illness, and an overrepresentation of pathology among the left-handers.

Carter-Saltzman's data (1979) suggest that mirror imaging of ectodermally derived tissue is related to a genetic abnormality in the control of asymmetry development, but in some of Boklage's cases, both twins were left-handed. Possibly, like the iv/iv mice, asymmetry determination is random in certain individuals, resulting in associations with mirror imaging in twin pairs, but also, on occasion, leading to concordance for an atypical asymmetry. There is no understanding at present regarding the nature of environmental effects that, acting on a genotype in whom asymmetry specification by the genes is either absent or weak, lead to one or the other direction of lateralization.

The Boklage (1976) observation that pathology was more common in left-handers when either one or both members of the pair were left-handed implies that the direction of lateralization has consequences for psychological function. If so, then a reversal of the usual pattern of brain laterality, with linguistic, analytic, and temporal functions specialized to the left, and nonverbal, synthetic, and spatial functions specialized to the right, does not mean mirror identity. Quite possibly, either the level or quality of verbal and associated capacities differs depending on which hemisphere is specialized for these processes, and similarly for nonverbal and associated abilities.

In any case, the available evidence strongly suggests that, although genetic information guides asymmetry development in the majority of individuals, in certain rare cases, the critical genes do not gain expression, or are expressed only weakly, and unknown environmental factors in early development determine the direction, and possibly the degree, of cerebral lateralization and other ectodermally derived asymmetries.

Once the basic embryological pattern has been set, either by genes or by unknown environmental events, it is probable that the asymmetries of most traits (e.g., fingerprint asymmetries) are quite implastic to subsequent environmental factors acting late in fetal development or in the perinatal, neonatal, and early childhood period. Such implasticity, however, would not be expected to hold with respect to the brain and the behaviors it controls. The human brain is built to respond to environmental contingencies and does so throughout the life of the individual. Learning itself may be presumed to reflect micro-reorganizations of neural relationships, and until senescence sets in, this plastic reorganizational capacity is maintained.

It is therefore reasonable to inquire about the possibility of macro-reorganizations, involving the direction of lateralization of the hemispheres, in young organisms. In the majority of adults, damage to the language zones results in a permanent aphasia, and the intact hemisphere is unable to assume the functions that have been disordered. Plasticity of reorganizational capacity in most adults is evidently insufficient to adapt to such major trauma.

Earlier researchers (Lenneberg, 1966) believed that the neonatal brain was perfectly symmetric in manifest function and that there was a progressive growth of lateral differentiation. It was claimed that total hemispherectomy performed in infancy had no consequences for the structure of cognitive abilities in the adult as a function of which hemisphere had been removed. In other words, each hemisphere of the infant brain was said to be equipotential with respect to all cognitive process. The basis for this conclusion was the report that young children develop transient aphasic disorders following cortical damage, regardless of which hemisphere suffers the lesion (Basser, 1962). While it is true that most aphasias in young children are transient, and that, if the lesion occurs prior to the onset of speech, speech will usually be acquired, independently of the side of lesion, Basser's own data clearly show the two hemispheres are not equally involved in language in young children. Of 30 patients studied in whom unilateral injury was acquired after the onset of speech, but well before puberty, half with left- and half with right-side damage, 13 of 15 of the former lost speech entirely consequent to the lesion, but only 7 of 15 of the latter had any aphasic symptoms at all (Exact $p = .0251$).

Nevertheless, the general view was that, perhaps, prior to the onset of speech the hemispheres had been equipotential, even if a degree of asymmetry was present after speech had been acquired. Recent studies rule out this possibility. That the hemispheres do not have equal potential for the acquistion of all cognitive abilities has been shown by investigations of Kohn and Dennis (1974) and Dennis and Kohn (1975) of adult infantile hemiplegics with hemispherectomy performed in infancy or childhood. Patients having only a right hemisphere manifest severe difficulty in assigning correct meaning to passive voice sentences that are semantically unconstrained, whereas those having only a left hemisphere are incapable of performing visuo-spatial tasks dependent on cognitive skills that normally develop after 10 years of age.

Asymmetric potential, of course, does not rule out the possibility of manifest symmetry in the infant brain, but, in fact, the neonatal and infant hemispheres are differentiated with respect to both. Entus (1977) and Best and Glanville (1978) found that babies discriminate words better with the right ear and music better with the left ear with dichotic listening procedures. Gardiner and Walter (1977) observed asymmetric blocking of the infant's homologue of the alpha rhythm over the left hemisphere in response to verbal stimuli and over the right hemisphere in response to musical stimuli. Molfese (1977) found larger evoked responses to musical chords over the infant right hemisphere and larger evoked responses to consonant-vowel syllables over the left hemisphere.

The functional asymmetries observed are congruent with anatomical asymmetries apparent in the neonatal brain (Wada, Clarke, & Hamm, 1975; Witelson & Pallie, 1973), and given the evidence that asymmetry determination begins in the early zygotic stage, it would have been surprising if, some 9 months later, the brain reflected no sign of this.

Thus, the infant brain is functionally and anatomically asymmetric at birth, presumably manifesting the effects of early asymmetry development for the body as a whole. As noted, however, the asymmetry seen at birth is plastic, though the plasticity is incomplete. Only subtle deficits are found in adult infantile hemiplegics; the intact remaining hemisphere can subsume a great part of the functions that would normally have been integrated by the missing hemisphere. The ability of a hemisphere to take over the functions of the other decreases with age, and this developmental change has led some researchers to propose that the *degree* laterlization increases developmentally. There is no satisfactory evidence to support this contention; observations render it unlikely.

When overall level of performance is controlled, no developmental changes are observed in the degree of perceptual asymmetry manifested on dichotic listening or on lateralized tachistoscopic tests (Witelson, 1977). Children of certain ages may fail altogether to display any perceptual asymmetry on a particular test designed to measure hemispheric lateralization, but if they show lateral differences, they are typically as large as those seen in adults.

Clearly, the properties of the cerebral hemispheres change with age, progressively losing the capacity for reorganization in the face of cerebral injury, but a decrease in plasticity does not necessarily imply an increase in lateralization. With cognitive growth, each hemisphere becomes increasingly competent to handle specialized functions within its domain, but it would hardly be valid or reasonable to claim that the failure of 3-year-olds to manifest a hemispheric difference in the capacity to interpret a passage from Shakespeare implies that adults are more verbally lateralized than children. In some instances, children fail to show asymmetries observed in adults, but yet perform above chance. A careful analysis will reveal, however, that the strategies applied by children in such cases are primitive and are laterally specialized in neither children nor adults. The absence of an asymmetry reflects the utilization of processes that are symmetrically organized in all age groups, processes that are bypassed in adults in favor of superior and laterally differentiated cognitive operations that are immature in children.

If the degree of lateralization does not increase with development, what mechanisms underlie the progressive loss of plasticity for lateral reorganization? How does a single hemisphere, deprived of its partner on the other side, come to have such a high degree of bifunctional competence if hemispherectomy is performed in childhood? Bifunctional competence of *both* hemispheres, similar to that seen in cases of infantile hemiplegia, is found in individuals with agenesis of the corpus callosum (Jeeves, 1969, 1972; Netley, 1977; Saul & Sperry, 1968), suggesting that each hemisphere, in the normal course of development, actively inhibits development of its own specialized functions within the other side of the brain: Loss of a hemisphere or absence of the major communicating pathway between hemispheres during critical maturational stages allows secondary programs in each hemisphere to mature and become elaborated, as we suppose usually happens only for the primary program of specialization.

It is being suggested that the infant is born with a primary program of specialization, as well as a less developed and less competent secondary program for the other hemisphere's specialties; that in the normal course of maturation, the secondary program fails to mature and undergoes functional regression as neural organization matures in service of the major specialized functions. If the language zones of the left hemisphere are lost at a sufficiently early age, homologous regions in the right hemisphere, freed from the postulated inhibition of linguistic organization, begin to mature in this direction, coming to assume language processes, though limited in the level that may be reached. When injury occurs relatively late in childhood, maturation of nondominant language functions begins from an infant baseline and has very few years before the onset of puberty, when, for most people, lateral plasticity is lost, to grow in functional competence.

On this model, the developmental loss of plasticity is not a passive consequence of aging. It reflects an active inhibition exercised by the other side of the brain, in the absence of which, as in cases of callosal agenesis, speech and other higher cognitive processes would develop bilaterally, but with differing degrees of functional competence. We appear to be built as a set of very wise contingency statements that allow and impel full elaboration of the specialized abilities of each hemisphere if and only if those on the other side are intact and in communication, but that demand a sharing of function within a hemisphere if those on the other side are lost or out of communication.

Although there is no evidence that trauma to the brain during the developmental period can reverse the direction of asymmetry, there is much evidence to suggest that either hemisphere can assume control of functions normally specialized to the other side. Particular lateralized behavioral traits such as handedness or eye dominance for sighting, may be shifted from their normal asymmetry by perinatal damage, but this does not imply that the asymmetry of the brain has been reversed. It merely suggests that damaged regions important for the control of the hand or for controlling sighting have yielded a command they otherwise would have maintained. A particular command function may be shifted, or even all functions in the case of hemispherectomy, but the intact hemisphere does not, thereby, shift its *own* specialized processes to the damaged side.

It appears that with the usual and benign intrauterine and postnatal environment, asymmetry development of the human hemispheres proceeds on a well-defined course, probably under control of information contained in the genes, but that under unusual circumstances, genetic control in early zygotic development may be lost and replaced by environmental information sources, and that in the face of trauma to the brain, the developing organism responds with neural reorganization that serves adaptive function.

Finally, some consideration should be given to the possible effects of the sociocultural environment on cerebral lateralization. At various times during the last century, investigators have suggested that culturally trained hand dominance, forced adoption of the less preferred hand by injury to the other for the control of writing and fine manual activities, or illiteracy changed the lateralization patterns of the brain as compared to what would have been seen in the absence of social pressure for hand usage, in the absence of unimanual injury, or if the individual were literate.

Gloning, Gloning, Haub, and Quatember (1969) obtained data that make it extremely unlikely that hand usage has any effects on cerebral lateralization. They examined 57 non-right-handed patients with unilateral injury to the right or left hemisphere for the presence of various linguistic disorders consequent to the lesion. These patients, all European and reaching puberty around 1918-1920, had been forced to use the right hand for writing during school years. After completing their education, however, 17 of these patients (30%) shifted to left-hand writing. Any effects of forced dextral usage during the entire developmental period would be expected to be identical for all patients in the sample, irrespective of whether they continued to use the right hand or shifted to the left hand after reaching puberty (WR and WL patients, respectively).

In contrast to predictions following from the hypothesis that hand usage conditions cerebral lateralization, Gloning et al. (1969) found a very strong writing hand × side-of-lesion interaction for the development of dysgraphia ($p < .00003$, 1-tailed) and dyslexia ($p < .0002$, 1-tailed), and weak interactions between writing hand and side-of-lesion for the development of long-lasting Broca's aphasia ($p = .0583$, 1-tailed) and long-lasting anomia ($p = .0548$, 1-tailed).

The authors suggest that their results may either be interpreted as meaning that the use of one hand for writing favors the development of reading and writing in the contralateral hemisphere (and, to some extent, speech functions also) or as meaning that the hand eventually chosen for writing is due to a greater underlying degree of dominance in the opposite hemisphere for linguistic processes. There are a number of considerations that rule out the first interpretation and that support the second. First, all the patients used the right hand for writing during their school years; the average period of schooling was 7.5 years, meaning that patients used the right hand for writing until an average age of 14 years. Even unilateral brain damage suffered at this age or later is normally incapable of shifting lateral dominance, and it would be remarkable if a shift in writing hand after this age could shift lateral functions. Second, if the hand used for writing can affect lateral dominance, one would expect that WL patients should be less laterally asymmetric than WR patients since the former had at least 7.5 years of experience in using the opposite hand for writing. Right-handed writing up to the age of puberty, if writing hand affects brain laterality, would be expected to ameliorate the effects of right-hemisphere damage in the WL group compared to the effects of left-hemisphere damage in the WR group. Similarly, it would be expected to exaggerate effects of left-hemisphere damage in the WL group compared to the effects of right-hemisphere damage in the WR group.

In fact, however, lesions contralateral to the writing hand produced dysgraphia in 91% of the WL group and 73% of the WR group, the difference being nonsignificant and in the opposite direction from what would be expected if writing hand affects lateral dominance. Lesions ipsilateral to the writing hand produced dysgraphia in 2 of 6 of the WL group and 1 of 14 in the WR group, and though, here, the expected "exaggeration effect" goes in the appropriate direction, the difference, by an exact probability test, is not significant ($p = .2017$).

Given these considerations, we are led to the conclusion that patients who eventually adopted the left hand for writing, in spite of being forced to use the right hand throughout school years, did so because reading and writing were specialized to the right hemisphere, at least in the majority of cases, while most who continued to write with the right hand did so because reading and writing were specialized to the left hemisphere. It should be noted, also, that cultural pressures exerted on non-right-handers for dextral writing have greatly decreased over time, that the proportion of the population who are left-handed writers has increased so that, currently, the fraction appears to be close to an asymptote of about 12%, and recent estimates of the proportion of left-handers having linguistic functions specialized in the right hemisphere are near 30%, the same percentage found by Gloning et al. (Levy, 1976).

Thus, the fact that the Gloning et al. patients were compelled to use the right hand for writing during school years had no effects on the distribution of right- versus left-hemisphere language functions compared to modern estimates, nor on the cerebral laterality patterns eventually manifested by the patients within the sample.

There are insufficient data in the literature to allow any conclusions to be drawn regarding the effect of illiteracy on cerebral laterality. Though individual case reports have appeared of right cerebral dominance for language in illiterate right-handed patients (e.g., Wechsler, 1976), there is no evidence that illiteracy increases the likelihood of right-hemisphere language compared to what is found in literate people. Additionally, the capacity to read is not an evolved characteristic in the sense of speech or the comprehension of speech. The evolution of human brain characteristics critical for true language resulted in organizational properties making the acquisition of reading possible, but skill in reading could not have been specifically selected since written languages are of such recent origin and never developed in certain primitive extant cultures. On a priori grounds, therefore, it would be improbable that the developmental organization of the human brain would be contingent on the presence or absence of reading skills.

In contrast, the developing child may well be dependent on experience with spoken language for normal maturation and dynamic usage of specialized hemispheric capacities. Neville (1977) examined the visual evoked potentials of congenitally deaf chidlren who were either conversant in American Sign Language or not. In normal children, the visual evoked response (VEP) was larger over the right than left hemisphere, but in nonsigning deaf children, evoked response magnitude was small over both hemispheres and little hemispheric asymmetry was seen. In signing deaf children, the VEP was larger over the left hemisphere. These results suggest that in the absence of either oral or sign language, the brain becomes underreactive and little lateral differentiation of response develops. In normal children, the visuo-spatial right hemisphere develops a greater response to visual signals than the left, but in deaf signing children, for whom language is purely visual, it is the left linguistic hemisphere that becomes asymmetrically responsive to visual signals. Whether the differences observed by Neville among hearing, nonsigning deaf, and signing deaf children reflect actual changes in the direction or degree of cerebral asymmetry of function, or, alternatively, changes in dynamic activational properties of the hemispheres, cannot be decided, though the latter interpretation seems intuitively plausible. If the dynamic hypothesis is correct, this would not necessarily mean that if hearing were suddenly restored, the evoked response patterns would come to be similar to those of children with no congenital hearing defects. Particular activational patterns, gained during development, may be resistant to change once the developmental period is completed.

Profound environmental deprivation of social interaction and experience with language can, evidently, preclude the development of normal left-hemisphere function that cannot be overcome when a normal environment is provided after puberty. Genie, the child of a blind mother and a psychotic father, was found at the age of 13 years 7 months after having been isolated from all human contact in a small locked room from the age of 20 months (Curtiss, 1977). At the age of 20 years, dichotic listening and tachistoscopic tests, as well as studies of Genie's auditory and visual evoked responses, indicated that her response to language, as well as to other stimuli, is mediated almost entirely by her right hemisphere. On cognitive tests depending on left-hemisphere abilities, Genie functions in the extreme defective range, but on tests depending solely on right-hemisphere processes, she scores well above the normal adult level; on some tests, she scored at the 98th percentile.

Though conclusions drawn from single case reports are necessarily speculative, the fact that Genie is unique in her cerebral organization pattern strongly suggests that social interaction and experience with language during prepubertal years is essential for left-hemisphere development, and that, in its absence, there may be hyperdevelopment of the right hemisphere.

Whether variations in environmental enrichment or deprivation within the normal range of environments have effects on lateral development cannot be decided on the basis of available evidence. Dorman and Geffner (1974) found no differences between low- and middle-SES children on ear asymmetry for a verbal dichotic listening test, and Geffner and Dorman (1976) observed no SES differences at 4 years of age, similar to results of Knox and Kimura (1970). However, Bever (1971) examined two groups of low-SES children, one of which had been exposed to an enrichment program, and observed a right-ear superiority on verbal dichotic listening only in this latter group, with no ear asymmetry being seen in the nonenriched group. The two groups of children did not differ in overall performance or in IQ, but the enriched group manifested greater manual asymmetry.

Since the Bever results are not consistent with those reported by others, they are difficult to interpret. They may reflect different attentional strategies that had been developed in the two groups of children, and the failure of others to find SES effects on ear asymmetry may be due to sampling differences, low-SES children in these latter investigations being equivalent to Bever's enriched group. Clearly, more data are needed to establish whether variations in SES background affect the rate at which asymmetric processes are manifested in behavior or the ultimate degree of differential hemispheric usage. As noted earlier, there is no evidence that lateralization *per se* undergoes progressive development, but the age at which asymmetric hemispheric skills are applied to cognitive problems almost surely does, and attentional strategies that affect indices of hemispheric asymmetry would similarly be expected to change with age.

To summarize this section, the early intrauterine environment can affect asymmetry development, including that of the brain; trauma to the brain in prepubertal years can induce adaptive reorganizations of laterally specialized functions; and extreme environmental deprivations can, at the least, change activational properties of the two sides of the brain to sensory stimulation and, with radically inadequate social and linguistic environments, can almost totally preclude left-hemispheric development, while permitting, and perhaps promoting, development of the right hemisphere. There is evidence against the view that hand usage affects hemispheric lateralization, and the evidence is mixed with respect to the question of whether normal variations in the level of sociocultural environments affect any aspects of lateralized hemispheric usage. Behavioral traits such as handedness are easily susceptible to cultural conditioning, but may revert to preferred manual dominance when pressures are removed, and social conditioning of dextrality for one manual skill does not generalize to others (Teng, Lee, Yang, & Chang, 1979), showing the superficial effects of social conditioning on manual dominance, and adding weight to the conclusion that central neural organization is not typically affected by manual usage.

Genetic Determinants of Cerebral Lateralization

The studies of monozygotic twins, discussed previously, strongly suggest that there are normally asymmetry determinants in the genes that guide morphogenesis leading to species-invariant asymmetries of the viscera, and, probably, to species-typical asymmetries of the brain. The prevalence of left-handedness and monozygotic twinning in the families of twin pairs showing mirror imaging of ectodermally derived traits (Carter-Saltzman, 1979), taken in conjunction with the evidence offered by Boklage, Elston, and Potter (1979) that monozygotic twinning is associated with disharmonies in development, both within and between sides of the body, and given animal studies establishing that in certain mutant strains of mice, genetic determination of asymmetry is repressed (Layton, 1976), one is led to the conclusion that genetic information, in the majority of individuals (but not all), is responsible for the direction in which asymmetry develops.

This conclusion, however, does not answer the question as to whether variations in the patterns of brain asymmetry seen in the human population are due to variations in genetic information guiding development. Although unusual genotypes (e.g., the *iv/iv* mice and certain classes of monozygotic twins and their relatives) are, evidently, associated with a prevalence of unusual laterality phenotypes, this is certainly due to genetic repression of asymmetry-determining genes in mice, and probably due to the same cause in the case of certain monozygotic twins and their relatives. In a certain sense, of course, genetic variation in these instances can be said to control phenotypic variation, but, for unusual phenotypes, this is the consequence of loss of genetic guidance in asymmetry development, and the resultant random determination of lateral organization.

Another issue is whether unusual cerebral asymmetry patterns and associated traits such as handedness and eye dominance for sighting, can develop under direct control of unusual asymmetry-determining genes, rather than as a consequence of random environmental factors when asymmetry-controlling genes are repressed. It could be, for example, that all instances of "nonstandard" brain asymmetry patterns and associated traits are due either to repression of genes for asymmetry determination, and control by random environmental effects, to brain trauma suffered in prepubertal years, with consequent reorganization of lateralized functions, or to extreme sociocultural environments experienced during development. Only the latter possible etiology is easy to establish or rule out in particular cases, since perinatal stress on the brain is always possible, and its presence, level, or effects not necessarily known, and since, in the absence of considerably more advanced knowledge of human developmental genetics than we currently possess, it is extremely difficult to know whether an unusual laterality genotype represents repression of asymmetry-determining genes or, alternatively, genetically controlled lateral development in an unusual direction. In both cases, the frequency of unusual lateral phenotypes would be increased in families of probands manifesting the atypical asymmetry, and in the absence of detailed knowledge of the mechanisms of transmission of asymmetry-determining genes, it is impossible to calculate what the expected familial increase should be for the two cases.

However, there clearly exist left-handers whose manual dominance is due to different factors from those having such major effects in monozygotic twins. As discussed, lack of concordance for right-handedness in monozygotic twins, at least one of whom is schizophrenic, is associated with discordance for schizophrenia, nonnuclear schizophreniform subtypes in twins with pathology, and more frequent pathology in left-handers than in right-handers, associations that are not observed in dizygotic twins, at least one of whom is schizophrenic (Boklage, 1977). Further, though ear asymmetry on dichotic listening tests is strongly associated with handedness in singletons (Satz, Achenbach, & Fennell, 1967), there is no association between ear asymmetry and handedness in monozygotic twins (Satz, Fennell, & Reilly cited by Boklage, 1977), possibly reflecting the disharmony of development in the latter, noted previously. Carter-Saltzman, Scarr-Salapatek, Barker, and Katz (1976) found very different performance patterns in monozygotic and dizygotic twins as a function of handedness on cognitive tests.

Among discordant pairs, left-handers performed better than their right-handed co-twins for the monozygotic pairs, but worse than their right-handed co-twins for the dizygotic pairs. However, for pairs concordant for left-handedness, disygotic twins had higher scores than monozygotic twins, these latter scoring worse than any other group, the former scoring better than any other group. No significant differences were reported between zygosities when twins were concordant for right-handedness.

It is apparent, based on the foregoing observations, that left-handedness in monozygotic twins is a different phenotype than left-handedness in dizygotic twins or in singletons, that its relationship with cerebral asymmetry differs, that its relationship with cognitive function differs, and that it almost certainly has a different etiology. It therefore becomes extremely unlikely that, if left-handedness in monozygotic twins is frequently due to random development of asymmetry, and failure of genes guiding lateral development to be expressed, that left-handedness in other groups could, except in rare instances, be due to the same cause. Assuming the validity of this conclusion, it then becomes possible to investigate whether genetic variations in cerebral asymmetry and its associated traits in non-twin or dizygotic twin populations exist, with some confidence that, if genetic factors are implicated, this is not merely a consequence of random determination of asymmetry in unusual phenotypes. If evidence for genetic variation is found in singleton populations, the most reasonable inference would be that different genotypes within such groups guide lateral development in different directions.

There is little argument that people vary in many lateralized traits, including handedness, eyedness, dermatoglyphic asymmetries, and cerebral asymmetry, and that there are significant correlations among these lateralized traits (Levy, 1976). Nor is there any disagreement that handedness runs in families (Nagylaki & Levy, 1973). However, many have argued that right-handedness and left cerebral dominance for language are species-invariant traits, and that any deviations from this pattern are due to embryonic anomalies, trauma to the brain, extreme environmental factors resulting in incapacity to use the right hand or, as in the case of Genie, pathological development, or social conditioning resulting from having a left-handed parent or sibling. As for many behavioral or psychological traits, there seems to be a very strong resistance toward admitting the possibility of genetic variation.

Nevertheless, the evidence is now considerable, and, I believe, compelling, that genetic factors are implicated in variations of cerebral laterality patterns and their associated traits. Hand preference for particular activities, though susceptible to cultural conditioning, is clearly correlated with cerebral asymmetry which, as discussed, is not. Almost all right-handers (over 95%) have the major speech zones in the left hemisphere; left-handers, in contrast, are highly heterogeneous in cerebral lateralization, some having speech on the left, some having speech on the right, and a considerable fraction being, to some extent, bilateralized with respect to some language functions. Handedness is also associated with neuroanatomical asymmetries (Galaburda, LeMay, Kemper, & Geschwind, 1978; Hochberg & LeMay, 1975; LeMay & Culebras, 1972; McRae, Branch, & Milner, 1968), at least some of which are present at birth, and, possibly, all.

Sinistrality in the families of right-handers is associated with an increased probability of recovery from aphasia following a left hemisphere lesion (Luria, 1970) and with smaller and more variable perceptual asymmetries on behavioral tests of lateralization (Hines & Satz, 1971). Eye dominance, unlike handedness, is expected to be unaffected by specific training since most people are unaware of the fact of eye dominance. Significant family correlations between parents and offspring are observed (Merrell, 1957).

Morgan (1977) argues that all such associations are mediated by maternal factors in the oöcyte (though, even if true, this would imply genetic variation since the cytoplasm of maternal cells contains the mother's gene products and other biological molecules formed under their enzymatic control). The Morgan argument predicts that lateralized traits in the father could have no effects on the progeny. In matings discordant for eyedness, Merrell (1957) found just as strong an effect of paternal as of maternal sinistrality.

Liederman and Kinsbourne (1980) have provided definitive proof of the transmission of laterlized traits via nuclear DNA. The large majority of infants display rightward turning biases at birth (Gesell & Amatruda, 1945; Turkewitz, 1977; Turkewitz, Birch, Moreau, Levy, & Cornwell, 1966), biases that are correlated with subsequent handedness (Churchill, Inga, & Senf, 1962; Gesell & Ames, 1947), left-handedness being overrepresented among the minority of children who displayed leftward turning biases at birth. Liederman and Kinsbourne found that rightward biases occurred only in infants having two right-handed parents; when either the father or mother was left-handed, no bias, either to the left or right, was seen. (There were no cases in which both parents were left-handed.) Since the paternal effect was as strong as the maternal effect, since the infants examined had had no opportunity to be conditioned by postnatal environmental factors, and since the only means by which a paternal trait can have effects on the neonate is via nuclear DNA, these observations establish that neonatal turning biases are, at least in part, under genetic control, and by implication, handedness and cerebral laterality.

Thus, one may not only conclude that the most common pattern of cerebral, sensory, and motoric lateralization develops under control of asymmetry-determining genes, the environment acting to modulate their typical expression, but that there are variations in genetic directives, some, particularly in monozygotic twins and their families, apparently resulting in repression of asymmetry determinants, others, usually occurring in non-twin families, specifying an unusual pattern of lateral development. The identification of factors underlying the maturation of certain trait characteristics does not, however, necessarily provide an understanding of their mechanistic organization.

Although most neurologists feel they have a good understanding of the basis of the association between handedness and cerebral asymmetry, the relationship is actually highly complex and presents a number of unsolved puzzles of relevance to developmental issues. In the next section, the neurology of the hand-brain relationship is discussed in the context of neural maturation.

Development of the Hand-Brain Relationship

The relationship between manual and cerebral dominance seems to be straightforward in the large majority of right-handers. The left hemisphere is specialized for language, analytic processes, and the comprehension and expression of temporally ordered events, and the right hemisphere is specialized for nonverbal functions, synthetic processes, and for the comprehension and expression of spatially ordered relationships. The majority of fibers in the pyramidal tracts, controlling fine distal movement, decussate in the lower medulla to descend as the lateral pyramidal tracts. Manual activities of the hands predominantly dependent on the specialities on one hemisphere are, in consequence, better performed by the hand contralateral to the specialized hemisphere. In right-handers, this means that writing and all ordered sequential movements will be better performed by the right hand, and that static spatial finger positioning will be better performed by the left hand (Ingram, 1975; Kimura & Vanderwolf, 1970). If all right-handers displayed this pattern, and if all left-handers displayed the mirror organization, there would be little to discuss.

But a few percent of dextrals and a majority of sinistrals have their dominant writing hand ipsilateral to the hemisphere specialized for speech (Rasmussen & Milner, 1977), and it is not understood how the dominant writing hand is controlled. In the discussion of the Gloning et al. (1969) observations, it was noted that there were extremely strong interactions between writing hand and side-of-lesion for both dysgraphia and dyslexia in their non-right-handed patients, but only a weak association for long-lasting Broca's aphasia and anomia. In other words, there was a significant dissociation in the lateralization of reading and writing versus speech. Evidently, a considerable fraction of the patients had reading and writing specialized to one hemisphere and speech specialized to the other. Hécaen and Sauguet

(1971) have also noted the peculiar pattern of language disorders in non-right-handed patients following unilateral cerebral injury.

The Gloning et al. data seem to suggest a simple solution to the unusual cerebral-manual associations observed in left-handers (and a small minority of right-handers): Those non-right-handers who write with the right hand have graphic skills specialized in the left hemisphere, regardless of the localization of speech, and those who write with the left hand (though adopted only after social pressures for dextral writing were reduced) have graphic skills specialized to the right hemisphere, regardless of speech localization. There is a serious problem with this solution in considering left-handers who developed in a more permissive environment, where left-handed writing is allowed.

The fraction of left-handed writing has increased in the United States from about 2% in 1932 to about 11% at present, and appears to be close to an asymptotic proportion (Levy, 1976). The patients studied by Gloning et al. were in their mid-50s when the data were collected (between 1950 and 1965) and would have completed their schooling in Germany around 1918. Had these individuals matured in an unbiased environment where left-handed writing was permitted, we may expect that a majority would have written with the left hand, including those with graphic skills specialized to the left hemisphere, an expectation that is given added weight by the fact that all patients were left-handed for activities not socially conditioned. Further, tachistoscopic and dichotic listening studies of normal left-handers who use the left hand for writing show that a small majority have left-hemisphere specialization for reading (tachistoscopic studies) and a small majority have left-hemisphere specialization for speech comprehension (dichotic listening), findings that cannot be easily reconciled with the postulate that the dominant writing hand is contralateral to graphic skills (Levy & Reid, 1976, 1978; McKeever, 1979; Smith & Moscovitch, 1979). Although the Gloning et al. observations suggest that speech laterality in left-handers cannot predict with any degree of accuracy the localization of graphic abilities, so that the dichotic studies may be of little relevance to the issue, they also show that specialization for reading and writing are strongly related. If this association generalizes to other left-handers, then asymmetries for reading in favor of the left hemisphere imply left-hemisphere specialization for writing also.

We are left with the strong inference that in many, and perhaps a small majority of, left-handers, specialized programs for the control of writing are ipsilateral to the writing hand. Via what pathways is writing controlled? Heilman, Coyle, Gonyea, and Geschwind (1973) suggest that interhemispheric mediation is necessary. They described a left-handed patient who wrote with the right hand. Following a right hemisphere lesion and left hemiplegia, this patient was totally free of aphasic symptoms (indicating speech localization in the left hemisphere), but had a profound agraphia of the right hand. Heilman et al. suggest that the right hemisphere was specialized for praxis, that praxic signals were transcommissurally relayed from the right hemisphere to the left to be integrated with linguistic information, and that the final common path to the right hand was from the left hemisphere. Presumably, this patient would have written with the left hand except for cultural pressure for dextrality, and in that event, by the Heilman et al. hypothesis, linguistic information would have been relayed from the left to right hemisphere, and motoric signals sent from there to the left hand. Gloning et al. observed one patient who wrote with the right hand and developed dysgraphia following a right hemisphere lesion, but the other 13 WR patients with right-side lesions did not.

If it is true that most of the WR patients would have written with the left hand in the absence of culture pressure, just as they used the left hand for most skilled activities, it is apparent that, in most instances, praxic-graphic skills would *not* be specialized in the right hemisphere. There were a total of 25 patients in the Gloning et al. sample with right-hemisphere lesions, of whom 11 became dysgraphic and 14 did not. Thus, it appears that only 44% of non-right-handers have praxic skills specialized on the right.

It is conceivable, of course, that even if writing is specialized in the left hemisphere of left-handed writers, that signals are relayed from the left hemisphere to motor centers on the right for control of the left hand. If so, however, right-hemisphere lesions should produce agraphia just as frequently in left-handers as left-hemisphere lesions do in right-handers, since, in the former, the right hemisphere is the final common pathway to the left hand regardless of whether speech and/or praxis is specialized to the left or right. Yet, Hécaen and Sauguet (1971) found a very low frequency of agraphia after right-side lesions in left-handers, much lower than the agraphia rate after left-side lesions in either right- or left-handers.

This is not consistent with the transcommissural-relay hypothesis, but the only obvious alternatives are equally unsatisfactory. First, it might be that some left-handers have a predominance of pyramidal fibers running in the ventral (uncrossed) pyramidal tract, but Milner (personal communication, 1979) has never observed an instance of ipsilateral hemiplegia following unilateral hemispheric inactivation in left-handers. On the predominance-of-uncrossed-fibers hypothesis, a substantial proportion of left-handers should manifest ipsilateral hemiplegia, either after amytal inactivation or after a unilateral lesion. In fact, homolateral pyramidal syndromes are exceedingly rare, too rare to account for asymmetry patterns of brain and hand in left-handers. Second, it might be that, in spite of larger crossed than uncrossed pyramidal tracts, left-handers with ipsilateral specialization of reading and writing rely on the uncrossed pathway for control of hand and finger movements. This hypothesis entails the *a priori* unlikely postulate that a transmission pathway having fewer fibers than another, and, consequently, a lower information-carrying capacity, can, nevertheless, exercise more precise and refined control than the larger pathway.

There is a nonobvious possibility to account for what may be called "ipsilateral left-handers" (or, in a few instances, "ipsilateral right-handers"). The pyramidal tracts have their cells of origin predominantly in the precentral cortex, with some neurons also lying in the postcentral cortex. They are spatially organized to map the various bodily regions they control in the form of a distorted homunculous, with some areas of the body greatly expanded (e.g., the hands and fingers), and others reduced (e.g., proximal regions of the body axis). Input to the pyramidal neurons comes from all regions of the brain, either directly or indirectly, and there is a two-way feedback between these neurons and those that feed into them. Visual, auditory, somesthetic, kinesthetic, olfactory, and taste information can guide selective activation of the pyramidal pool, and, additionally, communication between nonneocortical and cortical motor neurons modulates and refines the programmed output. Thus, brainstem, cerebellar, basal ganglia, and limbic integrations all serve in guiding the final output programs.

There is no reason to suppose that the maturation rates of pyramidal axons destined to control different body regions, or selectively responsive to input from different sensory integrations and subcortical systems are identical, and good theoretical justifications for supposing that they are not. First, developmental control of the body proceeds in a cephalo-caudal direction, from organization of closely associated muscle systems to larger integrations across distantly related regions, from gross to precise control, from axial to distal musculature, and from controlled reactions to ambient and subcortical regulations to focal and neocortical regulations (Trevarthen, 1973). Second, the systems organizing these regulatory inputs to cortical motor neurons mature at different rates. It is therefore not unreasonable to suppose that the embryogenesis of various pools of pyramidal neurons proceeds at different rates, that their outgrowing axons reach the lower medulla at different times in development, and that, dependent on the embryonic gradients controlling directional growth of axons that change in an organized manner over the course of prenatal development, the axons will either decussate to form the crossed pyramidal tracts, or will continue their growth ipsilaterally down the length of the spinal cord to form the uncrossed pyramidal tracts.

From the behavioral development of the fetus and infant, there is good reason to suppose that regulated control of finger and refined hand movements is one of the last motor systems to develop, and control of the fingers and hands in accordance with focal visual information, either in the external world or represented internally, may be the last sensorimotor system to attain maturation. Based on these considerations, it may be that pyramidal axons serving as the final common path of this integration are the last to reach the point of decussation at the medullary-cordal border. As Kuypers (1973) has shown from studies of monkeys, the uncrossed pathways exercise much better control over gross movements of the axial musculature than over refined movements of the distal musculature; the crossed pyramidal pathways appear to be essential for individuation of finger movements in response to visual stimuli. The same distinction relating uncrossed pathways to axial control and crossed pathways to distal control obtains for people (Trevarthen, 1980).

Though we have little information regarding the embryonic gradients controlling midline development and determining whether midline systems such as the pyramidal decussation or the corpus callosum will develop normally or not, there are certain observations suggesting that midline development may be blocked at certain stages of morphogenesis or, at the least, differentially permitted. First, there is the indirect evidence that early developing axial control is capable of mediation by the uncrossed pyramidal tracts, whereas late-developing distal control, in the large majority of people, is critically dependent on the crossed pyramidal tracts, an observation consistent with the possibility that during the stage when axial control systems are maturing, there may be a partial blocking of midline development, forcing ipsilateral growth of a substantial proportion of pyramidal axons. At the stage when distal control systems are maturing, the postulated midline block may be absent, and the outgrowing axons will cross to the other side of the nervous system.

It should be noted that crossed control is the primitive condition of the chordate nervous system, being present in amphioxus, and it has been suggested (Sarnat & Netsky, 1974) that the cross-wired vertebrate condition arose from behavioral contraints in the vertebrate ancestor. The cephalochordates, probably quite similar to the progenitors of the vertebrates, are filter feeders, extrude their sperm and ova into the water, and display no approach reflexes. They show, however, an escape-avoidance reflex such that, if touched on one side of the body, a coiling reflex away from the source of stimulation is elicited. Thus, stimulation of receptors on one side of the body must trigger muscle contractions on the other side of the body, implying a cross-wired system. Sarnat and Netsky postulate that we, and all other vertebrates, are the inheritors of a cross-connected control system (likely to horrify a design engineer) that arose due to the behavioral limitations of our ancestors. They also suggest that the development of commissural systems was important in inhibiting the primitive escape-avoidance reflex and in permitting approach reflexes to emerge.

On the basis of such considerations, it seems more reasonable to posit that crossing of axons will occur naturally, unless there is an active blocking of midline development at certain stages of morphogenesis, than to propose that uncrossed growth is the more passive condition, with crossed growth occurring only via specific elicitation. Thus, in the course of normal development, there may be an early period when midline maturation is relatively inhibited, and a later period when it is permitted.

The major portion of the pyramidal axons reach the medullary-cordal border at the end of the fourth gestational month in the human fetus (Humphrey, 1960), at the same time that the corpus callosum is just beginning its morphogenesis (Davies, 1963), and well after the anterior commissure can be clearly delineated. Evidently, in normal development, this is an "open" period for midline genesis, and apparently, given the well-developed anterior commissure, there were other "open" periods much earlier in development. However, at the end of the fourth gestational month, there must be very critical controls operating regarding the opening of the developing organism to midline development and that, under certain circumstances, can go awry. Though I am aware of no cases of agenesis of the anterior commissure, there are a large number of reports in the literature of disorders of midline development for systems undergoing their major development starting at the end of the fourth gestational month.

There are many descriptions of patients with callosal agenesis (Ettlinger, Blakemore, Milner, & Wilson, 1972, 1974; Ferriss & Dorsen, 1975; Jeeves, 1972; Lehman & Lampe, 1970; Saul & Gott, 1976), and, additionally, agenesis of the corpus callosum has been identified in other instances only at autopsy. Further, various disorders of midline development, though not seen in all patients with callosal agenesis, are represented at a much higher frequency than normal in such patients (Francois, Eggermont, Evens, Logghe, & DeBock, 1973), and is sometimes associated with failure of pyramidal decussation (Lagger, 1979).

In the Dandy-Walker syndrome (Dandy & Blackfan, 1914; Taggert & Walker, 1942), many disorders of midline neural development appear in association: agenesis of the vermis of the cerebellum, fusion of the anterior frontal lobe into a monomorphic structure, midline cyst of the fourth ventricle, cleft palate, nondecussation of the cortico-spinal tracts, and, most prevalently, agenesis of the corpus callosum (Lagger, 1979).

Midline developmental disorders are therefore seen to be represented for structures undergoing development just at the time when pyramidal decussation begins, and various degrees of midline dysgenesis are observed, ranging from the usually fatal Dandy-Walker syndrome, in which many different midline systems are disordered, to minor midline anomalies in which a greater proportion than usual of corticospinal fibers fail to decussate, with no other evidence of midline dysgenesis (Nyberg-Hansen & Rinvik, 1963;

Verhaart & Kramer, 1952; Yakovlev & Radic, 1966), or in which there is partial dysgenesis of the corpus callosum in the absence of other symptoms.

It is not difficult to imagine that there might be some individuals in whom the timing of midline development, or the timing of the postulated midline blockage, is abnormal, and that, in some cases, this results in a normal *proportion* of corticospinal fibers decussating, but in an abnormal representation of the *types* of fibers running in the crossed and uncrossed bundles, since, as discussed, there is reason to suppose that those controlling large bodily movements and the axial musculature reach the decussation point before those involved in controlling fine distal reactions of the fingers and hands in focal sensorimotor integration. If a blockage were to be present (abnormally) at the later period, certain regulated finger movements and/or certain subclasses of sensorimotor integrations involving precision movements of the fingers or hands may be mediated via the uncrossed pathways; in contrast, large movements of the arms, legs, axial regions of the body, sensorimotor integrations regulating posture and movement in space, would, as in most individuals, have a predominant reliance on the crossed pathways, with capacity for control by the uncrossed pathways possibly normal or possibly diminished. In this case, the normal contralateral hemiplegia would be observed following unilateral damage or inactivation of a hemisphere, but for certain restricted classes of fine distal movements, there would be a greater-than-normal number of fibers running ipsilaterally and controlling the ipsilateral side, and a consequent reliance on these pathways for the restricted class of distal movements or sensorimotor integrations.

Given the Gloning *et al.* (1969) observations of a strong association in left-handers between the control of writing and literate capacity, and the weak association of these with speech, "ipsilateral" individuals may manifest a specific reliance on uncrossed pathways for visuo-motor reactions and for fine distal movements relating to graphic-praxic skills, but reliance on the crossed pathways for other manual responses and activities. Of course, a disordering of timing in development that entailed a blocking of decussation of corticospinal axons specific for late-developing neuronal systems might also have effects on callosal development which is occurring at the same time. Any late-appearing blockage having the effects postulated on pyramidal decussation would have to have a brief duration; otherwise, one would observe major developmental anomalies, like those of the Dandy-Walker syndrome, occurring with frequency in left-handers. Though left-handedness is unusually prevalent in cases of callosal agenesis, there is no evidence of an increase in any obvious midline disorders in normal sinistrals. It might be, however, that among some fraction of ipsilateral left-handers, callosal development is less complete than in others. If so, then like patients with complete callosal agenesis, in whom greater-than-normal bifunctional competence develops in both hemispheres, these individuals might manifest signs of weak lateralization of function. Whether this would be true or not depends on what portions of the corpus callosum are critically involved in inhibiting maturation of secondary programs in the other side of the brain.

The callosum develops in an anterior-posterior direction, so that frontal lobe connections develop first, and occipital lobe connections develop last. The hypothesized brief and late-occurring midline blockage would occur during the initial stages of callosal development, since the pyramidal axons are at the medullary-cordal border only for a short time before proceeding ipsilaterally down the spinal cord or decussating prior to descent. In contrast, the corpus callosum only begins development around the end of the fourth gestational month and must continue to form posteriorly until the splenium is complete. One would guess that if any dysgenesis of the callosum occurs in ipsilateral writers, this would involve predominantly the more anterior portions. Destruction of the anterior commissure and of the genu and body of the corpus callosum, leaving the splenium intact, in neurosurgical patients results in no split-brain syndromes: there is transfer of sensory information between the hemispheres, ability of each hand to write or draw to verbal direction, and transfer of tactile learning from one hand to the other (Gordon, Bogen, & Sperry, 1971). Gordon et al. note that these observations "leave unsolved the problem of what functions are mediated by the large frontal sector of the corpus callosum. (p. 334)." Possibly, the anterior portion of the corpus callosum, though evidently not necessary for either sensory or motor cross-integration in the adult patient, plays an important role in organizing interhemispheric relationships during development and in governing certain attentional and regulatory functions in the adult. Gordon (1973) found that the partially split-brain patients displayed one peculiar symptom as compared to control subjects. Subjects were given word lists through one or both ears that they were required to repeat, with a 200 msec delayed feedback of the subject's response to the opposite ear (in the one-ear condition) and to both ears (in the binaural condition). The binaural condition generated higher error rates than the monaural condition in control subjects, and a similarly high error rate was seen for the patients. However, with the monaural condition (with delayed feedback to the other ear), patients made as many errors as with the binaural condition, and about twice that of control subjects.

Gordon also noted that, unlike control subjects, the partially split-brain patients, when given two simultaneous verbal messages to the two ears, reported their contents in intermixed order instead of reporting first the complete message to one ear and then the complete message to the other ear. It appears, as Gordon suggests, that absence of the anterior portion of the corpus callosum produces deficits in inhibitory regulation between the two sides of the brain. If so, a consistent inference would be that it is the anterior callosum that also regulates interhemispheric maturation. In this event, and based on the speculations offered regarding the development of control pathways in ipsilateral writers, such individuals should display no deficits in interhemispheric integration, but might be unusually weakly lateralized compared to other groups.

Studies of normal left-handers have bearing on the various issues. Levy and Reid (1976, 1978) gave two tachistoscopic tests, one verbal, one nonverbal, to right-handers having the typical noninverted handwriting posture (Group RN, the hand held below the line of writing, with the pen pointing toward the top of the page), to left-handers having the noninverted handwriting posture (Group LN), to one right-hander with an inverted handwriting posture (Subject RI, the hand held above the line of writing, with the pen pointing toward the bottom of the page), and to left-handers with the inverted handwriting posture (Group LI). A right-visual-field/left-hemisphere (RVF/LH) superiority was seen in RN and LI subjects for the verbal test and a left-visual-field/right-hemisphere (LVF/RH) superiority was seen for these groups on the nonverbal test. The reverse field asymmetries appeared in subject RI and in group LN. Thus, reading (and by inference, writing) was left-hemisphere specialized in LI and RN subjects and right-hemisphere specialized in LN subjects and the RI subject. In addition to the apparent ipsilateral relationship between writing hand and specialization for reading, subjects with the inverted hand posture displayed extremely small perceptual asymmetries between visual fields as compared to others.

Smith and Moscovitch (1979) confirmed the relationship between hand posture and the direction of cerebral asymmetry on tachistoscopic tests similar to those of Levy and Reid (but did not observe any differences in *degree* of asymmetry between the two hand-posture groups). They found no relationship between hand posture and a dichotic listening measure of cerebral asymmetry (assessing speech comprehension) and suggested that the hand-posture variable was specific for visual functions. In a similar vein, Herron, Galin, Johnstone, and Ornstein (1979), using asymmetric suppression of the EEG alpha rhythm during performance on cognitive tests as an index of cerebral lateralization, saw no association of hand posture with cerebral asymmetry from central or parietal leads, but the expected relationship from occipital leads. They concurred with Smith and Moscovitch that hand posture is specifically associated with visual functions. Herron *et al.* also examined alpha suppression while subjects were writing, finding greater suppression over the left hemisphere (as expected) for RN subjects at all leads and greater suppression over the right hemisphere for LN subjects at all leads. Subjects in group LI had greater right-hemisphere suppression at central and parietal leads, but slightly greater (almost symmetric) suppression over the left hemisphere at occipital leads.

McKeever and van Deventer (1980) observed no association between hand posture and cerebral asymmetry on either a verbal tachistoscopic test or on a dichotic listening test. McKeever (1979) found no association between hand posture and cerebral asymmetry on one verbal tachistoscopic test, but did find the association reported by Levy and Reid on another verbal tachistoscopic test, though they attributed these latter results to familial sinistrality which they found to be increased in LI subjects compared to LN subjects. Lawson (1978), using a nonverbal visual test assessing attentional biases in looking at faces (toward the left half of space for right-handers), observed relationships between hand posture and bias in the predicted direction for male subjects and in the opposite direction for female subjects! Possibly, there are male-female differences in strategies employed in face perception, as some have reported (Rizzolatti & Buchtel, 1977), and females may rely, more than males, on the verbal hemisphere.

Dabbs and Choo (1980) indexed hemispheric blood flow by measuring temperature of the left and right ophthalmic arteries, since others had found with direct blood-flow measures that right-handers have more flow to the right. Dabbs and Choo confirmed this by finding higher blood temperature of the right ophthalmic artery in RN subjects. Subjects in the LI group displayed the same right-side asymmetry, but LN subjects manifested the reverse. It is difficult to relate the Dabbs and Choo observations to possible modality-specific asymmetries, but, conceivably, blood-flow asymmetries are related to visuo-spatial specialization.

Parlow (1978) investigated manual asymmetries in static finger positioning, a task for which the nondominant left hand is superior in right-handers (Kimura and Vanderwolf, 1970). Subjects in the LI group, like those in the RN group, had a left-hand superiority, but LN subjects were mirror reversed with a right-hand superiority. Parlow and Kinsbourne (submitted) examined a variety of manual activities. On all measures eliciting hand asymmetries, LN subjects were mirror images of right-handers: a right-hand superiority for static finger positioning, a left-hand superiority for tapping rate, a greater decrement in tapping rate induced by concurrent verbal activity for the left hand. No manual asymmetries were observed on any measures for LI subjects, similar to the weak lateralization observed by Levy and Reid. Gregory and Paul (1980) compared RN, LN, and LI subjects on a large neuropsychological battery, and in addition to finding a general inferiority of LI individuals compared to others on certain of the measures, observed differences in hand asymmetry indices. Whereas, in a speed-of-writing test, RN and LN subjects were, as expected, superior with the dominant hand (the right and left hands, respectively, for the two groups), LI subjects were superior with the nondominant right hand.

The Gregory and Paul data are particularly interesting in showing that, in spite of a preference for the left hand and adoption of the left hand for writing, LI people can actually write faster with the right hand. It is difficult to identify possible reasons for the left-hand preference in such people, especially because the greater right-hand skill is not consistent with the speculation that ipsilateral individuals may have a predominance of uncrossed motor fibers for fine visuo-motor control of the fingers and hand if it is assumed, as the data from RN and LN subjects confirm, that the writing of a word depends on verbally specialized writing skills. It is possible, of course, that word writing can be transformed into a drawing task, where verbal content is ignored and spatial form becomes the critical identifying aspect. In this case, the right-hand superiority of LI subjects would reflect spatial superiority of the right hemisphere, but one must question why only LI subjects would apply a spatial strategy, with other subjects applying a verbal strategy. Of possible relevance, in the Gregory and Paul sample, LI subjects were significantly inferior to others on Part B of the Trail-Making Test, but not on Part A, the former typically showing disruptions after lesions to the language hemisphere in right-handers, the latter showing disruptions after lesions to the mute hemisphere in right-handers (Reitan & Tarshes, 1959). It may be that LI subjects were selectively deficient in language-related functions and, when possible, applied nonverbal strategies to cognitive and/or motoric tasks.

Todor (1980) measured speed of alternating movements of the left and right hands in LI and LN subjects, finding that, for both groups, the mean time per movement was less for the left than right hand, the hand difference becoming magnified as task difficulty increased. However, for both hands, LN subjects were faster than LI subjects, the difference between groups increasing as task difficulty increased. The task difficulty × group interaction did not approach significance for the left hand, but was significant for the right hand: the rate of increase in time-per-movement with increases in task difficulty was larger for LI than for LN subjects for right-hand performance. These data seem to go in an opposite direction from those of Gregory and Paul (1980) with respect to right-hand proficiency. However, difficult alternating movements may necessarily require the specialized programs guiding temporally ordered sequences, usually associated with linguistic functions, and may not be susceptible to translation into a spatial task. This would not explain, however, why, under the assumptions that praxic sequential skills are specialized to the left in LI people and to the right in LN people and dependent on uncrossed pathways in the former and crossed pathways in the latter, the left hemisphere of LI people should be selectively deficient in controlling the right hand on difficult tasks as compared to the right hemisphere of LN people. For both groups, under the assumptions stated, transcommissural relay would be required. Possibly, the uncrossed bundles contain fewer fibers in LI people than do the crossed bundles in LN people for the critical subset of motor fibers involved in the task, a speculation that would account for the general inferiority of LI subjects in speed of movement.

More direct evidence regarding motor control pathways has been obtained by Moscovitch and Smith (1979) and by McKeever and Hoff (1979), utilizing manual reaction time measures to laterally presented stimuli. Moscovitch and Smith (1979) confirmed the expected superiority for homolateral hand-stimulus combinations in RN and LN subjects for the visual, tactile, and auditory modalities, and for LI subjects in the tactile and auditory modalities. However, for visuo-motor responses, LI subjects responded faster with the left hand to RVF signals and faster with the right hand to LVF signals, indicating reliance on the uncrossed motor pathways for manual reactions to visual stimuli. McKeever and Hoff (1979) only investigated the visual modality, confirming Moscovitch and Smith with respect to LN subjects. However, in LI subjects, they found a homolateral superiority for left-hand responses, as in LN subjects, but a heterolateral superiority for right-hand responses. They suggest that in LI subjects, there is a functional disconnection between visual and motor regions within the left hemisphere, requiring two transcommissural relays to allow a right-hand response to a RVF signal, but only one transcommissural relay to allow a left-hand response to a RVF signal. This suggestion is not consistent with their finding that right-hand responses of LI subjects were some 20 msec *faster* than for LN subjects; they attribute this difference to one LN subject with unusually long reaction times, but even if this subject is excluded, the LI subjects still respond 12 msec faster. Since, by the McKeever and Hoff hypothesis, one additional transcommissural relay is entailed for LI subjects as compared to LN subjects, their reaction time should be *slower*. The discrepancy between the Moscovitch and Smith and the McKeever and Hoff findings will have to be resolved by future research. However, it should be noted that Day (personal communication, 1977), testing reaction times only for the visual modality, found results identical to those of Moscovitch and Smith: heterolateral superiorities for both hands in LI subjects.

A final observation relevant to this issue is Milner's (personal communication, 1978) failure to find any association between hand posture and speech localization in left-handers, assessed by unilateral inactivation of the brain by amytal injection, a result totally congruent with the absence of hand-posture effects on dichotic listening tests, the absence of hand-posture effects on EEG alpha suppression measured from central and parietal leads, and the very weak associations observed by Gloning et al. (1969) between lateralization for reading and writing, on the one hand, and speech, on the other, within left-handed patients.

Clearly, there are many questions that remain to be answered, and discrepancies among different studies are not understood, but the body of available evidence suggests that (a) individuals using the inverted hand posture have reading, and probably writing, specialized in the hemisphere ipsilateral to the preferred writing hand, whereas those using the noninverted posture have these skills specialized to the contralateral hemisphere; (b) ipsilateral individuals rely on uncrossed motor pathways for visuo-motor responses of the hands, but contralateral individuals rely on the normal crossed pathways; (c) left-handers using the noninverted posture appear to be mirror images of right-handers with respect to manual asymmetry on a variety of tasks and with respect to visually lateralized functions; (d) left-handers and right-handers using the inverted posture have an unusual neural organization, with visually lateralized language functions and visuo-motor lateralized language functions ipsilateral to the writing hand, lesser manual asymmetry for some tasks, but not all, and selective advantages of right-hand function in some circumstances and selective disadvantages in others.

Whether the peculiar pattern of relationships seen in individuals using the inverted hand posture actually derives from certain anomalous timing events during fetal development that affect midline development cannot be decided on the basis of current data, but there are no observations that would rule out the possibility. As discussed, observations establish that left-handedness in those with critical language functions in the left hemisphere cannot be due to an overall failure of normal decussation for, in this case, ipsilateral hemiplegia would be observed. Nor does it appear likely that such individuals rely on a relay through the right hemisphere for manual control since agraphia with right-hemisphere lesions in left-handers would be as common as agraphia with left-hemsphere lesions in right-handers, which it is not. Nor does it seem reasonable to suppose that, in spite of having higher information-carrying capacities in the crossed pathway for the manual activities involved, individuals rely on an uncrossed pathway. If the three possibilities just listed have little probability of being correct explanations for the observed ipsilateral hand-brain relationship, there are few alternatives left.

The observations that hand posture appears to have specific relationships with visually lateralized functions and with manual activities that may reflect internal visual representations, that heterolateral superiorities seen only in LI and RI subjects are restricted to visuo-motor reactions, and that studies of neurological patients show dissociations between visuo-motor lateralization (reading and writing) and speech functions in left-handers, are all congruent with the possibility that there may be a highly selective nondecussation of a small class of corticospinal fibers that are crossed in most individuals. If, also, the two studies showing weak lateralization in LI people provide a valid characteristic of this group, partial callosal dysgenesis of the anterior callosum may be implicated. There is only one certainty with respect to these issues: Future research will either confirm one or more of the various hypotheses that have been raised or will refute them all in favor of an alternative shown to represent reality.

Sex Differences in Cerebral Asymmetry

There is still considerable controversy concerning whether male-female differences in lateralization exist, the nature of sex-related variations, the relations of these to psychological sex differences, and the causes of any differences that may obtain. McGlone (1980) has reviewed studies pertinent to these issues, concluding that the evidence points to more functional bilateralization and a lesser degree of asymmetry in females as compared to males. Various commentaries following her article offer support for or disagreement with this inference, the latter being in the minority and arguing that there is no good evidence to suggest a distinction between the sexes in hemispheric lateralization. In brief, clinical studies of neurological patients and behavioral dichotic listening or tachistoscopic studies of normal populations find, respectively, less symptom differentiation as a function of side-of-lesion in women and smaller asymmetries between ears or visual fields in women. The lack of concensus regarding what the observations mean arises because in many studies, no sex differences are observed. The possible reasons for these discrepancies would require a small volume to analyze, and it seems likely that greater insight might be gained by addressing other relevant points and discussing selected studies in detail.

There are good conceptual grounds for supposing that the sexes should differ in brain organization. Experimental animal studies leave no doubt that the sex hormones, acting during critical periods of development, have major organizing effects on the brains of mammals and other vertebrates (Goy & McEwen, 1980), and studies of people exposed prenatally to unusual levels or kinds of sex hormones or suffering from various anomalies of sexual development manifest behavioral and psychological differences as compared to control subjects (see Peterson, 1979, and Reinisch, Gandelman, & Spiegel, 1979, for reviews). It would be surprising if sex hormones during critical periods affected neural organization and behavior in other mammals, but not in people.

The question, of course, is whether, even given that human brains, like other mammalian brains, react to sex hormones in critical stages, the effects could be expected to be manifested in asymmetry patterns. Diamond (1980) has made the remarkable discovery that in male rats from birth to old age, the right cerebral cortex was thicker than the left, whereas in female rats, studied only at 90 days of age, the left cerebral cortex was thicker than the right in all regions examined, but, for females, the asymmetry was not significant. Of equal importance, females subjected to neonatal ovariectomy, developed asymmetry patterns indistinguishable from those of males. Rogers and Chen (as cited by Denenberg, in press) have also shown that disruption of the right hemisphere in rats depresses maze-learning ability, whereas disruption of the left hemisphere does not (only male animals were studied). Male rats are generally superior to females in maze learning.

If Diamond's observations are not merely specific to rats, but are generalizable to other mammals and, in particular, to people, they suggest a relatively greater right-hemisphere development in males and a relatively greater left-hemisphere development in females and, as McGlone suggests, a lesser degree of asymmetry in the latter. Many have noted the superior capacities of human males at map reading, three-dimensional visualization, and other visuo-spatial abilities associated with specialized functions of the right hemisphere and the superior capacities of females in certain aspects of verbal communication (Maccoby & Jacklin, 1974). Additionally, Reid (1980) observed right-hand (left-hemisphere) advantages in 5-year-old girls for discrimination of tactile rhythms among RN and LI children and for discrimination of palpated nonsense shapes in LN children, asymmetries seen in both sexes within these handedness/hand-posture groups at 8 and 10 years of age, but not observed in 5-year-old boys. In contrast, 5-year-old boys manifested left-hand (right-hemisphere) advantages for discrimination of palpated nonsense shapes in RN and LI children and for discrimination of tactile rhythms in LN children, asymmetries found in both sexes at 8 and 10 years of age, but not present in 5-year-old girls.

In other words, irrespective of the nature of its specialized capacities, the left hemisphere of girls and the right hemisphere of boys had reached a sufficient level of maturation by 5 years of age so that asymmetric specialized strategies could be applied to the tasks that Reid administered, resulting in behavioral asymmetries not observed in the other sex until a later age. However, neither the male superiority in certain visuo-spatial functions and the female superiority in certain verbal communication functions, nor the developmental sex differences allow the conclusion that there is an overall right-hemisphere advantage in males or an overall left-hemisphere advantage in females.

Though Witelson (1976) found a left-hand superiority for nonsense-shape identification in boys, but not in girls, and Rudel, Denckla, and Spalten (1974) and Rudel, Denckla, and Hirsch (1977) observed a left-hand advantage for Braille letter learning earlier in boys than girls, Ghent (1961) reported greater thumb sensitivity on the left hand for single-point pressure emerging earlier in girls than boys, Cioffi and Kandel (1979) found no developmental sex differences for nonsense-shape identification or for tactile word identification, and there are a variety of right-hemisphere specialized functions for which there are no sex differences or a female superiority and, similarly, left-hemisphere specialized functions for which there are no sex differences or a male superiority.

Cioffi and Kandel (1979) did observe one sex difference at all ages studied (the youngest being 7.5 years of age), and no change with development. For two-letter tactile bigrams, boys had a left-hand superiority and girls, a right-hand superiority, suggesting that the former treated the stimuli as nonlinguistic, spatial information and that the latter treated the stimuli as linguistic. The two sexes apparently applied different strategies to the tasks, males relying on the right hemisphere and females relying on the left hemisphere. The discrepancy between Witelson's (1976) observations with respect to nonsense-shape identification in girls and those of Cioffi and Kandel (1979) and Reid (1980) are possibly due to subtle methodological differences that, in the former investigation, led to bihemispheric participation in girls and that, in the latter investigations, encouraged reliance on right-hemisphere processing, at least by 7.5 or 8 years of age.

The Ghent (1961) study points to the possibility that certain types of right-hemisphere processes are at a maturational advantage in girls, while others, measuring understanding of spatial relationships, are at a maturational advantage in boys. This inference is consistent with observations in adults showing females to be superior in particular subsets of right-hemisphere functions. Thus, females surpass males in the perception of fine visual detail (Andrew & Paterson, 1946; Bennett, Seashore, & Wesman, 1959; Schneidler & Paterson, 1942), in understanding the meaning of facial expression (Buck, Savin, Miller, & Caul, 1972), and in identifying the affective implications of tone of voice (Soloman & Ali, 1972). Males surpass females in mathematical reasoning (as distinct from arithmetical computation) (Droege, 1967; Hilton & Berglund, 1971; Svensson, 1971; Very, 1967), a formal language of logic, expected to be asymmetrically dependent on the left hemisphere, and particular subsets of left-hemisphere functions may be at a maturational advantage in males.

There is, however, a certain discrepancy between the proposition that hemispheric maturation rates are the same in the two sexes, but for different aspects of verbal and nonverbal functions, and the evidence showing that human females have a bias to rely on left-hemisphere processes, the Diamond (1980) findings that female rats, in contrast to males, have greater cortical symmetry with a bias toward greater left-hemisphere development, and the Reid data (1980) showing earlier emergence of left-hemisphere asymmetries in girls independent of whether specialization is for verbal or nonverbal processes.

One possibility is that, in both sexes, the communicative aspects of language mature earlier than the capacity to utilize language in service of abstract, logical representations, consistent with Piagetian theory, and that spatial schemata mature earlier than the capacity to apply nonverbal integrative skills in apprehending the contextual, social implications of nonverbal cues. If there is a maturational advantage of the left hemisphere in girls and of the right hemisphere in boys, then, dependent on whether a hemisphere is specialized in the verbal or nonverbal domain, the earlier emerging functions will be verbal-communicative in girls with verbally specialized left hemispheres and in boys with verbally specialized right hemispheres, and spatial-schematic in boys with nonverbally specialized right hemispheres and in girls with nonverbally specialized left hemispheres.

The earlier developing processes may gain a guiding role over subsequent maturation of functions both within the same hemisphere and of those in the other hemisphere, either because of inbuilt maturational programs, because the earlier developing capacities have prior access to environmental reinforcement and encouragement, or, more likely, both. The typical individual having verbal processes specialized in the left hemisphere and nonverbal processes specialized in the right hemisphere would become progressively more sex-differentiated with respect to hemispheric function. In females, both sides of the brain would become organized for communicative skills and for interhemispheric integration of their verbal and nonverbal aspects, with a left-hemisphere bias, in certain circumstances, in approaching problems. In males, both sides of the brain would become organized for highly structured schemata and for interhemispheric integration of their verbal and nonverbal aspects, with a right-hemisphere bias in approaching problems in certain circumstances.

Basic sensory or motoric asymmetries may mature earlier in girls as a general reflection of their faster physiological maturation.

If there is any validity to these speculations, the question arises as to how they are related to evidence suggesting greater hemispheric asymmetry in males than females. Highly structured and tightly organized logico-verbal and physico-spatial representations, posited to be specialized to the left and right hemispheres, respectively, in males, can, in principle, be integrated between the two halves of the brain by well-specified mapping operations not too dissimilar from the types of mapping operations applied by the analytic geometer in transforming an algebraic equation into a geometric picture or vice-versa. The radically different natures of an equation, on the one hand, and a pictorial representation of a set of spatial relationships, on the other, do not preclude the mapping of one onto the other. Both contain well-defined sets of relationships that can be transformed, by abstract rules, from their symbolic to their spatial form or reverse.

It is being suggested that the representational structures of the male hemispheres are highly organized and well defined, and in consequence, can be radically asymmetric in form, while yet permitting interhemispheric integration and communication via postulated abstract mapping rules. This would imply that for diffuse, loosely organized representations, the male might have serious difficulty in interhemispheric communication and integration: these would be resistant to formal mapping from one hemisphere to the other, and, if the forms of representation differ greatly, the "languages" of the two sides of the brain would be too dissimilar to permit easy cross-hemisphere communication.

If the female hemispheres are organized for communication, both receptive and expressive, for apprehending a broad spectrum of rich contextual information, this means that a great deal of information, even if relatively insusceptible to highly organized structuring, must be incorporated, remembered, and integrated. Linguistic information must be interpreted in accordance with social context, with the environmental framework, much of which, to a male, might be considered incidental and irrelevant, with tone of voice, with facial expression, and with experientially gained knowledge that, though bearing no obvious and specifiable relationship with current input, can be integrated to reach valid inferences. The so-called field-dependency of females (Witkin, Goodenough, & Oltman, 1977) might more accurately be described as context sensitivity.

Context sensitivity, and the ability to integrate information from many simultaneous sources having no clearly defined structural relationships among themselves, means that integration must occur in the absence of formal mapping rules. The "languages" of the two sides of the brain must be sufficiently similar to permit communication of representations that are information-rich, not abstracted to a high level of encoding, and, of necessity, only loosely organized and weakly structured. Understanding the implicative differences when a statement is made in one context versus another, with one tone of voice rather than another, with one facial expression rather than another, requires that specialized representations of the two hemispheres be accessible to the other side of the brain in relatively direct form, bypassing the need by formal mapping rules applicable only to highly structured sets of relationships.

The female brain may be more functionally bilateralized than the male brain in order to permit interhemispheric communication and integration at a more intimate and less abstracted level, and more similar to the integrations that occur within hemispheres. This would suggest that the greater is context sensitivity and the better are skills for integrating events and information sources bearing few obvious formal relationships, the more functional bilateralization of the hemispheres. Further, the probability that both hemispheres participate in a task or that shifts occur, depending on circumstances, from predominant reliance on one hemisphere to predominant reliance on the other, would be closely associated with the degree of bilateralization of function.

There has been a great deal of argument in the literature regarding whether small asymmetries in tachistoscopic or dichotic listening tests displayed by women are due to bihemispheric participation or, alternatively, to functional bilateralization. I suggest that these are generically related, that individuals having lesser functional lateralization are more likely, in certain circumstances, to engage both hemispheres in the task. Also, there should be some conditions in which individuals having strongly asymmetric brains would display little asymmetry on a lateralization task. In particular, if the task requires interhemispheric communication of representations not susceptible to formal mapping rules, then both hemispheres should perform poorly in strongly lateralized people, one hemisphere because it lacks information required from the other, the other hemisphere because its specializations are not well designed for the task presented. The consequence would be that the left and right hemispheres, for different reasons, would perform poorly, generating little or no asymmetry on the task.

There is evidence to support the suggestion that field-dependency (context sensitivity) is associated with greater bilateralization of function. Oltman, Ehrlichman, and Cox (1977) had subjects judge whether a composite face composed from the left half of a poser's face and its mirror image or a composite composed from the right half of a poser's face and its mirror image more resembled the normal face from which the composites were composed. Gilbert and Bakan (1973) had demonstrated that right-handed subjects attend more to the left half of face photographs (containing the right half of the poser's face) and, therefore, judge composites made from two right half-faces as being more similar to the original. This bias for attending to the left half of space in looking at faces derives from right-hemisphere specialization for face recognition. The degree of bias can be interpreted to reflect the degree of right-hemisphere specialization for face recognition and/or the extent to which reliance is placed on the right hemisphere for face recognition.

Oltman et al. (1977) found that the degree of bias was significantly correlated, for both males and females, with measures of field independence, field-independency being positively correlated with degree of asymmetric bias, field-dependency being negatively correlated with degree of asymmetric bias.

In a tachistoscopic study with male subjects, Zoccolotti and Oltman (1978) found field-independent subjects to have a RVF superiority on a letter-discrimination test and a LVF superiority on a face-discrimination test, but field-dependent subjects displayed no significant field asymmetries on either test. Congruent with the inference that field-dependent subjects have greater functional similarity between the hemispheres than field-independent subjects, Oltman (1979) found the EEG amplitudes, correlated over 5-sec epochs, had significantly greater covariation between hemispheres for field-dependent than for field-independent subjects irrespective of whether they were viewing abstract shapes or faces or were engaged in a word recognition or arithmetical task. Similarly, O'Connor and Shaw (1978) found greater coherence between hemispheres for alpha activity in field-dependent as compared to field-independent subjects.

Rapaczynski and Ehrlichman (1979) compared field-dependent and field-independent women on a tachistoscopic face-recognition test. Subjects had memorized a set of target faces and responded to laterally flashed faces with a discriminated reaction time response designating whether the stimulus was a target or not. Field-independent women had a reaction-time advantage for the LVF of 52 msec, whereas field-dependent women had a reaction-time advantage for the RVF of 45 msec and were slower in both visual fields than field-independent women. The field-dependent women in Rapaczynski and Ehrlichman's study were considerably more field-dependent than the field-dependent men in Zoccolotti and Oltman's investigation (1978). The RVF superiority for field-dependent women observed by Rapaczynski and Ehrlichman may indicate that when field-dependency becomes extreme, bihemispheric participation becomes reduced in favor of dominance by the verbal hemisphere even for a face-recognition task where performance level becomes diminished in consequence. Under the developmental hypothesis discussed previously, this might occur when verbal–communicative skills of the left hemisphere are greatly advanced maturationally over spatial–schematic skills of the right hemisphere, allowing a long developmental period when left- but not right-hemisphere skills have the benefit of excercise and environmental encouragement. A strong bias for reliance on left-hemisphere processes would develop and persist throughout adulthood.

The investigations relating field-dependency to degree of cerebral lateralization, and suggesting that the nonverbal hemisphere is selectively engaged in certain nonverbal tasks when field-dependency becomes extreme, offer a reinterpretation of the Lawson study (1978), discussed in an earlier section. Lawson employed a task identical to that of Oltman et al. (1977), where subjects had to judge whether face composites made from two right half-faces or from two left half-faces more resembled the original face. Dextral noninverted males and sinistral inverted males had a bias for the left half of space, the bias for the latter being of smaller magnitude than for the former, and male inverted dextrals and noninverted sinistrals had a bias for the right half of space, the bias for the former being of smaller magnitude than for the latter. Thus, the data from males are consistent with inferences of Levy and Reid (1976, 1978) with respect to the direction of lateralization and further suggest, based on the Oltman et al. (1977) results, that inversion of hand posture is associated with increased field-dependency and less extreme lateral differentiation.

In females, Lawson found that noninverted dextrals had a bias for the left half of space (of smaller magnitude than for comparable males), but that inverted sinistrals had a bias for the right half of space. Noninverted sinistrals displayed no bias and inverted dextrals had a very strong bias for the left half of space. The right-bias for inverted sinistrals may reflect strong field-dependency, and *not* specialization of the left hemisphere for nonverbal processes, and the left-bias for inverted dextrals may indicate similarly strong field-dependency, and *not* specialization of the right hemisphere for nonverbal functions, and suggesting again, as for males, that hand inversion is associated with field-dependency and weak lateralization. Under this interpretation, the Lawson observations become perfectly congruent with other studies relating hand-posture to the direction and degree of cerebral asymmetry, with research showing that field-dependency is associated with bilateralization of function, and with suggestions, derived from other investigations and from theoretical considerations, that inversion of hand posture is correlated with weak lateralization. If this is a valid explanation of the Lawson (1978) data, it will be necessary in future studies to assess field-dependency before inferences can be drawn regarding the direction and degree of cerebral asymmetry in male and female left- and right-handers with and without inversion of hand posture, based on asymmetries observed in laterality tasks.

If females, better than males, can integrate the verbal with the nonverbal aspects of social communication, they would have, by definition, better verbal access to the meaning of facial emotion and better capacity for visualizating facial expressions representing verbally described emotions. Safer (in press) compared men and women on a task requiring matching of emotional expressions seen on a study photograph presented for 8 sec in the center of the visual field followed, 1 sec later, by a briefly exposed lateral face displaying the same or a different emotion. Subjects were either run with labeling instructions, being told to think of the name of the emotion shown on the study face and to repeat this to themselves until the lateral face appeared, or with

6. LATERALIZATION AND ITS IMPLICATIONS FOR VARIATION IN DEVELOPMENT 217

empathy instructions, being told to attempt to feel the emotion shown on the study face. Three possible strategies were available to subjects on this task: (a) they could make the same/different judgment on the basis of certain physical invariants extracted from the center and lateral face; (b) they could gain a visual representation of a generic expression representing the emotion; or (c) they could gain a verbal representation of the emotion sufficient to allow a match to the lateral face. A pure face-matching task was given in a second experiment where the sole basis of matching was physical identity: The same poser with different expressions or the same expression with different posers was to be judged as "different" and only if the same poser with the same expression occurred for the center and lateral face was a judgment of "same" to be given. By comparing the results of the first and second experiments, Safer could infer the strategy applied to the first.

In the emotion-judgment task, males displayed a large LVF advantage, unaffected by instructional condition, and of the same magnitude as in the face-matching task, suggesting that they were relying on the extraction of physical invariants in both. Females displayed a small LVF advantage with empathy instructions and a small RVF advantage with labeling instructions. On the face-matching task, they manifested a LVF superiority as large as that of males. On the emotion-judgment task, women were superior to males in both visual fields and under both instructional conditions. Thus, even with labeling instructions where men had a LVF advantage and women had a RVF advantage, the female LVF performance was still superior to that of males.

These data strongly suggest that, unlike men, women did not rely on the extraction of physical invariants in the emotion-judgment task. Instead, with empathy instructions, they appeared to generate a representation in the right hemisphere of a generic expression carrying the emotion of the study face, but, given the small LVF advantage, engaging both hemispheres in the task to some extent. With labeling instructions, again, both hemispheres appeared to participate in the task, with a bias toward a left-hemisphere verbal strategy.

It is an interesting question as to what males might have displayed if the paradigm employed had not allowed reliance on the extraction of physical invariants in judging emotions, if, instead, facial emotions had to be identified as matching or not some designated verbal label. By hypothesis, men should have great difficulty in generating a refined facial representation in the right hemisphere from a verbal description decoded by the left hemisphere. The right hemisphere would have only a crude representation and the matching process would be lengthy and/or often in error. The left hemisphere would have to apply an analytical strategy to the matching task, ill-designed for the recognition of the meaning of facial expression. Women, under these conditions, would be expected to be able to decode the full meaning of the verbal label, would be able to apply their own special left-hemisphere strategies to the task, with participation from the right, could generate a refined right-hemisphere representation from the left hemisphere's decoding, and would have excellent right-hemisphere performance. Whether a left-field, right-field, or no-asymmetry would emerge for women under these conditions would depend on particular methodological aspects of the task that would selectively bias one hemisphere or the other or encourage bihemispheric engagement.

Ladavas, Umilta, and Ricci-Bitti (1980) employed precisely this paradigm using a go-no-go reaction time response. On every trial, subjects were given the name of the target emotion (happiness, sadness, surprise, fear, disgust, or anger) and a face expressing one of these six emotions was flashed to one or the other visual field. If the expression matched the named target, subjects were to respond; otherwise, they were to refrain from responding. The go-no-go paradigm entails a right-hemisphere bias since other studies have shown that "same" responses (the only type allowed in the Ladavas et al. study) tend to yield right-hemisphere advantages, whereas "different" responses tend to yield left-hemisphere advantages (Egeth & Epstein, 1972). Based on the theoretical considerations discussed previously, women would be predicted to have a LVF superiority, but better performance in both visual fields than men.

Reaction times of females were faster than those of males by a large amount in both visual fields (47 msec for the RVF; 77 msec for the LVF); women had a substantial LVF superiority (29 msec), and men had no field asymmetry (a 1 msec nonsignificant advantage for the RVF). Women had a LVF superiority for all six emotional expressions and were superior to men in both visual fields for all six emotional expressions.

Ladavas et al. (1980) point out that Buchtel, Campari, De Risio, and Rota (1978) and Landis, Assal, and Perret (1979) have found LVF asymmetries in males for assessing emotional expression, but in these studies, only two or three emotional expressions were used and even crude right-hemisphere representations would be expected to be sufficient for differentiating sadness from happiness. Thus, though the various studies are superficially discrepant, they are completely concordant when methodological differences are considered.

All suggest that women have a better capacity than men for recognizing emotional expressions with both hemispheres, that they are at an advantage in integrating the verbal with the nonverbal representations of emotion, in apprehending, both verbally and nonverbally, the meaning of facial emotion, and in integrating the functions of the two hemispheres in judging facial expressions. They are flexible in using one or the other hemisphere, depending on conditions, being biased for the right hemisphere when either empathy instructions are given, "same" responses are required, or other aspects of the task encourage reliance on the right hemisphere, but being able to shift to a predominant reliance on the left hemisphere when verbalization is encouraged, and probably engaging both hemispheres to some extent under all conditions.

Although considerably more research will be required to establish the nature and genesis of sex differences in hemispheric organization, the available evidence is concordant with the view that females are more functionally bilateralized, that there is a maturational advantage of the left hemisphere in females, conditioning subsequent development, and of the right hemisphere in males, conditioning subsequent development, and that these asymmetries of development lead to a capacity for interhemispheric integration and flexible hemispheric usage in women for processes that, by necessity, can only be loosely structured, and, in males, to highly structured representations that are communicated between hemispheres by a set of formal mapping operations.

Given the functional asymmetries observed between hemispheres for discrimination of words and musical sounds in infants, it would not be expected that differing hemispheric maturation rates for boys and girls would be revealed in dichotic listening tests, nor for other lateralized functions that are manifested asymmetrically in babies. Many, if not most, of the discrepancies in the developmental or adult literature on lateral differences between the sexes are apparent only, and could be resolved by careful considerations of methodological variations and differences among processes that are assessed. Different lateralized functions are likely to have dissimilar developmental courses in boys and girls and to be manifested in adult men and women in different ways. Subtle methodological differences between studies can lead to radically different strategies in subjects, selectively encouraging unihemispheric reliance on either the left or the right hemispheric or bihemispheric participation, with differing consequences for the two sexes.

Human beings are sexually dimorphic along many dimensions of anatomical structure and physiological function, the dimorphisms controlled, to a major extent, by fetal hormones acting during critical periods of development. In other mammals, sex differences in brain organization are observed, in some instances with respect to lateral differentiation, and the organizational differences have been demonstrated to be under control of hormonal status during critical periods. There is no reason, on *a priori* grounds, to believe that human sexual dimorphisms occur for all portions of the body except the brain, or that sexually dimorphic brains occur in all mammals except people. Indeed, given that the most characteristically human distinction we possess is our brain and the behaviors and functions it controls, that the human culture and social system is centered around the family structure, whether monogamous, polygynous, or polyandrous, it would be remarkable if the organ in our heads were totally sexually undifferentiated, and if we were the single mammalian species for which this held.

Why Variation?

A final question that deserves at least brief attention concerns the reasons for variations in cerebral asymmetry patterns observed in the human population. Many are induced by environmental factors beginning in the early prenatal period and continuing, through the mechanisms of learning, throughout life. They represent organismic adaptations and reorganizations that incorporate environmental information and tend to maximize integrity of behavior and adjustment to conditions even, and perhaps especially, when those conditions do damage to the adjusting organ itself. The number of contingencies that the developing organism is prepared to meet is astonishing, and the organizational responses available give proof of the potentialities inherent in a single fertilized ovum. The adjustments that are made, even in cases as extreme as that of Genie where the left hemisphere gives little evidence of functional competence, serve to assure the development of a social being, a human person who, though perhaps deviant in many ways, is part of our social world.

But in addition to those variations representing responses to varying environmental conditions, the available evidence also strongly suggests that there are genetic differences among people in the programs guiding cerebral development and organization, differences both within and between sexes. Why are we not as genetically invariant in our patterns of cerebral asymmetry as we are in our bipedalism? We may safely presume that genes for bipedalism were under directional selection and were fixed in the human population at some distant period in our evolutionary past. An upright posture served hominid adaption better than any other and those controlling deviations were lost. Were there some ideal pattern of cerebral asymmetry, one that, under all circumstances, enhanced fitness, the currently observed variations would not exist.

An evident conclusion is that no single pattern of lateralization confers invariant advantages, but there are two possibilities that might account for variability. First, the asymmetry organization of the brain might have been selectively neutral and differences among people would merely reflect random factors such as genetic drift. This is mathematically possible for within-sex variations, but not for any genetically controlled differences between sexes. It is biologically extremely unlikely, however, that, even for within-sex variations, fundamental properties of brain organization in the human species could have been selectively neutral. The second possibility is, therefore, that active variance-preserving mechanisms operated throughout our evolution to establish a balanced polymorphism of brain laterality types.

If neurological variations have psychological consequences, this would mean that what has been perserved are differences among people in skills, propensities, personality, information-processing strategies, and in the nature of social roles that bring fulfillment, differences that would surely be of value for a socially specialized species where members give to the culture the benefits of their special abilities and ways of seeing the world and receive, in turn, the benefits of the special skills of others.

I would suggest that we were self creating, that the major selective forces operating in our evolution derived from the social system itself, conditioning the structure of the human gene pool that had consequences for subsequent social evolution. When our ancestors were still too ignorant to realize that stones could be manufactured to suit social needs, those with skills in stone finding were valued and protected. When, by social development, some apish creature invented stone chipping, the entire system underwent a radical change, the invention spreading by learning throughout the group. But now, those skilled in stone chipping came to be valued and socially protected. New selective forces, emerging from the social structure, came into being through a social invention. Social change eventuated in changes in the gene pool and these changed future social evolution. By means of frequency-dependent selection, whereby rare and valuable members of the society received fitness-conferring social benefits, thus increasing their numbers and decreasing their rarity and special value, individuals were tailored to social needs and the social structure was simultaneously tailored to the needs of individuals. A balanced polymorphism was achieved providing for social stability.

Thus, it is not only that the human environment determines which genes are to be expressed, which organizational adaptations are to be made, in the developing organism, but also determined which genetic structures, which social organizations, were evolved in our species.

We vary because we evolved as human social beings and variation aided in making possible a stable society, and we vary because, developing as human beings, we are designed to adjust to our conditions so as to be as human and as competent as possible. The potential paths that may be followed by a single-celled human zygote are enormous, and those open to our species are close to infinite. There is a wisdom in our evolution and a wisdom in our ontogenesis and their directives are the same. Our variations are valuable and probably essential to our survival. We must permit and encourage individuals to follow their own bents, whether it is a man who seeks to care for and educate our infants or a woman who seeks to discover the secrets of the cosmos, and if we have the courage to grant such freedom, both individual and social needs will be met.

References

Andrew, D. M., O Paterson, D. G. *Minnesota clerical test: Manual.* New York: The Psychological Corporation, 1946.

Basser, L. S. Hemiplegia of early onset and the faculty of speech with special reference to the effects of hemispherectomy. *Brain,* 1962, *85,* 427-460.

Bennett, G. K., Seashore, H. G., & Wesman, A. G. *Differential aptitude tests* (3rd ed.). New York: The Psychological Corporation, 1959.

Best, C. T., & Glanville, B. B. Cerebral asymmetries in speech and timbre discrimination by 2-, 3-, and 4-month-old infants. Paper presented at The First International Conference on Infant Studies, Providence, R. I., March, 1978.

Bever, T. G. The nature of cerebral dominance in speech behavior of the child and adult. In R. Huxley & E. Ingram (Eds.), *Language acquisition: models and methods.* New York: Academic Press, 1971.

Boklage, C. E. Schizophrenia, brain asymmetry development, and twinning: A cellular relationship with etiologic and possibily prognostic implications. *Biological Psychiatry,* 1976, *12,* 19-35.

Boklage, C. E. Embryonic determination of brain programming asymmety—A caution concerning the use of data on twins in genetic inferences about mental development. In J. Dimond & D. A. Blizard (Eds.), *Evolution and lateralization of the brain* (Vol. 299), *Annals of the New York Academy of Sciences.* New York: The New York Academy of Sciences, 1977.

Boklage, C. E. On timing of monzygotic twinning events. Paper presented at The Third International Congress on Twin Studies, Jerusalem, June, 1980.

Boklage, C. E., Elston, R. C., & Potter, R. H. Cellular origins of functional asymmetries: Evidence from schizophrenia, handedness, fetal membranes, and teeth in twins. In J. H. Gruzelier & P. Flor-Henry (Eds.), *Hemisphere asymmetries of function in psychopathology.* Amsterdam: Elsevier/North-Holland, 1979.

Buchtel, H. A., Campari, F., De Risio, C., & Rota, R. Hemispheric differences in discriminative reaction time to facial expressions. *Italian Journal of Psychology,* 1978, *5,* 159-169.

Buck, R. W., Savin, V. J., Miller, R. E., & Caul, W. F. Communication of affect through facial expression in humans. *Journal of Personality and Social Psychology,* 1972, *23,* 362-371.

Bulmer, M. G. *The biology of twinning in man.* Oxford: Clarendon Press, 1970.

Carter-Saltzman, L. Mirror twinning: Reflection of a genetically mediated embryological event? *Behavior Genetics,* 1979, *9,* 442-443.

Carter-Saltzman, L., Scarr-Salapatek, S., Barker, W. B., & Katz, S. Lefthandedness in twins: Incidence and patterns of performance in an adolescent sample. *Behavior Genetics,* 1976, *6,* 189-203.

Churchill, J. A., Igna, E., & Senf, R. The association of position at birth and handedness. *Pediatrics,* 1962, *29,* 307-309.

Cioffi, J., & Kandel, G. Laterality of stereognostic accuracy of children for words, shapes, and bigrams: A sex difference for bigrams. *Science,* 1979, *204,* 1432-1434.

Corballis, M. C., & Beale, I. L. *The psychology of left and right.* New York: John Wiley and Sons, 1976.

Curtiss, S. *The case of Genie: A modern day wild child.* New York: Academic Press, 1977.

Dabbs, J. M., Jr., & Choo, G. Left-right carotid blood flow predicts specialized mental ability. *Neuropsychologia,* 1980, *18,* 711-713.

Dandy, W. E., & Blackfan, K. D. Internal hydrocephalus: An experimental, clinical and pathological study. *American Journal of Diseases of Children*, 1914, *8*, 406-482.

Davies, J. *Human developmental anatomy.* New York: The Ronald Press Company, 1963.

Denenberg, V. H. Hemispheric laterality in animals and the effects of early experience. *The Brain and Behavioral Sciences*, in press.

Dennis, M. & Kohn, B. Comprehension of syntax in infantile hemiplegics after cerebral hemi-decortication: Left-hemisphere superiority. *Brain and Language*, 1975, *2*, 472-482.

Diamond, M. New data supporting cortical asymmetry differences in males and females. *The Brain and Behavioral Sciences*, 1980, *3*, 233-234.

Dorman, M. F., & Geffner, D. S. Hemispheric specialization for speech perception in six-year-old black and white children from low and middle socioeconomic classes. *Cortex*, 1974, *10*, 171-176.

Droege, R. C. Sex differences in aptitude maturation during high school. *Journal of Conseling Psychology*, 1967, *14*, 407-411.

Egeth, H., & Epstein, J. Differential specialization of the cerebral hemispheres for the perception of sameness and differences. *Perception and Psychophysics*, 1972, *12*, 218-220.

Entus, A. K. Hemispheric asymmetry in processing of dichotically presented speech and non-speech stimuli by infants. In S. J. Segalowitz & F. A. Gruber (Eds.), *Language development and neurological theory.* Ney York: Academic Press, 1977.

Ettlinger, G., Blakemore, C. B., Milner, A. D., & Wilson, J. Agenesis of the corpus callosum: A behavioural investigation. *Brain*, 1972, *95*, 327-346.

Ettlinger, G., Blakemore, C. B., Milner, A. D., & Wilson, J. Agenesis of the corpus callosum: A further behavioural investigastion. *Brain*, 1974, *97*, 225-234.

Ferriss, G. S., & Dorsen, M. M. Agenesis of the corpus callosum: I. Neuropsychological studies. *Cortex*, 1975, *11*, 95-122.

Flor-Henry, P. Lateralized temporal-limbic dysfunction and psychopathology. In S. Harnad, H. Steklis, & J. Lancaster (Eds.), *Origins and evolution of language and speech* (Vol. 280), *Annals of the New York Academy of Sciences.* New York: The New York Academy of Sciences, 1976.

François, J., Eggermont, E., Evens, L., Logghe, N., & De Bock, F. Agenesis of the corpus callosum, the median facial cleft syndrome and associated ocular malformations. *American Journal of Ophthalmology*, 1973, *76*, 241-245.

Galaburda, A. M., LeMay, M., Kemper, T. L., & Geschwind, N. Left-right asymmetries in the brain. *Science*, 1978, *199*, 852-856.

Gardiner, M. F., & Walter, D. O. Evidence of hemispheric specialization from EEG. In S. Harnad, R. W. Doty, L. Goldstein, J. Jaynes & G. Krauthamer (Eds.), *Lateralization in the nervous system.* New York: Academic Press, 1977.

Geffner, D. S., & Dorman, M. F. Hemispheric specialization for speech perception in four-year-old children from low and middle socio-economic classes. *Cortex*, 1976, *12*, 71-73.

Gesell, A., & Amatruda, C. S. *The embryology of behavior.* New York: Harper, 1945.

Gesell, A., & Ames, L. B. The development of handedness. *Journal of Genetic Psychology*, 1947, *70*, 155-175.

Ghent, L. Developmental changes in tactual thresholds on dominant and nondominant sides. *Journal of Comparative and Physiological Psychology*, 1961, *54*, 670-673.

Gilbert, C., & Bakan, P. Visual asymmetry in the perception of faces. *Neuropsychologia*, 1973, *11*, 355-362.

Gloning, I., Gloning, K., Haub, G., & Quatember, R. Comparison of verbal behavior in right-handed and non right-handed patients with anatomically verified lesion of one hemisphere. *Cortex*, 1969, *5*, 43-52.

Gordon, H. W. Verbal and non-verbal cerebral processing in man for audition. (Doctoral dissertation, California Institute of Technology, 1973).

Gordon, H. W., Sperry, R. W., & Bogen, J. E. Absence of deconnexion syndrome in two patients with partial section of the neo-commissures. *Brain*, 1971, *94*, 327-336.

Goy, R. W., & McEwen, B. S. *Sexual differentiation of the brain.* Cambridge, Mass.: The MIT Press, 1980.

Gregory, R., & Paul, J. The effects of handedness and writing posture on neuropsychological test results. *Neuropsychologia*, 1980, *18*, 231-235.

Gruzelier, J. H., & Venables, P. H. Skin conductance responses to tones with and without attentional significance in schizophrenic and nonschizophrenic psychiatric patients. *Neuropsychologia*, 1973, *11*, 221-230.

Guttmacher, A. F. An analysis of 521 cases of twin pregnancy: I. Differences in single and double ovum twinning. *American Journal of Obstetrics and Gynecology*, 1937, *34*. 76-84.

Hécaen, H., & Sauguet, J. Cerebral dominance in left-handed subjects. *Cortex*, 1971, *7*, 19-48.

Heilman, K. M., Coyle, J. M., Gonyea, E. F., & Geschwind, N. Apraxia and agraphia in a left-hander. *Brain*, 1973, *96*, 21-28.

Herron, J., Galin, D., Johnstone, J., & Ornstein, R. E. Cerebral specialization, writing posture, and motor control of writing in left-handers. *Science*, 1979, *205*, 1285-1289.

Hibbard, E. Orientation and directed growth of Mauthner's cell axons from duplicated vestibular nerve roots. *Experimental Neurology*, 1965, *13*, 289-310.

Hilton, T. J., & Berglund, G. W. Sex differences in mathematics achievement—A longitudinal study. *Educational Testing Service Research Bulletin*, 1971.

Hines, D., & Satz, P. Superiority of right visual half-fields in right-handers for recall of digits presented at varying rates. *Neuropsychologia*, 1971, *9*, 21-25.

Hochberg, F. H., & LeMay, M. Arteriographic correlates of handedness. *Neurology*, 1975, *25*, 218-222.

Humphrey, T. The development of the pyramidal tracts in human fetuses: correlated with cortical differentiation. In D. B. Tower & J. P. Schade (Eds.), *Proceedings of the Second International Meeting of Neurobiologists, 1959. Structure and function of the cerebral cortex.* Amsterdam: Elsevier, 1960.

Ingram, D. Motor asymmetries in young children. *Neuropsychologia*, 1975, *13*, 95-102.

Jeeves, M. A. A comparison of interhemispheric transmission times in acallosals and normals. *Psychonomic Science*, 1969, *16*, 245-246.

Jeeves, M. A. Further studies of the effects of agenesis of the corpus callosum in man and neonatal sectioning of the corpus callosum in animals. In J. Cernacek & F. Podivinsky (Eds.), *Cerebral interhemispheric relations.* Bratislava;: Vydatelstvo Slovenskej Akademie Vied, 1972.

Kimura, D., & Vanderwolf, C. H. The relation between hand preference and the performance in individual finger movements by left and right hands. *Brain*, 1970, *93*, 769-774.

Knox, C., & Kimura, D. Cerebral processing of nonverbal sounds in boys and girls. *Neuropsychologia*, 1970, *8*, 227-237.

Kohn, B., & Dennis, M. Selective impairments of visuo-spatial abilities in infantile hemiplegics after right cerebral hemidecortication. *Neuropsychologia*, 1974, *12*, 505-512.

Kuypers, H. G. J. M. The anatomical organisation of the descending pathways and their contributions to motor control, especially in primates. In T. E. Desmedt (Ed.), *New developments in E. M. G. and clinical neurophysiology* (Vol. 3). Basel: Kerger, 1973.

Ladavas, E., Umilta, C., & Ricci-Bitti, P. E. Evidence for sex differences in right-hemisphere dominance for emotions. *Neuropsychologia*, 1980, *18*, 361-366.

Lagger, R. L. Failure of pyramidal tract decussation in the Dandy-Walker syndrome. *Journal of Neurosurgery*, 1979, *50*, 382-387.

Landis, T., Assal, G., & Perret, E. Opposite cerebral hemispheric superiorities for visual associative processing of emotional facial expressions and objects. *Nature*, 1979, *178*, 739-740.

Lawson, N. C. Inverted writing in right- and left-handers in relation to lateralization of face recognition. *Cortex*, 1978, *14*, 207-211.

Layton, W. M., Jr. Random determination of a developmental process. *The Journal of Heredity*, 1976, *67*, 336-338.

Lehmann, H. J., & Lampe, H. Observations on the interhemispheric transmission of information in 9 patients with corpus callosum defect. *European Neurology*, 1970, *4*, 129-147.

LeMay, M., & Culebras, A. Human brain—Morphologic differences in the hemispheres demonstrable by carotid arteriography. *New England Journal of Medicine*, 1972, *287*, 168-170.

Lenneberg, E. H. Speech development: Its anatomical and physiological concomitants. In E. C. Carterette (Ed.), *Brain function* (Vol. 3), *Speech, language, and communication*. Los Angeles: University of California Press, 1966.

Levy, J. A review of evidence for a genetic component in the determination of handedness. *Behavior Genetics*, 1976, *6*, 429-453.

Levy, J., & Reid, M. Variations in writing posture and cerebral organization. *Science*, 1976, *194*, 337-339.

Levy, J., & Reid, M. Variations in cerebral organization as a function of handedness, hand posture in writing, and sex. *Journal of Experimental Psychology: General*, 1978, *107*, 119-144.

Liederman, J. & Kinsbourne, M. Rightward motor bias in newborns depends upon parental right-handedness. *Neuropsychologia*, 1980, *18*, 579-584.

Lindsley, D. L., & Grell, E. H. *Genetic variations of Drosophila melanogaster*. Washington: Carnegie Institute of Washington, Pub. No. 627, 1967.

Luria, A. R. *Traumatic aphasia*. The Hague: Mouton, 1970.

Maccoby, E. E., & Jacklin, C. N. *The psychology of sex differences*. Stanford, Calif.: Stanford University Press, 1974.

McGlone, J. Sex differences in human brain asymmetry: A critical survey. *The Brain and Behavioral Sciences*, 1980, *3*, 215-263.

McKeever, W. P. Handwriting posture in left-handers: Sex, familial sinistrality and language laterality correlates. *Neuropsychologia*, 1979, *17*, 429-444.

McKeever, W. F., & Hoff, A. L. Evidence of a possible isolation of left hemisphere visual and motor areas in sinistrals employing an inverted handwriting posture. *Neuropsychologia*, 1979, *17*, 445-454.

McKeever, W. F., & van Deventer, A. D. Inverted handwriting position, language laterality, and the Levy-Nagylaki genetic model of handedness and cerebral organization. *Neuropsychologia*, 1980, *18*, 99-102.

McRae, D. L., Branch, C. L., & Milner, B. The occipital horns and cerebral dominance. *Neurology*, 1968, *18*, 95-100.

Merrell, D. J. Dominance of hand and eye. *Human Biology*, 1957, *29*, 314-328.

Mittwoch, U., & Kirk, D. Superior growth of the right gonad in human foetuses. *Nature*, 1975, *257*, 791-792.

Molfese, D. Infant cerebral asymmetry. In S. J. Segalowitz & F. A. Gruber (Eds.), *Language development and neurological theory*. New York: Academic Press, 1977.

Morgan, M. J. Embryology and inheritance of asymmetry. In S. Harnad, R. w. Doty, L. Goldstein, J. Jaynes & G. Krauthamer (Eds.), *Lateralization in the nervous system*. New York: Academic Press, 1977.

Morgan, M. J., & Corballis, M. C. Scrotal asymmetry and Rodin's dyslexia. *Nature*, 1976, *264*, 295-296.

Moscovitch, M., & Smith, L. C. Differences in neural organization between individuals with inverted and non-inverted hand posture during writing. *Science*, 1979, *205*, 710-713.

Nagylaki, T., & Levy, J. The sound of one paw clapping is not sound. *Behavior Genetics*, 1973, *3*, 279-292.

Netley, C. Dichotic listening of callosal agenesis and Turner's Syndrome patients. In S. J. Segalowitz & F. A. Gruber (Eds.), *Language development and neurological theory*. New York: Academic Press, 1977.

Neville, H. Electroencephalographic testing of cerebral specialization in normal and congentitally deaf children: A preliminary report. In S. J. Segalowitz & F. A. Gruber (Eds.), *Language development and neurological theory*. New York: Academic Press, 1977.

Nyberg-Hansen, R., & Rinvik, E. Some comments on the pyramidal tract with special reference to its individual variations in man. *Acta Neurologica Scandinavica*, 1963, *39*, 1-30.

O'Connor, K. P., & Shaw, J. C. Field dependence, laterality, and the EEG. *Biological Psychology*, 1978, *6*, 93-109.

Oltman, P. K. Cognitive style and interhemispheric differentiation in the EEG. *Neuropsychologia*, 1979, *17*, 699-702.

Oltman, P. K., Ehrlichman, H., & Cox, P. W. Field independence and laterality in the perception of faces. *Perceptual and Motor Skills*, 1977, *45*, 255-260.

Parlow, S. Differential finger movements and hand preference. *Cortex*, 1978, *14*, 608-611.

Parlow, S. & Kinsbourne, M. Handwriting posture and manual motor asymmetry in sinistrals. Submitted.

Pernkopf, E. Asymmetrie, Inversion, und Vererbung. *Zeitschrift für menschliche Vererbungs und Konstitutionslehre*, 1937, *20*, 606-656.

Petersen, A. C. Hormones and cognitive functioning in normal development. In M. A. Wittig & A. C. Petersen (Eds.), *Sex-related differences in cognitive functioning*. New York: Academic Press, 1979.

Rapaczynski, W., & Ehrlichman, H. Opposite visual hemifield superiorities in face recognition as a function of cognitive style. *Neuropsychologia*, 1979, *17*, 645-652.

Rasmussen, T., & Milner, B. The role of early left-brain injury in determining lateralization of cerebral speech functions. In S. J. Diamond & D. A. Blizard (Eds.), *Evolution and lateralization of the brain*, (Vol. 299) Annals of the New York Academy of Sciences. New York: The New York Academy of Sciences, 1977.

Reid, M. Cerebral lateralization in children: An ontogenetic and organismic analysis. (Doctoral dissertation, in preparation, University of Colorado, 1980.)

Reinisch, J. M., Gandelman, R., & Spiegel, F. S. Prenatal influences on cognitive abilities: Data from experimental animals and human and endocrine syndromes. In M. A. Wittig & A. C. Petersen (Eds.), *Sex-related differences in cognitive functioning*. New York: Academic Press, 1979.

Reitan, R. M., & Tarshes, E. L. Differential effects of lateralized brain lesions on the Trail Making Test. *Journal of Nervous and Mental Diseases*, 1959, *129*, 257-262.

Rizzolatti, G., & Buchtel, H. A. Hemispheric superiority in reaction time to faces: A sex difference. *Cortex*, 1977, *13*, 300-305.

Rudel, R., Denckla, M., & Hirsch, S. The development of left-hand superiority for discriminating Braille configurations. *Neurology*, 1977, *27*, 160-164.

Rudel, R., Denckla, M., & Spalten, E. The functional asymmetry of Braille letter learning in normal sighted children. *Neurology*, 1974, *24*, 733-738.

Sarnat, H. B., & Netsky, M. G. *Evolution of the nervous system*. London and New York: Oxford University Press, 1974.

Satz, P., Achenbach, K., & Fennell, E. Correlations between assessed manual laterality and predicted speech laterality in a normal population. *Neuropsychologia*, 1967, *5*, 295-310.

Saul, R. E., & Gott, P. S. Language and speech lateralization by amytal and dichotic listening tests in agenesis of the corpus callosum. In D. O. Walter, L. Rogers, and J. M. Finzi-Fried (Eds.), *Conference on human brain function: Brain Information Service*. Los Angeles: BRI Publications, University of California, 1976.

Saul, R. E., & Sperry, R. W. Absence of commissurotomy symptoms with agenesis of the corpus callosum. *Neurology*, 1968, *18*, 307.

Schneidler, G. R., & Paterson, D. G. Sex differences in clerical aptitude. *Journal of Educational Psychology*, 1942, *33*, 303-309.

Smith, L. C., & Moscovitch, M. Writing posture, hemispheric control of movement and cerebral dominance in individuals with inverted and noninverted hand postures during writing. *Neuropsychologia*, 1979, *17*, 637-644.

Soloman, D., & Ali, F. A. Age trends in the perception of verbal reinforcers. *Developmental Psychology*, 1972, *7*, 238-243.

Spemann, H., & Falkenberg, H. Über asymmetrische Entwicklung and Situs inversus bei Zwillingen und Doppelbildungen. *Wilhelm Roux' Archiv*, 1919, *45*, 371-422.

Sperry, R. W. Restoration of vision after crossing of optic nerves and after contralateral transposition of the eye. *Journal of Neurophysiology*, 1945, *8*, 15-28.

Srb, A., Owen, R., & Edgar, R. *General genetics* (2nd ed.). San Francisco: W. H. Freeman, 1965.

Sutton, P. R. Handedness and facial asymmetry: Lateral position of the nose in two racial groups. *Nature*, 1967, *198*, 909.

Svensson, A. *Relative achievement. School performance in relation to intelligence, sex, and home environment.* Stockholm: Almquist & Wiksell, 1971.

Taggert, J. K., Jr., & Walker, A. E. Congenital atresia of the foramens of Luschka and Magendie. *Archives of Neurology and Psychiatry*, 1942, *48*, 583-612.

Teng, E. L., Lee, P., Yang, K., & Chang, P. C. Genetic, cultural, and neuropathologic factors in relation to laterality. In D. O. Walter, L. Rogers, & J. M. Finzi-Fried (Eds.), *Conference on human brain function: Brain Information Service*. Los Angeles: BRI Publications, University of California, 1976.

Todor, J. I. Sequential motor ability of left-handed inverted and non-inverted writers. *Acta Psychologica*, 1980, *44*, 165-173.

Torgersen, J. Familial transportation of viscera. *Acta Medica Scandinavica*, 1946, *126*, 319-322.

Torgersen, J. Genic factors in visceral asymmetry and in the development and pathologic changes of the lungs, heart, and abdominal organs. *Archives of Pathology*, 1949, *47*, 566-593.

Torgersen, J. Situs inversus, asymmetry, and twinning. *Americal Journal of Human Genetics*, 1950, *2*, 361-370.

Trevarthen, C. Behavioral embryology. In E. C. Carterette & M. P. Friedman (Eds.), *Handbook of perception* (Vol. 3). New York: Academic Press, 1973.

Trevarthen, C. Functional organization of the human brain. In M. C. Wittrock (Ed.), *The brain and psychology*. New York: Academic Press, 1980.

Turkewitz, G. The development of lateral differentiation in the human infant. In S. J. Dimond & D. A. Blizard (Eds.), *Evolution and lateralization of the brain*, (Vol. 299) *Annals of the New York Academy of Sciences*. New York: The New York Academy of Sciences, 1977.

Turkewitz, G., Birch, H. G., Moreau, T., Levy, L., & Cornwell, A. C. Effect of intensity of auditory stimulation on directional eye movements in the human neonate. *Animal Behavior*, 1966, *14*, 93-101.

Verhaart, W. J. C., & Kramer, W. The uncrossed pyramidal tract. *Acta Psychiatrica Scandinavica*, 1952, *21*, 181-200.

Very, P. S. Differential factor structures in mathematical abilities. *Genetic Psychology Monographs*, 1967, *75*, 169-207.

Wada, J. A., Clarke, R., & Hamm, A. Cerebral hemispheric asymmetry in humans. Cortical speech zones in 100 adult and 100 infant brains. *Archives of Neurology*, 1975, *32*, 239-246.

Wechsler, A. F. Crossed aphasia in an illiterate dextral. *Brain and Language*, 1976, *3*, 164-172.

Witelson, S. F. Sex and the single hemisphere: Right hemisphere specialization for spatial processing. *Science*, 1976, *193*, 425-427.

Witelson, S. F. Early hemisphere specialization and interhemispheric plasticity: An empirical and theoretical review. In S. J. Segalowitz & F. A. Gruber (Eds.), *Language development and neurological theory*. New York: Academic Press, 1977.

Witelson, S. F. & Pallie, W. Left hemisphere specialization for language in the newborn: Neuroanatomical evidence of asymmetry. *Brain*, 1973, *96*, 641-647.

Witkin, H. A., Goodenough, D. R., & Oltman, P. K. *Psychological differentiation: Current status* (ETS RB 77-17). Princeton, N.J.: Educational Testing Service, 1977.

Yakovlev, P. I., & Radic, P. Patterns of decussation of bulbar pyramids and distribution of pyramidal tracts on two sides on the spinal cord. *Transactions of the American Neurological Association*, 1966, *91*, 366-367.

Zoccolotti, P., & Oltman, P. K. Field dependence and lateralization of verbal and configurational processing. *Cortex*, 1978, *14*, 155-163.

V

EPISTEMOLOGY, THEORY, AND METHOD

In Chapter 7, Gollin discusses some major themes of developmental theory and research. These themes are cast within the framework of biological systems theory. Within this perspective, the notion that the meaning of behavior can be evaluated only within the context in which it occurs is underscored; the evaluation of behavior relative to some a priori criteria is considered inappropriate because such criteria often involve assumptions that cannot be justified. Gollin criticizes progressive and destinational definitions of development that describe universal end points for developmental processes. He does not, of course, deny that there is substantial similarity between individuals and cultures, but he stresses that the nature of developmental processes that lead to particular outcomes is stochastic. Thus, individual differences and cultural variations in the form of behaviors are to be expected and need not be regarded (as in destinational theories of development) as deviant or pathological. Variations in development can be understood as originating through gene-environment interactions that are multimodal (involving different behavioral and physiological substrates) and polyphasic (the nature of intermodal interactions change over time). This perspective is not without implications for research methods. An understanding of behavioral development must include a consideration of the contributions of the state of the organism, the history of the organism (both biological and experiential), and the adaptive context in which the behavior occurs.

Stephen Toulmin, Chapter 8, advocates a limited collaboration between philosophers and psychologists so that a better definition of the issues may be forthcoming. He rejects the traditional epistemological barriers that have interfered with collaboration in the past and makes clear why the development of human cognition and language requires collaborative efforts.

He then questions the assumption that mental development has a unique destination. He uses the theoretical work of Jean Piaget as a vehicle for dealing with the issue of destined development, and concludes that there are many reasons to be skeptical about the proposition that there is a universal human developmental destination. He distinguishes between the existence of substantive common elements across cultures and the supposition of an a priori necessity, pointing out that the existence of common elements still leaves room for considerable variation.

He then discusses the roles played by enculturation on human development. This leads to a discussion of methodological issues and suggestions for alternate procedures of inquiry.

7

Development and Plasticity

EUGENE S. GOLLIN

Introduction

Developmental plasticity refers to the possible range of variations that can occur in individual development. To understand the nature of the constraints upon and opportunities for variations in biological and behavioral development, it is worthwhile to consider one of the major unifying concepts of modern biology,[1] the concept of organization:

> This tells us that at every level from the molecule through the supra-molecular organelle, the cell, the tissue, the organism, the individual, and up to the population or the society, the properties of life depend only to a small degree upon the substances of which living matter is composed. To a much greater degree living things owe their nature to the way in which the components are organized into orderly patterns, which are far more permanent than the substances themselves [Stebbins, 1966, p. 1].

This concept of organization mandates that our focus of attention must include not only material and efficient causes in the social and physical environment but also the web of relationships that describe organismic capacity and environmental opportunity. The relationships between organisms and environments are not interactionist, as interaction implies that organism and environment are separate entities that come together at an interface. Organism and environment constitute a single life process (Freedman,

---

[1] The role assigned to biology in this chapter is that of a source for models. There is no implication of either biobehavioral isomorphism nor is it my intention to suggest that behavior is necessarily reducible to the biological substrate.

1974). For analytical convenience, we may treat various aspects of a living system and various external environmental and biological features as independently definable properties. Analytical excursions are an essential aspect of scientific inquiry, but they are hazardous if they are primarily reductive. An account of the *collective behavior* of the parts as an organized entity is a necessary complement to a reductive analytic program, and serves to restore the information content lost in the course of the reductive excursion (see Weiss, 1969, 1973).

In any event, the relationships that contain the sources of change are those between organized systems and environments, not between heredity and environment (Lehrman, 1953, p. 345). The kinds of adaptive behaviors that emerge are to a large extent dependent upon the adaptations already available. This applies to both the biological and the cultural spheres of change (Stebbins, 1969, pp. 29, 136).

Among the factors contributing to an individual's qualities of organization are its unique hereditary equipment, its position and orientation in space, and its history of activities. At any moment in time, an organism's prospects are dependent upon the operative status of all these factors, together with contemporaneous environmental events, which themselves must be at least partly defined by the developmental status of the organism, that is, its adaptive capacities. It follows, therefore, that what we typically designate as environment is unique to each individual. This relationship may be stated in another way: The unique properties of each individual in a population will operate differently upon a given set of designated environmental conditions. Thus, the ambient world is uniquely defined by each living creature. This applies to the newly fertilized egg, to the embryonic period, to the fetal period, to infancy, and to every period in the life sequence of an organism.

The determination of the successive qualities of living systems, given the web of relationships involved, is probabilistic. This is so because the number of factors operating conjointly in living systems is very great. Additionally, each factor and subsystem is capable of a greater or lesser degree of variability. Hence, the influence subsystems have upon each other, and upon the system as a whole, varies as a function of the varying states of the several concurrently operating subsystems. Thus, the very nature of living systems, both individual and collective, and of environments, assure the presumtive character of organismic change.

Living systems are organized systems with internal coherence. The properties of the parts are essentially dependent on relations between the parts and the whole (Waddington, 1957). The quality of the organization provides opportunities for change as well as constraints upon the extent and direction of change. Thus, while the determination of change is probabilistic, it is not chaotic.

Developmental Perspectives

The current perspectives for considering developmental plasticity are derived from concepts of development and evolution. Words such as *change, growth,* and *differentation* are frequently accompanied by modifiers such as *progressive* and *unidirectional*. Weiss (1939), reflecting an embryological approach, defines an effect as developmental if it leaves a permanent modification in the organism. He distinguishes between physiological and developmental events, characterizing the former as *conservative* and the latter as *progressive*. An example is secretory functioning: A physiological instance is salivation; a developmental instance is the secretion of the vitreous body of the eye, a one-time event.

Thoday (1953), operating from a phyletic perspective, further qualifies change in terms of its contribution to fitness for survival. If a change contributes to survival, it is *progressive;* if it decreases fitness for survival it is retrogressive; and if it has no effect on survival it is *mere* change. A limiting value of Thoday's definition lies in the fact that the determination of the quality of change may be possible only long after the change has occurred. Furthermore, a change that increases fitness for survival in one environmental setting may reduce it in others.

The terms *progressive* and unidirectional tend to be value laden in usage by behavioral as well as by biological scientists. In part, this reflects implicit and sometimes explicit assumptions about "developmental" and "evolutionary" destinations (see Toulmin, Chapter 8, this volume). The assumptions are quite clearly expressed in Piagetian psychology with regard to the emergence of intellectual psychological structures during ontogenesis. In Wernerian psychology (1957), whether or not a set of organismic events are to be characterized as developmental is dependent upon changes in the organismic state from relative globality to increasing differentiation and hierarchic integration. Werner (1957) has also separated the notion of development from ontogenesis and has applied developmental analyses to behavioral change outside the ontogenetic context.

The use of the term *progressive* also appears in definitions of development where there are no connotations of structural change but only the cumulation of response-produced changes in stimulation that in turn trigger new responses that in turn further modify stimulation, ad infinitum (Bijou & Baer, 1961). Within such a definitional system, cumulation becomes synonymous with progress.

The use of the adaptive criteria of Thoday and other biologists is not necessarily compatible with either the destinational or cumulative criteria employed by psychologists. Morphological and behavioral (Hodos & Campbell, 1969) secondary simplifications are clearly adaptive. Yet they would certainly appear to be retrogressive by the criteria of both destinational and cumulative theories of development. For example, parasitic flatworms have become well adapted to their environments and have lost many of the structures found in their free-living relatives (Ebert, Loewy, Miller, & Schneiderman, 1973). An analogue of secondary simplification on the behavioral level is the kind of behavioral adaption achieved by damaged organisms including brain-damaged humans (Goldstein, 1939). Frequently they are able to maintain goal directed behavior by adopting alternate and sometimes simplified means to achieve an end that is no longer attainable in other ways. The emergence of such behaviors permits organisms to continue to function, even though the means employed represent a "primitivization" or "simplification" by the definitional criteria of destinational and cumulative theories. The adaptive efficiency of these behaviors must be evaluated within the novel context imposed by the damaged nervous system. What the individual retains are means to accomplish ends wherein both means and ends may have undergone radical alteration.

Organic deficits are not the only conditions that produce simplified means-end procedures. There are likely to be a large variety of environmental arrangements and task demands wherein organisms opt for behaviors that achieve ends by simplified means. Sometimes the simplified means may be less efficient than alternate procedures in a variety of demand situations, but they may have the advantage for the actor of allaying stress or other forms of tension. In sum, such behavioral simplifications are not necessarily pathological, although sometimes they may arise consequent upon pathology. However, pathological factors need not play a role. Being able to deploy one's actions between complex and simple means-ends action sequences may entail a flexibility quite useful to organisms. It becomes, therefore, a matter of some importance to assess the adaptive value of behavior in the context in which it occurs rather than exclusively against some abstract definitional criteria.

A further problem, particularly for destinational definitions of development, entails the logical requirement of regarding sets of changes that do not meet destinational criteria as aberrational or pathological. Such a position creates not only scientific but also cultural and social problems. For example, there are societies where the use of Euclidean coordinates is not part of the intellectual repertoire (Olson, 1970). Yet such an outcome could hardly be classified as aberrational, as the people are able to function effectively in their environmental settings.

In a broader sense, the emphasis placed in Western culture on particular cognitive skills, abstract problem solving abilities, and the shift from heteronomous to autonomous attributions in the area of moral judgment would not necessarily constitute the developmental program of other societies. Members of such societies are best described as having followed alternate developmental paths and not ever as arrested in or deviant from some "true" developmental course. My aim is not to deny the legitimacy of aesthetic or moral or intellectual preferences for one or another kind of developmental course. Rather, it is to raise the possibility of alternate developmental forms that express different organismic histories, and at the same time avoid the pitfalls entailed in normative and culture-centered definitions of development.

Destinational theories in practice, if not in principle, detach developmental processes from both genetic and extragenetic influence except in the case of severe genetically based anomalies, or dysfunctions consequent upon environmental catastrophes, such as severe prolonged sensory deprivation. Between these extreme kinds of events, some proponents of destinational theories, for example, Piagetians, promulgate the notion that the developmental courses they have postulated unfold more or less independently of either genetic or experiential variation.

Available definitions of development and theoretical approaches to ontogenetic change all appear to contain restrictions on the extent to which they are able to contribute to the understanding of plasticity in development. Within the framework of existing formulations, ideas associated with progressive and unidirectional change and with morphogenetic and behavioral complexity need to be reconsidered.

Developmental Perspective and Plasticity

Although some destinational theories of development explicitly include mechanisms for transactions between organisms and environments, for example Piaget's assimilation and accommodation, they rarely contain statements relating to individual differences in development. There are some exceptions. First, it is held that the pace but not the structural and sequential character of development may vary between individuals. Second, it is suggested that there may be differences in how far individuals may progress along a hypothesized developmental course. Thus plasticity in development would necessarily consist of variations in the time of onset of the several phases or stages that constitute the particular developmental program, and/or the extent to which individuals realize the predetermined developmental program.

For cumulative models, uniqueness in development is no problem, as there are, in principle, as many forms of development as there are environmental programs. However, this definitional condition is achieved by either implicit or avowed (Skinner, 1963) denial of an organismic contribution to the course of behavior. The stimulus program constitutes development.

Many approaches to development are explicitly normative. For example, Waddington (1940, 1957, 1966, 1968, 1971), in presenting his developmental topography, has placed great emphasis on the normative aspects of development.

> Any organ such as a wing, a leg, or an eye, must always be regarded as expressing some sort of compromise or balance between the conflicting or competing activities of very many genes.
> It is particularly important to discover that these balances show a certain degree of stability, in the sense that it is quite difficult to persuade the developing system *not* to finish up by producing its *normal* end result [1966, pp. 46-47, italics added].

He conceives of the various regions of the developing embryo as having a number of possible developmental pathways. The course of development is well buffered in that abnormal environmental or genetic events activate self-regulatory mechanisms that tend to hold development to its normal course. He employs two illustrative metaphors that do not do justice to the dynamic properties of his developmental ideas, but that do expose some of the problems inherent in his "epigenetic landscape" model. The first metaphor is that of a ball rolling downhill through a series of branching valleys. If some event should push it one way or another, it will still "finish up in its normal place [1966, p. 49]." The second metaphor is that of a stream rushing down a hillside. Should it be blocked by a landslide, "it does not come back to the stream bed at the place where the diversion occurred, but some way farther down the slope [1969, p. 366]." This process of stabilized change over time is called homeorhesis. Despite occasional perturbations, the devleopmental course remains normal. The destinational character of Waddington's formulation is underscored by his introduction of the word *chreod* to stand for the kind of stabilized time trajectories he is interested in. The word chreod is constructed from parts of two Greek words meaning *fated* and *path*.

The Waddington system is conservative and normative, emphasizing processes that tend to maintain species and individual integrity. As a system of developmental principles, it may be more applicable to anatomical and physiological characteristics of organisms than to behavioral and sociocultural characteristics. However, even at the level of anatomical and physiological analysis, Waddington's system may be too restraining. Although his stream analogy might be adequate for a literal topographic landscape, it seems less appropriate for biological systems, given the dynamic properties that characterize them.

Thus we are confronted, on the one hand, with a set of conceptions that portray development as predestined, and, on the other hand, with conceptual systems that exclude organismic constraints and directions and vest variation exclusively in the experiential history of organisms. Plasticity in the former system is confined to the timing of the occurrence of change and/or to how far along the destined road an individual may proceed. Plasticity in the latter instance comes to stand for individual differences instigated by external events, although, paradoxically, individual differences have not been the central concern of researchers who espouse the cumulative developmental position. Finally, in some normative systems, such as Waddington's, plasticity is largely confined to means or courses of development but not to end states.

The Roles of Experience

To be clear about the character of plasticity, we must distinguish between those variations that lie on a continuum of variation around some hypothesized average value, and variations that entail structural and functional changes of a qualitative nature. In this regard, Gottlieb's distinctions between *maintaining*, *facilitating*, and *inducing* functions of experience on the development and maturation of the nervous system and of behavior are quite helpful. Maintaining functions or experiences preserve already developed states and keep "an immature system intact, going, and functional so that it is able to reach its full development [Gottlieb, 1976, p. 28]."

Maintaining functions appear to be very similar to Waddington's "tendencies toward normative development." In both Gottlieb's and Waddington's presentations there is the implication of some normal course of development that requires at least a low level of functional activity for its maintenance. Thus, one order of plasticity would involve quantitative variations around hypothesized mean values of the anatomical, physiological, and behavioral parameters that constitute development.

The effects of facilitation differ from the effects of maintenance in that the former assist in the developmental achievement of certain states or end points, whereas the latter operate to preserve already achieved states or end points (1976, p. 31). Experience is facilitative if it accelerates development, boosts terminal achievement, or produces a combination of these effects. Gottlieb writes that facilitation refers to "instances where development of the behavior (or neural structure and function) will eventually occur even in the absence of the experience[2] (though it may be subpar without experience)

---

[2] The use of the word *experience* in this context is somewhat confusing. It is inconceivable that a living system is not an experiencing system in the sense that it is by definition an active system. There can be no nonexperiential existence in that sense. The problem is resolvable if one distinguishes between specific experience (e.g., learning) and more general nonspecific experience (see Hebb, 1949, pp. 107-139, and also Hebb, 1958, pp. 454-455).

[p. 31]" An example of a facilitative effect is the earlier hatching of chick embryos after exposure to light.

Experience is inductive if it leads to the establishment of a behavioral or neural event. For example, Gottlieb described a study by Wiens (1970) that illustrated the inductive effects of experience. The behavior investigated was habitat selection of the red-legged frog (*Rana aurora*). Tadpoles were reared in one of three conditions: featureless and white; black and white stripes; and, black and white squares. The black and white striped rearing condition was the one most like the natural habitat of red-legged frogs, which contains willow branches, cattails, and so on, essentially linear structures casting linear shadows. The test chamber consisted of an area covered with black and white stripes and an area covered with black and white squares. Tadpoles reared in the featureless condition showed no preference for either of the patterned substrates. Animals reared in the square-patterned habitat also showed no significant substrate preference (there was a slight tendency to choose the squared substrate more often), whereas the tadpoles reared in the striped habitat demonstrated a significant preference for the striped area of the test chamber. Thus, without ecologically relevant experience, appropriate habitat choices were not forthcoming. Tadpoles raised in the featureless condition did not spontaneously develop appropriate choice behavior, and those reared in the black and white square condition did not develop a significant preference for the squared substrate. Rearing conditions at variance with ecologically valid substrates did not contribute to the development of strong alternate preferences in the tadpoles. Thus the Wiens experiment provides not only an instance of the inductive role of experience but also an example of a species-typical constraint upon plasticity.

An instance of induction in human behavior is provided by Leibowitz and Pick (1972), who presented various forms of the Ponzo perspective illusion (see Figure 7.1) to American and Ugandan college students, and to Ugandan villagers. Students at both Ugandan and American colleges overestimated the length of the horizontal line closer to the apex of the converging lines. The village dwellers showed no tendency to overestimate the length of the upper line. Leibowitz and Pick hypothesized that differences in susceptibility to the illusion were related to differences in the frequency of exposure to two-dimensional representations in photographs, books, television, and the like.

The three classes of experiential events, maintenance, facilitation, and induction are most useful as concepts if they are thought of as grading into each other and as providing contexts for the exploration of the constraints upon and opportunities for developmental variation.

It would appear that extant developmental theories are relatively compatible with the maintaining and facilitating functions of experience and less able

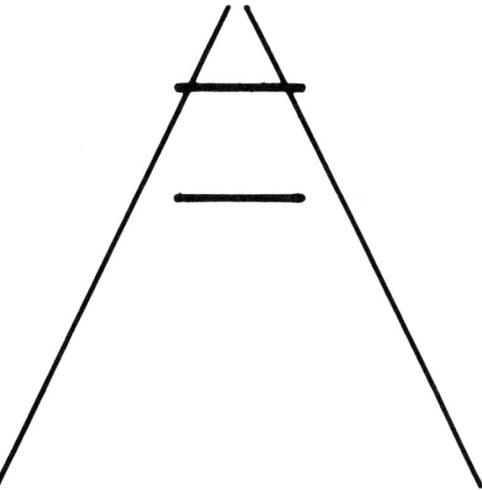

Figure 7.1. *The Ponzo illusion. In the Leibowitz & Pick (1972) study, in addition to the geometric display shown, there were textured photographs containing the illusion (e.g., railroad track) as well as a control condition that contained only the two horizontal lines.*

to deal with inductive variations. Thus their relevance to the study of developmental plasticity is curtailed to the extent that qualitative variations in ontogenesis are outside their scope.

Multimodal and Polyphasic Development

Development, whether viewed in the biochemical, anatomical, or behavioral realm consists of structural and functional events or aspects that change in differing ways during ontogenesis. For example, "The ontogenetic development of each of the different anatomical structures of the central nervous system proceeds at its own individual rate and during that development each structure shows quite different metabolic requirements [Davis, Himwich, & Agrawal, 1969, p. 37]." The growth of all tissues does not follow a linear course, but always involves one or more periods of rapid growth or growth spurts (Dobbing, 1975). Development is, thus, aptly described as multimodal and polyphasic, with different structures and subsystems undergoing different rates of change at different ontogenetic times.

Therefore, at any particular ontogenetic period, the living system is a configuration of many structures and functions. Because the various structures and functions change at different rates during ontogenesis, the configurational character of the system is different at successive periods in the life sequence of an organism. It follows that the consequences of perturbations to the system are dependent upon the configurational character of the system at the time of disturbance. The idea of *sensitive period* is a useful one in this context. In all organ systems there are three phases of growth: hyperplasia, hyperplasia and hypertrophy, and hypertrophy alone (Winick, 1974, 1975). The transition from phase to phase depends on the slowing down and finally the cessation of DNA synthesis. In the brain, for example, different regions manifest different patterns of timing of the successive phases. Malnutrition during proliferative cell growth (hyperplasia) produces a permanent reduction in brain cell number. Malnutrition during the period of cell enlargement (hypertrophy) leads to a temporary restriction of enlargement that is reversible by reestablishing adequate levels of nutrition. Winick (1974) also reports synergistic effects in both rats and human infants deprived of adequate nutrition (the humans putatively) both before and after birth. The number of cells contained in their brains is much smaller than the additive number found in either pre- or postnatally deprived organisms.

A comparable case against Waddington's relatively rigid self-corrective formulation is made by Altman (1971) and by Adinolfi (1971): The nervous system, as well as other organs, is more vulnerable to irreversible change during periods of cell proliferation than at other times during development. Additionally, different types of cells tend to differentiate sequentially in any brain region. The three basic cellular types are macroneurons, which constitute the "coarse wiring" of the nervous system, microneurons or interneurons, which constitute the "fine wiring" of the brain, and the neuroglia, which are responsible for sustenance, protection, and "fixation" of the developed brain circuitry. Furthermore, the maturation of these three types of cells is not synchronized across regions in the nervous system. The different regions of the nervous system and the several basic elements are differentially susceptible to the impact of various environmental and organismic events at different periods in ontogenesis. Chreodic processes need to be evaluated in the context of this complexity. A further complication is that there is as yet little specification of the consequences of disturbing one type of constituent or one region upon the morphogenetic course of other types of constituents or regions.

Amino acids in the developing rabbit visual system show comparable patterns of multimodal, polyphasic development (Davis, Himwich, & Agrawal, 1969). As is shown in Figure 7.2, there are marked changes in neural transmitter levels at successive periods during ontogenesis. The configuration of neural transmitter levels at successive periods of ontogenesis thus provides different organismic settings. For example (see Figure 7.2), at 21 days, there are relatively high levels of glutamine present and low levels of aspartic acid. At 90 days of age, the level of aspartic acid is at its highest, whereas the level of glutamine has dropped off.

The impact of stimulation upon the developing system will have different

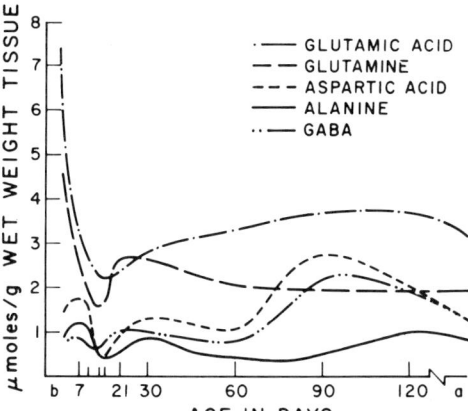

Figure 7.2. Concentration of amino acids in developing rabbit retina. (Adapted from J. M. Davis, W. A. Himwich, & H. C. Agrawal, Some amino acids in the developing visual system. Developmental Psychobiology, 1969, 2, 34-39. Reprinted by permission from John Wiley & Sons, Inc.)

consequences as a function of the configural character of the successive cross-sectional profiles that constitute the ontogenetic pattern. Additionally, when these substances are applied to neuronal membranes, they may either enhance or diminish the excitability of cells. The site, and perhaps the concentration, of the neurotransmitter plays a role in determining its effects (Kopin, 1975).

These principles are illustrated in a study by Greer & Alpern (1980). Each of two strains of mice were treated with either d-amphetamine or a saline solution and then exposed to a seizure-producing inhalant, flurothyl. The latency to the onset of seizure is plotted in Figure 7.3 for both strains at two age periods under each condition of treatment. Inspection of Figure 7.3 indicates that, at 15 days of age, the onset of seizure in *strain a* d-amphetamine treated mice is retarded relative to their saline treated controls, whereas it is accelerated for similarly treated *strain b* mice relative to *strain b* controls. In *strain a* mice, at 35 days of age, treatment with d-amphetamines prior to exposure to the seizure-producing inhalant produces a much shorter latency to onset of seizure than in their 15-day-old counterparts. This relationship is reversed in *strain b* mice. Furthermore, seizure onset time, although not differing between the controls of the two strains at 15 days of age, differs markedly between the control (NaC1) animals of the two strains at 35 days of age. Thus, latency to onset of seizure is dependent upon ontogenetic period, drug treatment, and genetic strain.

Multimodal, polyphasic patterns are not confined to physiological, anatomical, and biochemical events. They are also observable in behavioral domains such as concept development, as shown in Figure 7.4, which portrays data reported in a paper by Denney and Moulton (1976). In this study, children of different ages were asked to point to two pictures (from a large ar-

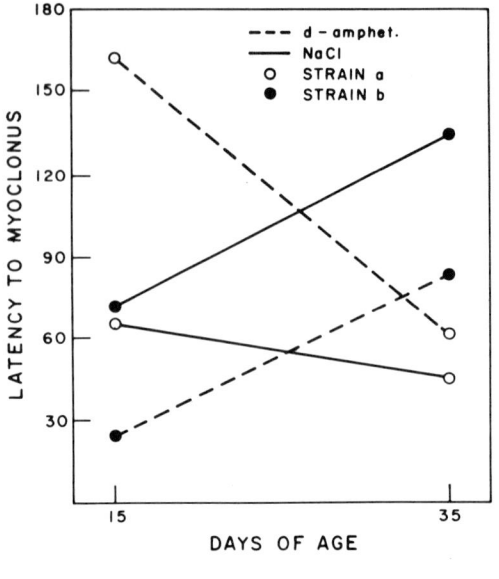

Figure 7.3. Latency to onset of seizure in two strains of mice. (The data were obtained from Figure 1 of C. A. Greer & H. P. Alpern. Paradoxical effects of d-amphetamines upon seizure susceptibility in two selectively bred lines of mice. Developmental Psychobiology, 1980, 13, 7-15. Reprinted by permission from John Wiley & Sons, Inc.)

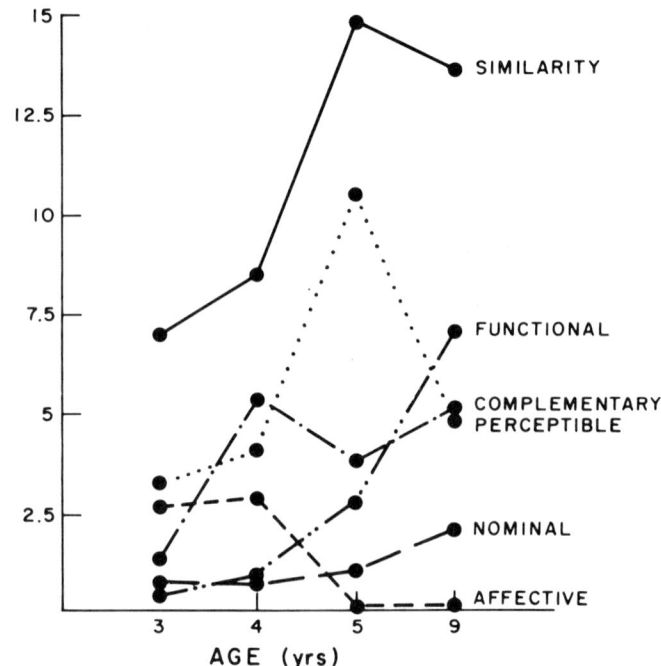

Figure 7.4. Means for concept category scores. (The data are from Table 1, Denney, D. R., & Moulton, P. A., Conceptual preferences among preschool children. Developmental Psychology, 1976, 12, 509-513. Adapted by permission.)

ray of pictures) that went together or were alike in some way. The children were then requested to explain their choices. There were a variety of categorization rules employed by the children. These rules may be thought of as modes of organization. As can be seen in Figure 7.4, these modes manifest different ontogentic courses. Other sets of materials, other experimental methods, and other task demands might very well produce alternate patterns of relationships among the inferred organizational modes. However, some manner of multimodal, polyphasic patterning is likely to emerge. For purposes of this discussion, consider that the diachronic pattern, the changing pattern developing over ontogenetic time, is constituted by a series of successive cross-sectional profiles, or short-term configurations. Each may be thought of as a *sensitive period* or *state* with somewhat different properties. A comparable description is applicable to the sequence of changing patterns shown by the variations in level of neural transmitters over the course of rabbit ontogenesis. Thus, at biochemical, anatomical, and behavioral levels of analysis, development is describable as multimodal and polyphasic.

The consequences of stimulation upon the behavior and subsequent development of an organism will be heavily dependent upon which of the temporally defined cross-sectional slices is the target of particular events or treatments. The problem will be further complicated if the stimulation persists over a sufficient duration to transact with changes in the pattern, that is, with successive cross-sectional configurations. Still another complication may arise if experience per se modifies the relationships between the several conceptual, physiological, anatomical, or biochemical modalities, that is, alters the configurational characteristics of subsequent ontogenetic periods.

Theoretical and Methodological Implications of a Multimodal-Polyphasic Model

A multimodal-polyphasic model of development shifts emphasis from the stimulus to the organism. This shift requires the construction of a central process theory and a methodology appropriate to a central process theory. Szentagothai and Arbib (1975), who are concerned with conceptual models of neural organization caution against "any naive stimulus-response or *unidirectional, information-flow* view of neural activity [p. 12, italics added]." They go on to remind us that there is continual interaction between different neural layers, and that each layer bears, "the marks of the animal's history, both as an individual and as a member of the species." The caution against *unidirectionality* places emphasis on the configural complexity of the central process system and upon the intricate reciprocities and feedback systems at every level of functioning. The concern with *information-flow* raises the question of what we mean by information. If, as indicated in earlier sections of this chapter, information means the signification imposed upon stimulation by the organism, then information must be defined in organismic as well as in extrinsic terms.

The need for tempering extrinsically defined events with organismic considerations becomes clear as soon as comparative or developmental work is undertaken. In the comparative sphere, it is essential to distinguish between chronological and developmental time, as chronological events coincide with different central events in different species. For example, to understand the effects of undernutrition on neonatal organisms, it is essential to know what is happening to the nervous system at the time of birth. What is happening differs from species to species (Dobbing, 1974; Dobbing & Sands, 1979). As shown in Figure 7.5, there is a brain growth spurt that occurs prenatally in guinea pigs, postnatally in rats, and perinatally in humans. According to Dobbing and Sands (1979), reasonable degrees of undernutrition either before or after the brain growth spurt produce no detectable effects that

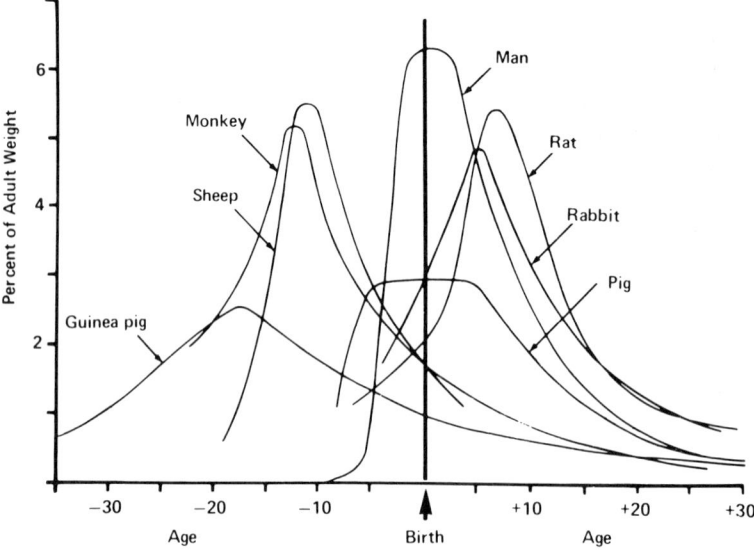

Figure 7.5. The brain growth spurts of mammalian species expressed as first-order velocity curves of the increase in weight with age. The units of time for each species are as follows: guinea pig—days; rhesus monkey—4 days; sheep—5 days; pig—weeks; man—months; rabbit—2 days; rat—days. Rates are expressed as weight gain as a percentage of adult weight for each unit of time. (From J. Dobbing & J. Sands, Comparative aspects of the brain growth spurt. Early Human Development, 1979, 3, 79-83. Reprinted by permission from Elsevier/North Holland Biomedical Press, Amsterdam.)

outlast nutritional rehabilitation. By contrast, nonreversible damage to the central nervous system is produced by even mild restriction during the brain growth spurt. Thus, the effect of identical nutritional deprivation varies within a species as a condition of the status of the developing system at the time of deprivation. Developmental time, used here as a measure of central state conditions, determines the significance of nutritional deprivation in comparative contexts, where there are species differences in the brain growth spurt relative to the time of birth.

In a similar fashion, at the level of behavioral analysis, the developing organism redefines and reorganizes environmental events as a function of the emergence of novel mediational systems or the refinement of already available mediational systems in the course of psychological ontogenesis (Gollin & Garrison, 1980; Gollin & Saravo, 1970). For example, the refinement of conceptual-verbal mediational systems provides alternative processing modes to young children who hitherto relied mainly upon perceptual-motor modes of processing (Gollin & Rosser, 1974). Once such a central state change occurs for children, there are shifts in environmental saliences, in causal attributions, in the ability to integrate events over time, as well as in the ability to shift behavior in the face of changing demands. Again, the organism determines what will and will not serve as a stimulus, and what the stimulus will signify.

The major methodological shift, whether in the laboratory, the therapeutic context, or the educational setting, is from scientific concern with the "rightness" or "wrongness" of a response or a behavior to the relevance of that response or that behavior to the organism involved. Such a shift in focus directs attention to the response as a sign of central state conditions rather than as a more or less direct copy of the stimulus, that is, a flow-through bit of information.

To be sure, how we characterize the performance of the organism is also a function of the method of inquiry, of the way in which we introduce stimulation (Goldstein, 1939). This is the other side of the coin of central state modification of stimulus input. In the Greer and Alpern (1980) study, cited earlier, the latency to seizure of the animals was a function of their genetic strain, their age, and the priming substance introduced into their systems prior to exposure to the seizure stimulant. Each of these factors affect the central state, and as it undergoes change the impact of the seizure stimulant also changes. A change in any one of the conditions would alter the data picture and change our characterizations of these animals. The methodological mandate is thus quite clear. Any reasonable account of the behavior of organisms, of their character, must involve a systematic exploration that includes both samplings of central state conditions and samplings of stimulus parameters. As theories become increasingly sophisticated, it should be possible to predict the significance of stimulation from a knowledge of central state conditions, and conversely, to make inferences about central state from a knowledge of stimulus-response patterns.

A psychological example of the interplay between intrinsic and extrinsic conditions is provided by an analysis of egocentric behavior in young children (Schachter & Gollin, 1979; Walker & Gollin, 1977). The task set for the children was to identify another's view of a scene before which the child was seated. The task was a variation of Piaget's classic three-mountain task. An observer faced the scene (in this case a house) from varying points around its periphery. The child was asked to identify the observer's view by selecting a photograph from a set that contained the child's own view (egocentric response), the observers view, and several other views. In an alternate format, the child was shown a pictorial view and was asked to generate that view for the observer by rotating the display.

Whether or not the children's responses were judged to be egocentric, in these contexts, was dependent on age (a rough index of developemental status), on the task arrangements, on the difficulty of the judgment as defined by angles of regard, on the presence or absence of a shield between the child and his own view, and so on. The same child, dependent upon the conditions of examination, may make egocentric errors, nonegocentric errors, or correct responses. Thus, any systematic characterization of children with respect to egocentricity becomes a complex function of central state and task conditions. Also important in how the respondents are characterized are the measures employed. Typically, children have been rated by the photograph they select. However, many 4-year-old children, after selecting the egocentric photograph to represent the observer's perspective, made spatially appropriate pointing motions on the photograph to indicate that they understood that the observers view was different from their own. This deictic response at once puts an inference of egocentricity into doubt, and also provides a basis for deducing something about young children's representational systems.

The multimodal-polyphasic model mandates a research methodology that seeks to access the central processes available to organisms, to plot their polyphasic relations over ontogenesis, and to specify the successive configural characters of organism-environment patterns. Neither stimulus copy theories nor simplistic stage theories are able to encompass the complexity revealed by research methodologies generated by the multimodal-polyphasic model. The essential feature of those methods is the recognition of the constructive, organizational character of central processes, their patterned inter-

7. DEVELOPMENT AND PLASTICITY   **247**

relations and variation during ontogenesis, and their species-specific qualitative and temporal properties.

Cultural and Biological Instances of Plasticity in Development

An example of induced changes in patterns of complex social behavior of monkeys has been reported by Japanese ethologists (e.g., Kawai, 1965). The animals (*Macaca fuscata*) living on the islet of Koshima were introduced to pieces of sweet potato placed on the sandy beach, and at a later time wheat grains were also thrown on the sand.

Before the introduction of food by man, the monkeys inhabited only the wooded mountainsides of the island. After the food was introduced, the monkeys extended their habitat to include the sandy beach and the sea. New behaviors associated with the novel food supply were initiated by a young female. These behaviors included washing and seasoning sweet potatoes in salt water, dropping handfuls of wheat and sand in the water, where only the wheat floated, thus becoming available for consumption, and swimming in the sea. Age, sex, kinship, and social organization all affected the likelihood that the new behaviors would be adopted. Almost all young monkeys adopted the behaviors, but adolescent and older males did not. This was due in part to the fact that adolescent males are driven to the periphery of the troop by troop leaders, away from feeding females and their young. As a result, adolescent males were excluded from the social interactions necessary to instate the behavior.

Initially, kinship propagation of the new behaviors was found to be from child to mother and from younger to older siblings. Once the behavior was established, learning it tended to be much easier, and propagation tended to be from mother to child.

An illustration of cognitive differences associated with diverse cultural experience has been provided by Moran and his co-workers (Moran, 1974; Moran & Huang, 1974; Moran & Murakawa, 1968): Word association tests were administered to North American, Japanese, and Chinese children, and to North American and Japanese adults. The predominant mode among the three linguistic groups of children was "action upon referent" responses (e.g., stool-sit, apple-eat). Despite the similarity among the children in mode of association, the two adult samples showed marked differences. North American adults responded primarily in a "logical" mode that involved the use of synonyms, superordinates, contrasts, or coordinates (e.g., crow-bird, black-white, apple-pear). Japanese adults, in contrast, gave responses that were primarily iconic (e.g., crow-black, black-hair, apple-red). Thus the cross-cultural commonality of childhood patterns of organization is replaced in adulthood by diverse modes of representation associated with diverse enculturation.

A dramatic example of biological and behavioral plasticity is found among the highland people of New Guinea (Gajdusek, 1970). They occupy a high altitude niche that provides very little water and a very low protein diet. Their growth rates are extremely slow. Early signs of puberty appear at about fifteen years and menarche between 18 and 20 years. This contrasts with menarchial ages of 12.5 and 13.5 years found in Chinese and English populations. The people typically show a pigmoid build. Schoolboys and young laborers who are given improved diets at schools and in labor compounds show dramatic growth acceleration. The changes are such that the young men are hardly recognizable even by their relatives. Such youths returning later to their home villages and no longer receiving the enriched diet are noticeably unfit and quickly sicken.

The highest serum growth levels in man have been found in these New Guineans. This may reflect a compensation for the low protein intake of their tranditional diet. (An analogous compensatory mechanism has been reported by Himwich and her collaborators (Davis, Himwich, & Agrawal, 1969; Himwich, 1967; Himwich, Davis, & Agrawal, 1967), who found an excessive accumulation of neural transmitters in the visual systems of visually deprived rabbits.)

The highland people live in a sodium-scarce environment. Gajdusek (1970) reports that there is a dramatic reversal of expected values for urinary sodium and potassium excretion compared to that found elsewhere: "Their daily intake of NaCl of 40-70 mg is less than one percent of what is considered normal elsewhere. . . . Urinary K/Na ratios are often 200-500:1, or 400 to 1,000 times the expected normal values [p. 33]." Daily urinary output values are so low that, in Europeans on normal diets, one would anticipate uremia within a few days.

The reaction to pain of these Melanesian people is also remarkable. Without wincing they tolerate scarification procedures produced by burning, digital amputation as a funerary rite, and slashing with bamboo rods during courtship. Yet they are just as able to detect minimal pain stimulation as are other people.

Here then is an ancient people embodying not only attitudes and values radically different from those familiar to Westerners but also manifesting a physical and biochemical composite beyond the reasonable expectations of the anthropologists who studied them. Their adaptation to the conditions in which their cultures evolved is an example of cultural and developmental plasticity, as are the adaptive accomplishments of other peoples including those we are most familiar with, namely, ourselves.

Conclusions

The physical, biochemical, and psychological variations just described are expressions of adaptation to particular sets of environmental circumstances and pose difficult problems for normative as well as destinational developmental approaches. If organism and environment are regarded as a unitary living system, the question of developmental plasticity becomes context dependent and must be explicated in those terms. Questions associated with end state recede in importance, and questions associated with normative values become paradigmatic guides rather than developmental imperatives. The concept of a *norm of development,* then, must be shifted from a value laden truism to a theoretically derived hypothesis that guides inquiry. Otherwise, we fall into the trap of regarding variations from a preferred or known context as deviant. Whether a particular variation is adaptive in a physiological or social sense will depend upon the context in which it operates. Principles governing growth and development are to be sought in the succession of organized matrices that constitute ontogenesis. The examples presented in this chapter suggest that we only know about a limited range of developmental courses. New cultural settings and environmental circumstances will very likely reveal unanticipated opportunities for and constraints upon developmental variations.

References

Adinolfi, A. M. The postnatal development of synaptic contacts in the cerebral cortex. In. M. B. Sterman, D. J. McGinty, & A. M. Adinolfi (Eds.), *Brain development and behavior.* New York: Academic Press, 1971.

Altman, J. Nutritional deprivation and neural development. In M. B. Sterman, D. J. McGinty, & A. M. Adinolfi (Eds.), *Brain development and behavior.* New York: Academic Press, 1971.

Bijou, S. W., & Baer, D. M. *Child development* (Vol. I). New York: Appleton-Century-Crofts, 1961.

Davis, J. M., Himwich, W. A., & Agrawal, H. C. Some amino acids in the developing visual system. *Developmental Psychobiology* 1969, *2,* 34-39.

Denney, D. R., & Moulton, P. A. Conceptual preferences among preschool children. *Developmental Psychology,* 1976, *12,* 509-513.

Dobbing, J. The later growth of the brain and its vulnerability. *Pediatrics,* 1974, *53,* 2-6.

Dobbing, J. Human brain development and its vulnerability. In *Biologic and clinical aspects of brain development*. Mead Johnson Symposium on Perinatal and Developmental Medicine, 6, 1975.

Dobbing, J., & Sands, J. Comparative aspects of the brain growth spurt. *Early Human Development*, 1979, *3*, 79-83.

Ebert, J. D., Loewy, A. G., Miller, R. S., & Schneiderman, H. H. *Biology*. New York: Holt, Rinehart & Winston, 1973.

Freedman, D. G. *Human infancy: An evolutionary perspective*. New York: Erlbaum, 1974 (distributed by John Wiley & Sons).

Gajdusek, D. C. Physiological and psychological characteristics of stone age man. *Engineering and Science*, 1970, *33*, 26-33, 56-62.

Goldstein, K. *The organism: A holistic approach to biology derived from pathological data in man*. New York: American Book Company, 1939.

Gollin, E. S., & Garrison, A. Relationships between perceptual and conceptual mediational systems in young children. *Journal of Experimental Child Psychology*, 1980, *30*, 325-335.

Gollin, E. S., & Rosser, M. On mediation. *Journal of Experimental Child Psychology*, 1974, *17*, 539-544.

Gollin, E. S., & Saravo, A. A developmental analysis of learning. In J. Hellmuth (Ed.), *Cognitive studies (Vol. 1)*. New York: Bruner/Mazel, 1970

Gottlieb, G. The roles of experience in the development of behavior and the nervous system. In G. Gottlieb (Ed.), *Neural and behavioral specificity: Studies on the Development of behavior and the nervous system* (Vol. 3). New York: Academic Press, 1976.

Greer, C. A., & Alpern, H. P. Paradoxical effects of d-amphetamines upon seizure susceptibility in two selectively bred lines of mice. *Developmental Psychobiology*, 1980, *13*, 7-15.

Hebb, D. O. *The organization of behavior: A neuropsychological theory*. New York: John Wiley, 1949.

Hebb, D. O. Alice in Wonderland, or psychology among the biological sciences. In H. F. Harlow & C. N. Woolsey (Eds.) *Biological and biochemical bases of behavior*. Madison: University of Wisconsin Press, 1958.

Himwich, W. A. Multi-disciplined studies of the visual system in developing rabbits. In L. Jilek & S. Trojan (Eds.), *Ontogenesis of the brain*. Proceedings of the International Symposium Neuroontofeneticum, Praga, 1967. Prague: Charles University.

Himwich, W. A., Davis, J. M., and Agrawal, H. C. Biochemical substrates of the development of the matured evoked potential. In J. Wortis (Ed.), *Recent Advances in Biological Psychiatry* (Vol. 9). New York: Plenum, 1967.

Hodos, W., & Campbell, C.B.G. Scala naturae: Why there is no theory in comparative psychology. *Psychological Review*, 1969, *76*, 337-350.

Kawai, M. Newly acquired pre-cultural behavior of the natural troop of Japanese monkeys on Koshima Islet. *Primates*, 1965, *6*, 1-30.

Kopin, I. J. Neurotransmitters: Basic concepts, development and role in neurological disorders. *Biologic and clinical aspects of brain development*, Mead Johnson Symposium on Perinatal and Developmental Medicine, 6, 1975.

Lehrman, D. S. A critique of Konrad Lorenz's theory of instinctive behavior. *The Quarterly Review of Biology*, 1953, *28*, 337-363.

Leibowitz, H. W., & Pick, H. A. Cross-cultural and educational aspects of the Ponzo perspective illusion. *Perception & Psychophysics*, 1972, *12*, 430-432.

Moran, L. J. Comparative growth of Japanese and North American cognitive dictionaries. *Child Development*, 1974, *44*, 862-865.

Moran, L. J., & Huang, I. Note on cognitive dictionary structure of Chinese children. *Psychological Reports*, 1974, *34*, 154.

Moran, L. J., & Marakawa, N. Japanese and American association structures. *Journal of Verbal Learning and Verbal Behavior,* 1968, *7,* 176-181.
Olson, D. R. *Cognitive development: The child's acquisition of diagonality.* New York: Academic Press, 1970.
Schachter, D., & Gollin, E. S. Spatial perspective taking in young children. *Journal of Experimental Child Psychology,* 1979, *27,* 467-478.
Skinner, B. F. Operant behavior. *American Psychologist,* 1963, *18,* 503-515.
Stebbins, G. L. *Processes of organic evolution.* Englewood Cliffs, N.J.: Prentice-Hall, 1966.
Stebbins, G. L. *The basis of progressive evolution.* Chapel Hill: University of North Carolina Press, 1969.
Szentagothai, J., & Arbib, M. A. *Conceptual models of neural organization,* Cambridge, Mass: The M.I.T. Press, 1975.
Thoday, J. M. Components of fitness. *Symposia of the Society of Experimental Biology,* 1953, *7,* 96-113.
Waddington, C. H. *Organisms and genes.* London: Cambridge University Press, 1940.
Waddington, C. H. *The Strategy of the genes: A discussion of some aspects of theoretical biology.* London: George Allen & Unwin., 1957.
Waddington, C. H. *Principles of development and differentiation.* New York: Macmillan, 1966.
Waddington, C. H. The basic ideas of biology. In Waddington, C. H. (Ed.), *Towards a theoretical biology.* Chicago: Aldine, 1968.
Waddington, C. H. The theory of evolution today. In A. Koestler & J. R. Smythies (Eds.), *Beyond reductionism: New perspectives in the life sciences.* Boston: Beacon Press, 1971.
Walker, L. D., & Gollin, E. S. Perspective role-taking in young children. *Journal of Experimental Child Psychology,* 1977, *24,* 343-357.
Weiss, P. *Principles of development: A text in experimental embryology.* New York: Henry Holt, 1939.
Weiss, P. The living system: determinism stratified. In A. Koestler & J. R. Smythies (Eds.), *Beyond reductionism: New perspectives in the life sciences.* Boston: Beacon Press, 1969.
Weiss, P. *The science of life: The living system—A system for living.* New York: Futura, 1973.
Werner, H. The concept of development from a comparative and organismic point of view. In D. B. Harris (Ed.), *The concept of development.* Minneapolis: University of Minnesota Press, 1957.
Wiens, J. A. Effects of early experience on substrate pattern selection in Rana aurora tadpoles. *Copeia,* 1970, *3,* 543-548.
Winick, M. Malnutrition and the developing brain. In F. Plum (Ed.), *Brain dysfunction in metabolic disorders.* Research Publication of the Association of Nervous and Mental Diseases (Vol. 53). New York: Raven Press, 1974.
Winick, M. Maternal nutrition and intrauterine growth failure. *Nutrition, growth and development.* Mod. Probl. Paediat. Basel: Karger, 1975.

8

Epistemology and Developmental Psychology

STEPHEN TOULMIN

Introduction

When Gottlob Frege resurveyed *The Foundations of Arithmetic* in the closing years of the nineteenth century, his primary concern was with the philosophy of mathematics. So he took good care to distinguish "logical" questions about mathematical concepts and statements which (in his view) arose in philosophical contexts, from all merely "empirical" questions. Philosophers, he argued, had no business concerning themselves either with historical questions about the cultural evolution of mathematical methods, or with psychological questions about children's progressive recognition of arithmetical and other formal relationships (Frege, 1884). Like Plato long before him, Frege was not interested in recording contingent truths about the ways in which human beings—whether carpenters, accountants or professional mathematicians—have *in fact* done their thinking about numbers. He wished to concern himself exclusively with the *necessary* structures supposedly underlying all correct arithmetical thought. For his purposes, it was crucial to avoid sliding across the boundary lines separating logic and arithmetic from history or psychology, and to distinguish the rules of correct thought from generalizations about actual thinking.

Was Frege justified in believing that it is possible, even in arithmetic, to carry through completely his Platonic ambition to "strip away" all the historical and psychological "accretions" that "veil the eye of the mind" from recognizing "concepts in their pure form"? That is a legitimate question. But it will be more urgent to pay attention to Frege here, because his arguments have been taken as defining the twin heresies that most twentieth century philosophers have striven so hard to avoid—namely, "psychologism" and "the genetic fallacy." For some 50 years, these bogies have largely prevented any effective alliance between epistemology or philosophy of science, on the one hand, and the developmental psychology or history of science, on the other (Achinstein & Barker, 1969; Suppe, 1974). The authority of Frege's word and of his example was so weighty that his successors transferred his methods and maxims wholesale from their original locus in the philosophy of arithmetic, first into the philosophy of natural science, and subsequently into general epistemology. As a result, they not only kept the philosophical analysis of scientific concepts forcibly separate from the historical scrutiny of scientific changes: More basically still, they insisted on separating the epistemological criticism of our concepts and everyday ways of thinking from all studies of concept acquisition and mental development. If the "context of discovery" was dismissed as irrelevant to questions of justification in the philosophy of science, the "context of learning" was seen as more irrelevant to questions about the broader validity of human knowledge and understanding.

Since the mid-1960s, the worst strictures against the genetic fallacy have been relaxed. By now, most philosophers of science acknowledge that the rationality of changes in scientific thought can be fruitfully characterized, only by paying attention to sample episodes from the historical development of science. Just as abstract codes of law can be expounded convincingly only by exemplifying their practical force with the help of sample cases, so too a critique of scientific rationality demands at least the "rational reconstruction" of paradigmatic examples from the history of science. Meanwhile, those of us who seek to interpret the rationality of science itself on a historically developing, or "common law" model find the history and sociology of science a helpful source of insight into the *evolution* of human rationality (i.e., the processes of "conceptual phylogeny" through which, bit by bit, human thinkers and agents have collectively established an intellectual and practical command over their dealings with the world). In consequence, the considerations that make conceptual innovations acceptable within collective rational enterprises can be understood fully and clearly only by looking to see how, in actual historical fact, such innovations *come to be perceived* as acceptable (Toulmin, 1977).

Hitherto, by contrast, the ban against psychologism has remained in full force. For instance, philosophers generally still regard investigations into the psychology of mathematical discovery as irrelevant to the philosophy of mathematics, and ignore empirical studies of the ways in which children develop their conceptions of causality, conservation, and the rest, as having nothing to teach us about the epistemological grounding of thought and perception. It is one thing to analyze and criticize the mature concepts and structures that are the presumed destinations of a child's mental develop-

8. EPISTEMOLOGY AND DEVELOPMENTAL PSYCHOLOGY

ment. It is quite another thing to investigate the sequences of steps, whether stumbling or unerring, by which individual children move toward that presumed destination. Unless the final destination of mental development were recongizable in advance, these philosophers would reply, how could we even frame meaningful empirical questions about those steps? (Hamlyn, 1971; Peters, 1967). For, surely, our prior understanding of the mature forms is an indispensable template for judging when any child has taken such a "step" at all. So permitting matters or developmental psychology to influence our arguments about the theory of knowledge would once again (it seems) infect the necessary truths of philosophy with the contingency of empirical generalizations about actual human beings, and so (in Kant's phrase) debase philosophy to the level of "mere anthropology."

The aim of this chapter is to reopen consideration of that issue. In the long run (I shall argue), the barriers separating epistemology from developmental psychology and psycholinguistics can be no more watertight than those separating the philosophy of science from the history and sociology of science. The central philosophical enterprise—of validating the basic concepts that make human knowledge and experience both intelligible and "sayable"—remains admirable and indispensable. But we can hardly carry this task through to completion, in advance of all psychological investigations into concept acquisition and the *development* of human rationality (i.e., the processes of "conceptual ontogeny" through which, bit by bit, individual human beings develop their command over the rational inheritance of their species and cultures). The traditional program for an a priori epistemology is probably unrealizable *in principle;* and, even if it did turn out to be realizable in fact, this would presuppose certain general truths about human beings that are as yet far from certain, and must be brought into the light of day.

On this view, philosophers are not, of course, obliged to wait patiently for empirical data from psychology and psycholinguistics before even *beginning* on their own analytical tasks. But neither philosophers alone nor developmental psychologists alone can hope to *complete* their inquiries in complete isolation from the others. Up to a point, the philosophical claims of our everyday basic concepts can be discussed without explicitly raising psychological questions about the manner in which these concepts are acquired or developed, and vice versa. But the mutual independence of these two sets of issues is limited; and the *claim* to independence involves, on each side, presuppositions about the other side. Piaget's vaunted rejection of philosophy, for instance, rests on—and disguises—his own philosophical commitments: Conversely; epistemology can be protected against adulteration from psychology only if one makes substantial assumptions of a psychological kind. Instead of fighting for mutual independence at all costs, it might therefore be better for philosophers and psychologists to accept a limited collaboration, in the hope of defining more exactly the issues that arise for both subjects at the final frontiers of obscurity.

Does Mental Development Have a Unique Destination?

To focus on the crucial issue that arises at the intersection of philosophy, history, and psychology:

How far is it possible to determine and define in advance the final goal or destination toward which improvements in our procedures and abilities are directed, *either* during the historical evolution of our collective procedures, *or* during the psychological development of our individual abilities?

Evidently, if we could take the existence of such a uniquely definable and universal "destination" for granted, that would greatly simplify matters for us. For, then, the only significant problems left to investigate would be—for the philosophers—how that unique destination is to be characterized, and, for the psychologists, how we in fact move toward it. Unfortunately, presupposing the existence of such a universal destination would also close off certain possibilities for both psychology and philosophy in a way for which there is no clear justification.

Let me take as my central example the theoretical project of Jean Piaget. Piaget has always been interested in exploring the implications for psychology of Kant's fundamental position: Namely, that the mature "reason" operates according to a unique, universal set of concepts, categories, and forms. Despite all his careful empirical studies of children's behavior, as a result, Piaget's overall intellectual goal is not open-ended. His primary concern (he has said) is to discover how growing children "come to *recognize the necessity of*" conforming to the intellectual structures of logic, Euclidean geometry, and the other basic Kantian forms. He barely pauses to enquire whether there really *is* such a "necessity": He does not ask, for instance, whether our current adult scheme of spatial concepts must *necessarily* be represented by a specifically Euclidean set of formal operations. In his eyes, the central question for any "genetic epistemology," whether for developmental psychology or for the history of thought, is simply how children and cultures alike come to recognize that Kantian destination, whether during their individual lifetimes or their collective histories (Toulmin, 1972).

To pose the problem of "developmental destinations" in these terms, however, begs the crucial question. If it had already been demonstrated in fact, on convincing grounds and to the general satisfaction, that such a unique and universal destination was available to all cultures and in all epochs, this procedure might be legitimate. Philosophers and psychologists could then go their separate ways: philosophers to map the region so defined, psychologists to look and see how children find their ways about it. Yet no such demonstration has in fact been given. So, rather than taking the existence of any simple, general destination for granted, we should be asking, rather:

In what respects, to what extent, and on what conditions, does the psychological development of human individuals—like the historical evolution of human cultures—proceed *in fact* toward some universal and uniquely definable destination?

The Grounds for Skepticism about Any Universal Destination

To start with, all we need here is a "burden of proof" argument. What little evidence we have undercuts any supposition that intellectual evolution and conceptual development have a single, unique, and universal destination. At the very least, that evidence falls short of what we should need to raise this supposition to a basic axiom either of epistemology, or of developmental psychology. Whether as philosophers or as psychologists, it is therefore desirable that we should consider how the theoretical ambitions of both fields need to be reformulated, once the burden of proof is shifted: That is to say, if we take as our starting point the contrary supposition, that the directions of conceptual ontogeny and phylogeny alike are to some significant degree indeterminate and open-ended.

To Piaget's credit, he has always emphasized the links between history and psychology, cultural evolution, and individual development. His use of the phrase "genetic epistemology," for instance, refers to both the historical analysis of our collective methods of thought and to the study of mental development in individuals. Yet his treatment of these topics oddly recalls Von Baer's hypothesis in early nineteenth century embryology, to the effect that "ontogeny recapitulates phylogeny." The collective experience of human cultures and the individual experience of human adolescents are both presented as leading to the same eventual destination; namely, to the adoption of the particular systems of "formal operations" that Piaget regards as characteristic of mature thought—whether about logic or arithmetic, spatial relations, causality, conservation (i.e., material substance) or whatever. Whereas concept acquisition in individuals is evidently, in part, a matter of socialization and enculturation—any particular child naturally ends up by reaching the "stage" of collective development typical of its own culture and epoch—all cultures are equally capable, in the long run, of progressing to the same ultimate stage of conceptual maturity; so that the individuals who grow up within them will thereafter be carried along to the same final goals.

About this position, certain naive but fundamental reservations are unavoidable. For instance: If Euclidean geometry provides the "mature" forms of thought about spatial relations in all cultures and epochs alike, why, then, did the Greek mathematicians (notably, Euclid himself) abstract those "mature spatial operations," and make them explicit, so late in the history of the human species? And, if the geometry of Euclid is really a shared, universal goal, why do we even now find peoples who understand, deal with, and answer spatial questions in terms that have nothing recognizably Euclidean about them? (Q: "How far away is that village?" A: "Two days' walk.") In nonindustrial cultures that make little use of formal mathematics, children evidently do not pursue the "natural sequence" of intellectual stages all the way to the "formal operational" stage of accepting Euclidean spatial concepts, and their counterparts in other spheres of activity.

Why is this? Is it because their impoverished cultural environment somehow *prevents* them from doing so? Must we assume that some active process of "blocking"—or *calage,* to use the French term—obstructs their natural development, until it is counteracted by a contrary process of "unblocking," or *décalage?* Or, alternatively, should we not rather suppose that young children are in themselves "neutral" toward alternative systems of spatial concepts (i.e., that they are subject to *no* uniquely "natural" tendency toward *Euclidean* geometry)? If that is so, the abstract geometry of the classical Greeks—in particular Euclid's own comprehensive synthesis of Greek geometry—was presumably a genuinely novel invention; and children in cultures historically unaffiliated to the classical Greek culture do not need to be actively *prevented* from "recognizing the necessity of" Euclidian geometry. If they grow up without acquiring either our sholastic grasp of the Euclidean formalism, or our pragmatic grasp of spatial operations, that is merely because their cultures have given them no *positive* occasions to acquire such a grasp.

Euclidean geometry no doubt serves with fair accuracy as an idealized formal representation of the "spatial relations" that are put to use in most practical contexts, and discussed more abstractly at school, within our own "modernized" societies and cultures; and the advantages of those particular spatial concepts, at least for everday purposes, can be explained in a pragmatic kind of way (Toulmin, 1972). Still, the implications of this fact must not be overinterpreted. Granted that Euclidean geometry adequately "represents" the spatial operations current in twentieth century industrial societies; and granted, also, that teen-agers in such societies learn at school to handle the abstract mathematical deductions of the Euclidean system; these two facts taken together still do not imply that the forms of Euclidean

8. EPISTEMOLOGY AND DEVELOPMENTAL PSYCHOLOGY

geometry are "necessary" or "inherent to" *all* spatial theory and praxis in any—still less, every—culture or epoch.

Even in our own culture, indeed, certain enterprises have set aside Euclidean geometry in favor of alternative systems. In certain parts of astrophysics, for instance, Euclidean geometry loses all practical possibility of application: How can we specify physically intelligible operations for identifying a "Euclidean plane surface" in the intergalactic regions? So, if we commit ourselves prematurely to the belief that all "mature" spatial thinking has a unique and universal form, we shall land ourselves with some paradoxical consequences. For then the question will arise: Which is the truly mature stage in spatial thinking—Euclidean or non-Euclidean? And neither answer will be satisfactory in all cases. Does the use of non-Euclidean geometry in astrophysics represent a *less* mature stage in spatial thinking than the Euclidean? In that case, twentieth century physics has undergone an odd historical regression, from fully mature Euclidean to immature, non-Euclidean thinking about geometry! Or are we to say that non-Euclidean thinking represents a new and more mature stage than the Euclidean? What are we, in that case, to say about spatial thinking in earlier cultures? Were children in those cultures, after all, "really" en route toward a Riemannian or Lobachevskian goal, rather than toward a Euclidean one? Was the goal of non-Euclidean "maturity" really available to earlier cultures all along? It is bad enough being required to treat the thinking of simple agricultural societies as "implicitly" Euclidean; but to be told that Euclid himself was really a "non-Euclidean" geometer *manqué* will surely be intolerable!

The Need for an Alternative Approach

Piaget's program of research in developmental psychology (I have argued) takes for granted the validity of a certain basic Kantian position. In Kant's arguments, the strict formal systematicity of deductions within Euclid's geometry and Newton's mechanics was concertina'd with their seemingly unrestricted pragmatic scope and range of practical application. Logical "necessity" (i.e., systematicity) was thus equated with pragmatic "necessity" (i.e., indispensability). Given this equation, Kant was open to the same temptations as Plato and Descartes before him. What the "intuited certainty" of geometrical principles had done for Plato and the "clarity and distinctness" of Euclidean concepts had done for Descartes, the systematic necessity *internal* to Euclidean geometry and Newtonian mechanics did for Kant: It seemingly guaranteed the universal relevance and applicability of those systems to practical situations *outside themselves* (Martin, 1955).

For ourselves, we must begin by insisting on the distinctions that Kant's equation conceals. Any formal system, simply as such, possesses the kind of internal systematicity associated with logical necessity; yet no such formal system, simply as such, can guarantee its own pragmatic necessity, or indispensability. Many formal systems have initially been given an unrestricted practical application, only to have limits subsequently placed on them in the light of actual experience. Euclidean geometry, for instance, was first arrived at by abstraction from practical operations that were familiar within classical antiquity, such as carpentry and land surveying: As a formal representation of such operations, its applicability remains as clear today as it was then. Still, this observation in no way implies, either that people in preclassical or nonindustrial cultures are "striving in vain" to think in Euclidean manner, or that the use of non-Euclidean geometry in astrophysics today calls in question the relevance of the Euclidean system to terrestrial practice.

Within our own epoch and culture, most young children grasp the basic practical operations from which the Euclidean formalism was abstracted in the course of mastering our common ways of dealing with, and talking about, spatial relations. But that is a comment, merely, on the technical subtlety of "general knowledge" in our culture: It is neither guaranteed by, nor does it reinforce, the "formal necessity" of the Euclidean system. (That is quite another matter.) In less technical cultures that have less occasion to employ those particular operations, correspondingly, the dominant ways of assessing spatial relations may well remain pre-Euclidean (e.g., "Two days' walk"). Meanwhile, the exigencies of space travel might yet give the children of some future intergalactic culture the same familiarity with Riemann or Lobachevsky that our own children have with Euclid.

In other spheres of thought and action, also, the claims for a unique, universal destination must be challenged in the same way. Thus, no single precise idea of "conservation," or "material substance," has clearly been shared by people in all epochs and cultures. On some sufficiently vague level, no doubt, people have long assumed a certain continuity of material substance through many kinds of changes. Some of the pre-Socratic philosophers even made the indestructibility of material substance a matter of abstract principle. But, even on the more sophisticated scientific level, the conservation of matter achieved an established position only in the eighteenth century, following the discovery of the first effective techniques for collecting and handling gases (Hales, 1727; Toulmin & Goodfield, 1962). Before that time, there was no practical way of demonstrating that evaporation and condensation did not involve the actual destruction and creation of liquids. And, by now, twentieth century physicists have been forced to recognize that, using sufficiently discriminating measures, matter is *not* in fact perfectly conserved in all physical transformations. Rather, it is sometimes annihilated or created by the transformation of electromagnetic radiation into electron-positron pairs, or vice versa. So we might once again ask: Which of these notions of conservation is it supposed that young children can discover or develop naturally, for themselves? Is it the pre-Newtonian idea, the classical nineteenth-century idea, or the sophisticated late twentieth-century idea?

There is no need to labor the point further. What we have available for study in developmental psychology today is not the spectacle of young children "coming to recognize the necessity of" certain universal formal schemata: Rather, it is the processes by which they are in fact socialized and enculturated into the actual praxis, thought, and speech current in any culture sharing our particular kinds of technical sophistication. As to the question, how far, for example, animistic cultures share our *façons de parler* and *façons de penser* in considering substance and causality: That is an issue we must approach with open minds. And as for the possibilities open to future, more complex cultures, there too we must be prepared to speculate open-mindedly. There, perhaps, people generally will take pride in having overcome the "illusions" of material conservation and Euclidean space alike, and may come to talk about everyday material objects with the same conceptual sophistication we ourselves display only toward such un-everyday things as electrons.

In certain limited respects, of course, our actual ways of dealing with, and thinking about, material objects, particularly in infancy, may yet prove, in fact, to be the same in all cultures and epochs. It may even prove that adult ways of speaking and thinking about material objects on the everyday level share a common "logical grammar" in all languages—even that specific perceptual and intellectual obstacles (embodied in our genetic inheritance, say) forever prevent us from dealing with material objects in any other terms. But, if this does turn out to be the case, it will be a significant *factual* discovery about the "cultural universals" of human thought and action. And, until that discovery is firmly established, taking the universality of our own modes of thought and praxis for granted is building on sand. Instead, all concerned, namely cultural anthropologists, developmental psychologists, philosophers, and anyone else with something relevant to say, would do better to sit down together and figure out just what differences of praxis, thought, and language would genuinely distinguish (e.g.), nonconserving or noncausal modes of life, thought, and speech: Only then shall we be in position to look and see just how *nearly* universal, in actual practice, our own forms of thought and language really are.

For the moment, then, we are left with several unanswered questions. To begin with: *Just how far* do the ostensibly universal elements in thought and praxis really take us? Just how far, and in what specific respects, will they really serve (e.g., to define a shared destination for mental development in different cultures)? All cultures having any real degree of modernization and technical sophistication most probably do share some substantial common elements: whether in their ways of handling spatial relations, or material substance, or causality, or whatever. But the existence of substantial common elements is one thing: The supposition of a priori necessity is something quite else. And the acknowledgment of substantial common elements, in some specific respects and to some recognizable extent, still leaves room for plenty of variety, both of kind and of degree. At this point, we must turn and consider the implications of that variety for both philosophy and developmental psychology.

Alternative Developmental Destinations and Trajectories

If there were a unique and universal destination toward which the mental development of children and adolescents was directed in all cultures and epochs alike, that would (as I remarked) have made the tasks of philosophy and developmental psychology much easier. If we set aside that hypothesis as oversimplified, even as unfounded, we must now introduce fresh complications into our enquiries on both the philosophical and the psychological fronts.

Philosophically speaking, we shall have to place severe restraints on the use of phrases like "the everyday conceptual framework" and "our ordinary way of speaking" (Strawson, 1959; 1966). The singular implications of these phrases (e.g., the use of the definite article "the") will now be misleading: How can we say *"the* framework" unless we have proved that there is *one and only one* such framework? Nor shall we be able to develop any clear philosophical contrast between "the actual world," as defined by the particular pattern of basic concepts and forms of thought we in fact use to structure our judgments and experience, and other "possible worlds" that would be defined by alternative patterns of forms and concepts. For, again, there will be a range of slightly different "actual worlds" to be discussed, and these will embrace a spectrum of "possible worlds," rather than one only.

Something of this complexity was recognized by Wittgenstein (1953), who used to challenge any assumption that our twentieth-century liberal industrial "everyday conceptual framework" is unique, inevitable, and authoritative for all language communities, regardless of their forms and conditions of life. Indeed, in his lectures, Wittgenstein used repeatedly to explore the various ways in which changes in those forms and conditions might make it—or might have made it—equally natural to develop alternative sets of basic concepts, "language games," and so on. However, Wittgenstein himself used to do this entirely in a speculative, hypothetical spirit. He was concerned to say only, "See how differently things *might well have* gone, had our forms and conditions of life been sufficiently different!," rather than, "See how different they actually *were, are, and may yet be* in the future, given sufficiently different forms and conditions of life!" His arguments were directed (that is to say) against any attempt to revive the Kantian project of providing a "transcendental deduction" of the necessary uniqueness of one particular set of basic forms of thought and experience: They were not designed to encourage any positive exploration into the *actual* diversity of those forms of thought—whether displayed in cultural variety or in historical variability. Least of all were Wittgenstein's discussions of our basic conceptual equipment offered as opening any new doors into developmental psychology. In his own way, he was as insistent as Kant, and also Frege (1892), had been on the need to avoid muddling philosophical analyses with psychological hypotheses, and he did not want anybody to suppose that he was offering speculations about how children *in fact* catch on to novel "language games."

Still: The philosophical tasks opened up for us by Wittgenstein's work are themselves worth pursuing for their own sakes. The assumption that, at their foundation, mature human thought and reasoning are structured by, or in accordance with, some single correct set of intellectual forms, is as old as philosophy itself. From the time of Plato on, it has been assumed that something or other defines that uniquely correct set of intellectual forms, whether an external, objective "world of forms," or our shared innate repertory of "clear and distinct ideas," or the basic "concepts, categories, and forms of intuition" involved in rational judgment, or (perhaps) the "everyday conceptual framework" embodied in the "logic of ordinary language". Now, however, we need to go behind that assumption, and to map the available spectrum of possible diversity represented by the actual forms of thought of human peoples in different cultures and epochs. For why should not the fundamental differences between, for example, animistic and mechanistic ways of thinking about causality be as great as they appear to be on the surface? The most that we are entitled to take for granted, surely, is that the practical demands of human life and experience in any particular sphere of thought and action will restrict the available destinations of mental development within a reasonably narrow compass. In this respect, the goal of philosophy from now on should no longer be to demonstrate the "logical necessity" of any particular conceptual framework, or set of "basic forms of thought." Rather, it should be to learn, from reflection on actual experience, how far and in what respects the "pragmatic necessities" of our society, cultural, and technical situations place constraints on the manual, intellectual, and other operations we employ in dealing with spatial relations, material commodities, and the rest: how far, and in what respects (on the contrary) they leave us with scope to choose, and to change, between one set of basic operations and another.

Psychologically, the new scope that is opened up for us in this way is considerable. To begin with, any adequate recognition of the part played in mental development by *enculturation* requires us to add a third dimension to studies of child behavior. Researchers in the Piagetian tradition of observation are trained to play down their own roles in the situations they study; and, as a result, they tend to present their findings as though the only significant elements in the situations they study were the child and the external situation he or she is dealing with. Yet, in fact, the researchers themselves are unavoidably parts of their own experiments in at least two distinct ways, and we cannot interpret their results satisfactorily without making allowance for this fact. In the first place, the experimental situations facing the child are themselves carefully selected in ways that reflect the conceptual presuppositions of the psychologist's own culture. So, the child is studied "catching on to", for instance, the twentieth-century Genevan notion of conservation, as represented in a twentieth-century Genevan laboratory situation: And the experiment itself becomes an element in the Genevan child's "enculturation." In the second place, the need for the child to be motivated to participate in the experiment at all carries with it the likelihood that it will be interested—like clever Hans, the "calculating horse"—in "psyching out" the experimenter, and getting the satisfaction of coming up to the standards reflected in the experimenter's own apparent hopes and expectations.

For our present purposes, accordingly, it is preferable to recognize explicitly the triangular character of the experimental situation in developmental psychology, and to design all observational studies with this in mind. Along with the child itself, and the specific problem situation facing it in any particular experiment, there is always "the culture," as embodied in the experimenter, the mother, and other significant and influential persons in the child's milieu—what we may call its "mentors." So, alongside studies of the respects in which a child comes naturally to "reinvent" aspects of the ambient culture within the psychologist's laboratory, by itself and for itself, we also need studies of the respects in which the entirety of a child's learning serves to "replicate" the ambient culture for a new generation, and of the processes by which that replication is achieved. Here we can learn from the work of L. S. Vygotsky, A. R. Luria, and their associates, who have focused attention helpfully on the ways in which speech—first, the speech of the mentors, but subsequently the child's own inner speech—serves as the "scaffolding" within which the forms of our thought and action are shaped and consolidated (Vygotsky, 1961, 1978). If the genetic coding of macromolecules in cell nuclei is responsible for the basic "hardware" of human development and ac-

8. EPISTEMOLOGY AND DEVELOPMENTAL PSYCHOLOGY 265

counts for the precision of replication displayed over the generations (e.g., the Hapsburg nose), the corresponding process of replication displayed in the exact transmission of cultural and social niceties from generation to generation is (on this view) encoded as "software" in the language games of the culture—which, when internalized, give shape to the child's "mental structures." (Notice: For Vygotsky, as for Wittgenstein, language is not seen as separable from other kinds of conduct. It has always to be mastered in conjunction with, and as part of, an associated larger constellation of actions and activities—i.e., a "language game.")

At this point, the philosophical and psychological aspects of present discussion flow together and reinforce one another. Only if we have seen just what a range of ideas of "material substance" and "conservation" are available to us as possible destinations of mental development, for instance, can we effectively go on to study the cues, and the processes, as an outcome of which a child in twentieth-century Geneva, in fact, ends up with one precise idea of "substance," a child in eighteenth-century Haiti a second, a child in twenty-second-century Tokyo or Akademogorsk (maybe) yet a third. Once we have developed ways of studying how, in actual fact, children come to pick up entire "language games"—linguistic and nonlinguistic elements together—we shall see where exactly within that process there is, and is not, scope for cultural diversity and historical variability in the alternative outcomes of that process. But, until we have done this, it will remain hard to counter the charge that our conservation studies, for example, are "culture bound," and our observations of moral development excessively focused on how a child learns to *talk* respectably in response to the ethical questions presented to it by the psychologist.

Certainly, the best way of avoiding the fragmentation between language and praxis that is apparent in much recent developmental psychology will be to incorporate the current two-dimensional studies of mental development into their natural three-dimensional matrix—with the child, the situation, and the mentors all incorporated fully into the script. Once this is done, we shall at once be more at home with the actual diversity to be found in cross-cultural studies. Consider, for instance, Kohlberg's (1966, 1969) claim that there is a standard sequence in which children come to recognize the personal, imaginary status of dreams. First, he says, they recognize the fantasy character of dream episodes: Only later do they discover their dreams are individual experiences, not collective ones. Checking out this result with Nigerian children, Shweder and LeVine (1975) have demonstrated that the sequences presented by Kohlberg are not in fact universal. Nigerian children come to recognize these features differently: Many of them recognize the individual nature of the experiences before their imaginary or fantasy character. Shweder and LeVine suggest that this difference is related to differences in children's sleeping arrangements in Nigeria and the Unites States. American children are accustomed to sleeping alone from infancy, and so have little chance to "compare notes" about their dreams with other people: Nigerian children usually share their sleeping quarters with others, and so discover early that their dreams are not shared.

More generally: A full acceptance of the diversity of developmental destinations will require us to rethink current assumptions in developmental psychology in three chief respects.

1. We shall have to question the whole familiar scheme of "stages" and "sequences" around which so much recent work has been structured. Rather than one single sequence of stages—from "sensorimotor" to "formal operational" or whatever—being accepted as relevant to cultural tasks of all kinds, we should now expect to find different "epigenetic sequences" appropriate to different kinds of tasks. No doubt, there will be certain specific and limited points at which we shall find correspondences between the sequences typical of different tasks in any given culture; but there will be equally specific and limited reasons why that proves to be the case. Either they will result from such evident facts as that 5-day-old infants do not speak and 5-year-old children do; or else they will reflect, for example, respects in which the sequences in both areas involve earlier tasks that are "prerequisites" of later ones in similar kinds of way (Feldman & Toulmin, 1976).

2. We shall have to make much more allowance for the possibility that different children may arrive at the same specific destinations by alternative routes—either in different cultures (as in the "dreaming" example) or within the same culture.

3. We should also set out to develop fresh analytical devices to replace the standard, oversimplified "stage" model (e.g., by developing a system of "populational" categories to replace the "typological" ones commonly used to define the phases through which children come to meet the demands of enculturation). Just as the apparent fixity of organic species has turned out to reflect, at most, a local and temporary stability in the flux of organic populations, so too the apparently static and universal character of developmental stages may be expected to reflect, at most, a statistical dominance of certain kinds of achievement among the children of a given age cohort in the culture concerned. This being the case, there will be quite as much to be learned by studying how and why certain children deviate from that pattern—being either precocious or backward in the relevant respect—as there is to be learned by concentrating on the general pattern itself. Similarly, there will be as much to be learned by looking to see how differently any given child develops in dealing with problems of different kinds—what that child's overall profile of achievements is, and how natural talents for, for instance, music or manual dexterity come to show themselves—as there is to be learned by

looking only at the common features evident in mental development in the course of dealing with varied and heterogeneous tasks.

References

Achinstein, P., Barker, S. F. *The legacy of logical positivism.* Baltimore, Md.: Johns Hopkins Press, 1969.
Feldman, C. F., & Toulmin, S. Logic and the theory of mind. In W. J. Arnold & J. K. Cole, (Eds.), *Nebraska Symposium on Motivation 1975.* Lincoln, Neb.: University of Nebraska Press, 1976.
Frege, G. *The foundations of arithmetic.* (J. L. Austin, Eng. trans.) Oxford, Eng.: Basil Blackwell, 1884.
Frege, G. On concept and object. *Vierteljahrsschrift für Wissenschaftliche Philosophie,* (1892) 16, 192-205.
Hales, S. *Vegetable staticks.* London, Eng.: W. & J. Innys, 1727.
Hamlyn, D. In T. Mischel, (Ed.), *Cognitive development and epistemology.* New York: Academic Press, 1971.
Kohlberg, L. Cognitive stages and pre-school development. *Human Development.* 9, 5-17, 1966.
Kohlberg, L. Stage and Sequence: The cognitive-developmental approach to socialization. In D. A. Goslin (Ed.), *Handbook of Socialization,* Chicago, Ill.: Rand McNally. pp 347-480, 1969.
Martin, G. *Kant's metaphysics and theory of science.* (P. G. Lucas, Eng. trans.) Manchester, Eng.: Manchester University Press, 1955.
Peters, R. *The concept of education.* London, Eng.: Routledge & Kegan Paul, 1967.
Shweder, R., & LeVine, R. Dream concepts of Hausa children: A critique of the doctrine of invariant sequence in cognitive development. *Ethos,* 1975, 3, 209-230.
Strawson, P. F. *Individuals.* London: Methuen, 1959.
Strawson, P. F. *The bounds of sense.* London: Methuen, 1966.
Suppe, F. *The structure of scientific theories.* Urbana, Ill.: University of Illinois Press, 1974.
Toulmin, S. *Human understanding,* (Vol. 1). Princeton, N.J.: Princeton University Press, 1972.
Toulmin, S. From form to function: Philosophy and history of science in the 1950s and now. *Daedalus,* 1977 106 (3), 143-162.
Toulmin, S., & Goodfield, G. J. *The architecture of matter.* London, Eng.: Hutchinson, and New York: Harper & Row, 1962.
Vygotsky, L. S., *Thought and language.* (G. Vakar & E. Hanfmann, Eng. trans.). Cambridge, Mass.: M.I.T. Press, 1961.
Vygotsky, L. S. *Mind in society.* (Eng. ed.: Cole, M. J. et al.) Cambridge, Mass.: Harvard University Press, 1978.
Wittgenstein, L. *Philosophical investigations.* Oxford, Eng.: Basil Blackwell, 1953.

Author Index

Numbers in italics refer to the pages on which the complete references are listed.

A

Abrahamson, D., 117, *129*
Abramov, I., 155, *166*, *167*
Abramson, A. S., 158, *169*
Achenbach, K., 193, *226*
Achinstein, P., 254, *267*
Adinolfi, A. M., 240, *249*
Agrawal, H. C., 239, 240, 248, *249*, *250*
Aiu, P., 38, *63*
Aldrich, C. A., 118, *129*
Ali, F. A., 211, *227*
Alpern, H. P., 241, 245, *250*
Altman, J., 240, *249*
Amatruda, C. S., 195, *223*
Ames, L. B., 195, *223*
Anderson, V. E., 27, *31*
Andrew, D. M., 211, *222*
Antonova, T. G., 123, 125, *129*
Appleton, T., 125, *129*
Arbib, M. A., 243, *251*
Ashmead, D. H., 111, *129*
Aslin, R. N., 60, *62*
Assal, G., 218, *224*
Atkinson, J., 56, 58, 59, *61*, *62*

B

Babkin, P. S., 116, *129*
Baer, D. M., 233, *249*
Baerends, G. P., 140, 164, *166*
Bakan, P., 214, *223*
Baker, F. H., 59, 60, *62*
Banks, M. S., 50, 58, 59, 60, *62*, *64*, *67*
Baptista, J., 147, *166*
Barber, A., 50, *62*
Barker, S. F., 254, *267*
Barker, W. B., 193, *222*
Barnet, A. B., 123, *129*
Bartoshuk, A. K., 48, 49, *62*
Basser, L. S., 185, *222*
Bateson, P. P. G., 142, *166*
Beale, I. L., 176, *222*
Bechtold, A. G., 56, *67*
Beecher, M. D., 151, 152, 153, *166*, *170*, *172*
Beer, C. G., 143, *166*
Bekoff, M., 83, *98*
Bennett, E. L., 104, *132*
Bennett, G. K., 211, *222*
Berglund, G. W., 211, *224*
Berlin, B., 154, 155, 156, *166*
Bernard, J., 36, *62*
Bernuth, H. V., 49, *65*
Best, C. T., 185, *222*
Bever, T. G., 191, *222*
Bijou, S. W., 233, *249*
Birch, H. G., 195, *227*
Blackfan, K. D., 201, *223*
Blakemore, C., 60, *66*
Blakemore, C. B., 201, *223*

Blomquist, A. F., 72, *98*
Bodmer, W. F., 18, 19, 20, *30*
Boettner, E. A., 51, 53, *62*
Bogen, J. E., 203, *224*
Boklage, C. E., 182, 183, 184, 192, 193, *222*
Bornstein, M. H., 55, *62*, 154, 155, 156, 157, *166*, *167*
Bosack, T. N., 116, *131*
Bouman, M. A., 58, *67*
Bower, T. G. R., 50, *62*, 113, 122, *129*
Bowes, W., 123, *129*
Boxerman, S. B., 118, *132*
Boyd, H. 142, *168*
Brackbill, Y., 117, 123, 125, *129*
Braddick, F., 58, *61*
Braddick, O., 56, 58, 59, *61*, *62*
Bradley, R. M., 36, *62*, *66*
Branch, C. L., 195, *225*
Brazelton, T. B., 123, *129*
Bredberg, G., 47, *62*
Bronshtein, A. T., 123, 125, *129*
Bronson, G., 51, *62*
Brown, J., 119, *129*
Buchtel, H. A., 205, 218, *222*, *226*
Buck, R. W., 211, *222*
Bullock, T. H., 136, *167*
Bulmer, M. G., 182, *222*
Burke, P., 127, *132*
Burke, P. M., 44, *62*
Bushnell, E. W., 56, *67*
Butcher, M. J., 43, *65*, 110, 111, *132*
Bystroletova, G. N., 117, *130*

C

Cable, C., 146, *168*
Caldwell, B. M., 46, *64*
Campari, F., 218, *222*
Campbell, C. B. G., 233, *250*
Campbell, D., 43, *63*
Campbell, H., 125, *131*
Capranica, R. R., 164, *167*
Carmichael, L., 61, *62*, 82, *98*
Caron, R. F., 122, *130*
Carpenter, G. C., 123, *130*
Carpenter, Y., 117, *129*
Carter-Saltzman, L., 182, 184, 192, 193, *222*
Caul, W. F., 211, *222*
Cauna, N., 47, *62*
Cavalli-Sforza, L. L., 18, 19, 20, *30*

Cavonius, C. R., 59, *63*
Chambers, B. E. I., 60, *66*
Chandler, M. J., 125, *132*
Chang, P. C., 192, *227*
Changeux, J.-P, 27, *30*
Chauvin, R., 72, 89, *98*
Chiszar, D., 72, *98*
Choo, G., 205, *223*
Churchill, J. A., 195, *222*
Cioffi, J., 211, *222*
Clarke, R., 186, *227*
Clifton, R., 125, *129*
Clifton, R. K., 48, *62*, 122, *130*
Cogan, D. G., 38, *64*
Cohen, L. B., 51, *62*
Conel, J. L., 50, *63*
Connolly, K., 116, *130*
Conway, E., 123, *129*
Cooke, F., 140, *167*
Cooper, F. S., 159, *169*, *171*
Cooper, L., 54, *63*
Corballis, M. C., 175, 176, *222*, *225*
Cornsweet, T. N., 57, 59, *63*
Cornwell, A. C., 195, *227*
Coss, R. G., 104, *130*
Coursin, D. B., 49, *63*
Cowan, P. J., 143, *167*
Cowett, R. M., 107, *130*
Cox, P. W., 214, 215, 216, *226*
Coyle, J. M., 197, *224*
Creutzfeldt, O. D., 60, *64*
Crook, C. K., 43, 44, *63*, 105, 112, *130*
Culebras, A., 195, *225*
Cummings, W., 155, *172*
Curtiss, S., 191, *222*
Cutting, J. E., 158, 159, *167*

D

Dabbs, J. M., Jr., 205, *223*
Dandy, W. E., 201, *223*
Darwin, C., 102, *130*
Daumer, K., 163, *167*
Davies, J., 201, *223*
Davis, J. M., 239, 240, 248, *249*, *250*
Day, R., 154, *172*
Day, R. H., 122, *132*
Dayton, G. O., 38, *63*
De Beer, G. R., 35, *63*
De Bock, F., 201, *223*
DeFries, J. C., 15, 25, 27, *31*
DeLucia, C. A., 120, *132*

Denckla, M., 211, *226*
Denenberg, V. H., 210, *223*
Denisova, M. P., 115, *130*
Denney, D. R., 241, *249*
Dennis, M., 185, *223, 224*
De Risio, C., 218, *222*
Desmond, A. J., 9, *30*
Desor, J. A., 43, 44, *63, 66*
DeValois, R., 154, 155, *167*
Diamond, M., 210, 212, *223*
Dixon, L. K., 27, *31*
Dobbing, J., 239, 244, *249, 250*
Dobson, V., 52, 53, 58, *63*
Dooling, R. J., 150, 159, *167, 170*
Doris, J., 54, *63*
Dorman, M. F., 191, *223*
Dorsen, M. M., 201, *223*
Drillien, C. M., 125, *130*
Droege, R. C., 211, *223*
Dubignon, J., 43, *63*

E

Ebert, J. D., 234, *250*
Edgar, R., 176, *227*
Egeth, H., 218, *223*
Eggermont, E., 201, *223*
Ehrlichman, H., 214, 215, 216, *226*
Eibl-Ebesfeldt, I., 82, *98*
Eilers, R. E., 160, 161, *167*
Eimas, P. D., 37, 49, *63*, 102, 121, *130, 131*, 158, 160, 161, *167*
Eisenberg, R. B., 48, 49, *63*
Elkonin, D. B., 101, 114, *130*
Ellingson, R. J., 36, *63*
Elston, R. C., 182, 192, *222*
Emde, R. N., 118, *130*
Emlen, J. T., 143, *170*
Engen, T., 39, 40, 42, *63, 65*, 110, 122, *130*
Entus, A. K., 185, *223*
Epstein, J., 218, *223*
Estévez, O., 59, *63*
Ettlinger, G., 201, *223*
Evans, R. M., 143, *167, 168, 170*
Evens, L., 201, *223*

F

Fabricius, E., 142, 143, *168*
Fagan, J. F., 50, *63*
Falkenberg, H., 178, 179, *227*
Falls, J. B., 146, *168*

Fantz, R. L., 50, 56, *63*, 123, *130*, 139, 168
Feind, C. R., 38, *65*
Feldman, C. F., 266, *267*
Fennell, E., 193, *226*
Ferriss, G. S., 201, *223*
Field, T., 125, *130*
Figurin, N. L. K., 115, *130*
Fiorentini, A., 58, *66*
Fitzgerald, H. E., 117, *129*
Flor-Henry, P., 183, *223*
Flory, C. D., 46, *67*
Francois, J., 201, *223*
Freedman, D. G., 231, *250*
Freeman, D. N., 59, *66*
Freeman, R. D., 61, *64, 66*
Frege, G., 253, 263, *267*
French, J., 56, 58, *62*
Freud, S., 45, *64*, 102, *130*
Friedman, S., 123, *130*
Frisch, K. von, 163, 164, *168*
Frishkopf, L., 164, *167*
Frisina, R. D., 47, *64*
Fulk, K., 54, *68*

G

Gajdusek, D. C., 248, *250*
Galaburda, A. M., 195, *223*
Galebsky, A., 38, *64*
Galin, D., 204, *224*
Gandelman, R., 209, *226*
Gardiner, M. F., 185, *223*
Garrison, A., 245, *250*
Geffner, D. S., 191, *223*
Gelber, E. R., 51, *62*
Gellis, S. S., 38, *64*
Gescheider, G. A., 47, *64*
Geschwind, N., 195, 197, *223, 224*
Gesell, A., 36, *64*, 104, *130*, 195, *223*
Ghent, L., 211, *223*
Gibson, E. J., 135, *168*
Gibson, J. J., 37, *64, 168*
Gilbert, C., 214, *223*
Glanville, B. B., 185, *222*
Globus, A., 104, *130*
Gloning, I., 188, 196, 202, 208, *223*
Gloning, K., 188, 196, 202, 208, *223*
Goddard, K. E., 43, *65*
Goff, W., 154, *172*
Gogel, W. C., 139, *168*
Gold, P. S., 143, *168*

Goldberg, S., 125, *129, 131*
Goldstein, K., 234, 245, *250*
Goldstein, P. J., 59, *66*
Gollin, E. S., 245, 246, *250, 251*
Gonyea, E. F., 197, *224*
Goodenough, D. R., 213, *228*
Goodfield, G. J., 260, *267*
Goodkin, F., 38, *64*
Goodwin, E. B., 139, *168*
Gordon, H. W., 203, *223, 224*
Gorman, J. J., 38, *64*
Gott, P. S., 201, *226*
Gottlieb, B., 104, *130*
Gottlieb, G., 35, 36, 37, *64,* 83, *98,* 142, *168,* 237, *250*
Goy, R. W., 209, *224*
Graham, F. K., 46, 48, *62, 64,* 123, *130*
Green, D. G., *64*
Green, S., 151, 153, *168, 172*
Greenberg, J. H., 159, *168*
Greenwood, M. M., 43, *65,* 110, 111, *132*
Greer, C. A., 241, 245, *250*
Gregory, R., 205, 206, *224*
Grell, E. H., 177, *225*
Griffin, D. R., 163, *168*
Griffin, E. J., 49, *63*
Griffith, B. C., 159, *169*
Grigg, P., 59, 60, *62*
Grohmann, J., 82, *98*
Gruzelier, J. H., 183, *224*
Gunther, M., 126, *130*
Guthrie, E. R., 79, *98*
Guttmacher, A. F., 182, *224*

H

Haegerstrom, G., 61, *66*
Hager, J. L., 80, *99*
Hailman, J. P., 94, *98,* 144, *168*
Haith, M. M., 51, *64,* 122, *131*
Hales, S., 260, *267*
Hall-Craggs, J., 146, *171*
Hamilton, W. D., 93, *98*
Hamilton, W. J. III, 164, *170*
Hamlyn, D., 255, *267*
Hamm, A., 186, *227*
Harlow, H. F., 72, *98*
Harlow, M. K., 72, *98*
Harmon, R. J., 118, *130*
Harris, K. S., 159, *169, 171*
Hatton, H. M., 48, *62*
Haub, G., 188, 196, 202, 208, *223*

Haynes, H., 56, *64*
Hebb, D. O., 237, *250*
Hécaen, H., 196, 198, *224*
Heck, W. E., 38, *64*
Hecox, K., 48, *64*
Heilman, K. M., 197, *224*
Held, R., 56, *64*
Hering, E., 53, *64*
Herrnstein, R. J., 146, *168*
Herron, J., 204, *224*
Hershenson, M., 54, *64*
Hess, E. H., 139, 140, *168*
Hibbard, E., 177, *224*
Hickey, T. L., 50, *64*
Hickox, J. E., 140, *169*
Hill, W. F., 77, 79, *98*
Hilton, T. J., 211, *224*
Himwich, W. A., 239, 240, 248, *249, 250*
Hinde, R. A. 80, *98*
Hines, D., 195, *224*
Hirsch, S., 211, *226*
Hochberg, F. H., 195, *224*
Hockett, C. F., 4, *31*
Hodos, W., 233, *250*
Hoff, A. L., 205, 207, *225*
Hoffman, H. S., 159, *169*
Hogan, J. A., 139, 140, *168*
Hohle, R. H., 48, *67*
Hohmann, A., 60, *64*
Hopkins, C. D., 163, *168*
Horowitz, F. D., 114, *131*
Hoversten, G. H., 48, 49, *65*
Howard, J., 25, *31*
Howland, H. C., 56, *62*
Huang, I., 247, *250*
Hubel, D. H., 59, *65, 68*
Hull, C. L., 80, *98*
Hulsebus, R. C., 114, *131*
Humphrey, T., 36, *65,* 201, *224*
Hunter, M. A., 49, *63*
Hurvich, C., 154, *168*
Hurvich, L. M., 55, *65*
Hutt, C., 49, *65*
Hutt, S. J., 49, *65*

I

Igna, E., 195, *222*
Ilg, F. L., 104, *130*
Immelman, K., 140, 141, 147, *168*
Impekoven, M., 143, *168*
Ingram, D., 196, *224*

AUTHOR INDEX

J

Jacklin, C. N., 210, 225
Jacob, F., 26, 31
Jacobs, G., 155, 167
Jacobson, S. W., 113, 131
Jaeckel, J. B., 142, 166
Jameson, D., 154, 168
Jeeves, M. A., 187, 201, 224
Jenkins, J. J., 159, 171
Jensen, K., 44, 65
Johansson, B., 36, 65
Johnstone, J., 204, 224
Jolly, A., 4, 31
Jones, M. H., 38, 63
Julesz, B., 60, 65
Jusczyk, P., 37, 49, 63, 121, 130

K

Kagan, J., 113, 131
Kakihana, R., 22, 31
Kamenetskaya, N. H., 123, 125, 129
Kandel, G., 211, 222
Karmel, B. Z., 50, 65
Kasatkin, N. I., 117, 131
Katz, S., 193, 222
Kawai, M., 247, 250
Kay, P., 154, 155, 156, 166
Kaye, H., 40, 46, 63, 65, 116, 117, 119, 131
Kear, J., 139, 169
Keeton, W. T., 163, 172
Kelly, W., 159, 170
Kemper, T. L., 195, 223
Kennell, J. H., 115, 131
Kenshalo, D. R., 47, 65
Kessen, W., 43, 66, 102, 122, 131, 155, 156, 167
Kimble, G. A., 104, 131
Kimura, D., 191, 196, 205, 224
Kinney, J. A., 159, 169
Kinsbourne, M., 195, 205, 225, 226
Kirk, D., 175, 225
Klaus, M. H., 115, 131
Klein, R., 160, 169
Klein, T. W., 25, 31
Kling, J., 104, 131
Knox, C., 191, 224
Kobre, K. R., 65, 107, 119, 131
Koenderink, J. J., 58, 67
Koenig, K. L., 118, 130
Kohlberg, L., 265, 267

Kohn, B., 185, 223, 224
Koltermann, R., 137, 169
Konishi, M., 147, 149, 150, 169
Kopin, I. J., 241, 250
Korner, A. F., 37, 65, 127, 131
Kovach, J. K., 140, 169
Krachkovskaia, M. V., 117, 131
Kramer, W., 202, 227
Kreithen, M. L., 163, 172
Kron, R. E., 43, 65
Kroodsma, D., 146, 169
Kruijt, J. P., 140, 164, 166
Kuhl, P. K., 154, 160, 161, 169
Kuypers, H. G. J. M., 200, 224

L

Ladavas, E., 218, 224
Lagger, R. L., 201, 224
Lamb, L. E., 48, 67
Lampe, H., 201, 225
Landis, T., 218, 224
Lane, H., 159, 169
Larsen, J. S., 50, 65
Lashley, K. S., 145, 169
Lasky, R., 160, 169
Lauer, J., 138, 169
Lawrence, M. M., 38, 65
Lawson, N. C., 205, 215, 216, 225
Layton, W. M., Jr., 181, 192, 225
Leavitt, L. A., 123, 130
Lee, P., 192, 227
Le Gros Clark, W. E., 4, 31
Lehmann, H. J., 201, 225
Lehrman, D. S., 82, 84, 99, 232, 250
Leibowitz, H. W., 238, 239, 250
Leigh, E. G., Jr., 92, 99
LeMay, M., 195, 223, 224, 225
Lenard, H. G., 49, 65
Lenneberg, E. H., 185, 225
Letson, R. D., 60, 62
LeVay, S., 59, 65
Leventhal, A. S., 48, 65
Levikova, A. M., 117, 131
LeVine, R., 265, 267
Levison, C. A., 122, 131
Levison, P. K., 122, 131
Levy, J., 189, 194, 197, 204, 216, 225
Levy, L., 195, 227
Levy, N., 46, 65
Lewin, K., 103, 131
Lewis, M., 123, 125, 131

Liberman, A. M., 159, *169, 171*
Liederman, J., 195, *225*
Lindauer, M., 137, 138, *169*
Lindsley, D. L., 177, *225*
Lipsitt, L. P., 40, 42, 43, 44, 46, 48, *63*, 65, 101, 102, 105, 107, 110, 111, 114, 116, 117, 118, 119, 120, 122, 126, 127, 128, 129, *130, 131, 132, 133*
Lipton, E. L., 48, *67*
Lisker, L., 158, *169*
Lissmann, H. W., 163, 169
Loewy, A. G., 234, *250*
Logghe, N., 201, *223*
Lorenz, K., 140, 144, *169*
Lorenz, K. Z., 77, 81, 86, 95, *99*
Loveland, D. H., 146, *168*
Luria, A. R., 195, *225*
Luppova, V. A., 123, 125, *129*

M

McClearn, G. E., 15, 22, 24, 27, *31*
Maccoby, E. E., 210, *225*
McDougall, W., 36, *66*
McEwen, B. S., 209, *224*
Macfarlane, A., 39, 40, *65*
McGlone, J., 209, *225*
McGraw, M. B., 38, *66*, 128, *132*
McKeever, W. F., 205, 207, *225*
McKeever, W. P., 197, *225*
McKenzie, B., 122, *132*
Mackintosh, N. J., 145, 157, 161, *169*
McRae, D. L., 195, *225*
Maffei, L., 58, *66*
Magoon, E. H., 50, *66*
Maisel, E. B., 50, *65*
Maller, O., 43, 44, *63, 66*
Mann, I. C., 50, *66*
Marakawa, N., 247, *251*
Maratos, O., 113, *132*
Marg, E., 59, *66*
Marler, P., 136, 146, 147, 148, 149, 150, 151, 153, 164, *169, 170, 172*
Marquis, D. P., 116, 117, *132*
Marshall, R. E., 118, *132*
Martin, G., 259, *267*
Matarazzo, R. G., 46, *64*
Mattson, M. E., 143, *168, 170*
Maurer, D., 50, *66*
Meltzoff, A. N., 113, *132*
Menzel, R., 137, *170*

Merrell, D. J., 195, *225*
Metcalf, D., 118, *130*
Milewski, A. E., 120, 121, *132*
Miller, D. E., 143, *170*
Miller, J. D. 159, 161, *169, 170*
Miller, R. E., 211, *222*
Miller, R. S., 234, *250*
Millodot, M., 61, *66*
Milner, A. D., 201, *223*
Milner, B., 195, 196, *225, 226*
Minifie, F. D., 160, 161, *167*
Miranda, S. B., 50, *63*
Mistretta, C. M., 36, *66*
Mitchell, D. E., 61, *66*
Mittwoch, U., 175, *225*
Moar, K., 56, 58, 59, *62*
Molfese, D., 185, *225*
Money, J., 27, *31*
Moncur, J. P., 48, 49, *65*
Monod, J., 26, *31*
Moody, D. B., 151, 152, 153, *166, 170, 172*
Moore, J. A., 118, *132*
Moore, J. M., 161, *167*
Moore, M. K., 113, *132*
Moran, L. J., 247, *250, 251*
Moreau, T., 195, *227*
Morgan, M. J., 175, 195, *225*
Moriarty, A. E., 118, *132*
Morse, P., 160, *170*
Moscovitch, M., 197, 204, 207, *225, 227*
Moskowitz-Cook, A., 52, 53, *66*
Moulton, P. A., 241, *249*
Movshon, J. A., 60, *66*
Mundinger, P., 146, 149, *170*
Muntjewerff, W. J., 49, *65*
Murphy, L. B., 118, *132*

N

Nagy, A. N., 123, *130*
Nagylaki, T., 194, *225*
Nas, H., 58, *67*
Nelson, A. K., 42, *66*
Nelson, M. N., 122, *130*
Netley, C., 187, *226*
Netsky, M. G., 200, *226*
Neville, H., 190, *226*
Nevo, E., 164, *167*
Newton, N., 126, *132*
Nicolai, J., 147, *170*

Nowlis, G. H., 43, 66
Nyberg-Hansen, R., 201, 226

O

O'Connor, K. P., 215, 226
Oh, W., 107, 130
Olrich, E. S., 123, 129
Olson, D. R., 234, 251
Oltman, P. L., 213, 214, 215, 216, 226, 228
Ordy, J. M., 56, 63
Orgel, L. E., 5, 31
Ornstein, R. E., 204, 224
Owen, R., 176, 227

P

Pallie, W., 186, 228
Papousek, H., 120, 132
Parlow, S., 205, 226
Pastore, R. E., 159, 170
Paterson, D. G., 211, 222, 227
Paul, J., 205, 206, 224
Pavlov, I. P., 77, 99
Peeples, D. R., 52, 53, 54, 55, 66, 67
Peiper, A., 42, 66, 114, 132
Peltzman, P., 59, 66
Perlmutter, M., 51, 68
Pernkopf, E., 181, 226
Perret, E., 218, 224
Peters, R., 255, 267
Peters, S., 150, 170
Petersen, A. C., 209, 226
Petersen, M. R., 151, 152, 153, 166, 170, 172
Peterson, J. L., 48, 67
Pfaffmann, C., 39, 66, 104, 112, 132
Phoenix, M. D., 43, 65
Piaget, J., 45, 66, 102, 113, 132
Pick, H. A., 238, 239, 250
Pick, H. L., 50, 68
Pirchio, M., 58, 66
Pisoni, D. B., 154, 159, 170
Pola, Y. V., 151, 171
Potter, R. H., 182, 192, 222
Powers, M. K., 56, 64
Pratt, K. C., 42, 66
Premack, D., 119, 132
Purpura, D. P., 104, 132

Q

Quatember, R., 188, 196, 202, 208, 223

R

Radic, P., 202, 228
Rakic, P., 50, 59, 66, 67
Rapaczynski, W., 215, 226
Rasmussen, T., 196, 226
Ratliff, F., 154, 170
Rawson, R. A., 38, 63
Reid, M., 197, 204, 210, 211, 212, 216, 225, 226
Reilly, B. M., 43, 65, 110, 111, 129, 132
Reimarus, H. S., 82, 99
Reinisch, J. M., 209, 226
Reitan, R. M., 206, 226
Ricci-Bitti, P. E., 218, 224
Richards, W., 60, 66
Richmond, J. B., 48, 67
Richter, C. P., 46, 67
Rieser, J., 39, 67
Riggs, L. A., 104, 131
Rinvik, E., 201, 226,
Rizzolatti, G., 205, 226
Robb, R. M., 50, 66
Robertson, D. R., 92, 99
Robertson, E. O., 48, 67
Romer, A. S., 9, 31
Rosch, E., 153, 154, 156, 170
Rosch-Heider, E., 154, 156, 171
Rose, M., 38, 63
Rosenzweig, M. R., 104, 132
Rosner, B., 159, 167
Rosser, M., 245, 250
Rota, R., 218, 222
Roth, L. M., 164, 171
Rozin, P., 95, 99
Rudel, R., 211, 226

S

Salapatek, P., 50, 56, 58, 59, 62, 67
Salapatek, P. H., 122, 131
Sameroff, A. J., 114, 116, 119, 125, 132, 133
Sands, J., 244, 250
Saravo, A., 245, 250
Sarnat, H. B., 200, 226
Satz, P., 193, 195, 224, 226
Sauguet, J., 196, 198, 224
Saul, R. E., 187, 201, 226, 227
Savin, V. J., 211, 222
Scammon, R. E., 50, 67
Scarr-Salapatek, S., 36, 67, 193, 222
Schachter, D., 246, 251

Schaie, K. W., 27, *31*
Schaller, M. J., 54, *67*
Scharf, B., 49, *67*
Schief, A., 164, *172*
Schiltz, K. A., 72, *98*
Schleidt, W., 164, *171*
Schneider, D., 164, *171*
Schneiderman, H. H., 234, *250*
Schneidler, G. R., 211, *227*
Schutz, F., 141, *171*
Searcy, M. A., 150, *167*
Seashore, H. G., 211, *222*
Sekel, M., 54, 55, *67*
Seligman, M. E. P., 80, *99*
Senf, R., 195, *222*
Shanks, B. L., 123, *129*
Shankweiler, D., 159, *169*
Shaw, J. C., 215, *226*
Sherman, I. C., 46, *67*
Sherman, M., 46, *67*
Shorey, H. H., 164, *171*
Shweder, R., 265, *267*
Sidman, R. L., 50, *67*
Siqueland, E. R., 36, 37, 49, *63*, *68*, 120, 121, *130*, *132*, *133*
Sjölander, S., 141, *171*
Skinner, B. F., 80, *99*, 235, *251*
Smith, L. C., 197, 204, 207, *225*, *227*
Snowdon, C., 151, *171*
Sokol, S., 59, *67*
Soloman, D., 211, *227*
Sonnemann, P., 141, *171*
Sontag, L. W., 36, *62*
Sostek, A., 116, *133*
Sostek, A. M., 116, *133*
Spalding, D. A., 82, *99*
Spalten, E., 211, *226*
Spears, W. C., 48, *67*
Spemann, H., 178, 179, *227*
Sperry, R. W., 178, 187, 203, *224*, *227*
Stevenson-Hinde, J., 80, *98*
Spiegel, F. S., 209, *226*
Spillmann, L., 54, *68*
Spinelli, D., 58, *66*
Springer, S., 154, *171*
Srb, A., 176, *227*
Stebbins, G. L., *30*, 231, 232, *251*
Stebbins, W., 151, 152, 153, *166*, *172*
Stebbins, W. C., 151, *170*
Steele, B., 38, *63*
Stein, M., 43, *65*

Steiner, J. E., 36, 39, 40, 44, *67*, 105, *133*
Steinschneider, A., 48, *67*, 123, 127, *129*, *133*
Stern, C., 18, *31*
Stevens, S. S., 48, *67*
Strange, W., 159, *171*
Strock, B. D., 123, *130*
Stratton, W. C., 118, *132*
Strawson, P. F., 262, *267*
Streeter, L. A., 160, *171*
Strock, B. D., 123, *130*
Struhsaker, T., 151, *171*
Studdert-Kennedy, M., 159, *169*, *171*
Sturner, W. Q., 127, *132*
Sun, K. H., 42, *66*
Suppe, F., 254, *267*
Sutherland, N. S., 145, 146, 150, *171*
Sutton, H. E., 20, *31*
Sutton, P. R., 175, *227*
Svensson, A., 211, *227*
Syrdal-Lasky, A., 160, *169*
Sytova, V. A., 123, 125, *129*
Szentagothai, J., 243, *251*

T

Taggert, J. K., Jr., 201, *227*
Tamura, M., 147, *170*
Tarshes, E. L., 206, *226*
Teller, D. Y., 52, 53, 54, 55, 58, *63*, *66*, *67*
Tenaza, R., 151, *170*
Teng, E. L., 192, *227*
Thibos, L. N., 61, *64*
Thoday, J. M., 233, *251*
Thoman, E., 37, *65*, 124, *133*
Thompson, C. I., 72, *98*
Thorndike, E. L., 77, 80, *99*, 103, 104, *133*
Thorpe, W. H., 146, 150, *171*
Thurlow, W. R., 48, *67*
Tibbling, L., 38, *67*
Tinbergen, N., 86, *99*, 137, 143, 144, 163, 165, *171*
Tischner, H., 164, *172*
Todor, J. I., 206, *227*
Tolman, E. C., 80, *99*
Torgersen, J., 181, *227*
Toulmin, S., 33, *33*, 254, 256, 258, 260, 266, *267*
Trevarthen, C., 199, 200, *227*

Turkewitz, G., 195, *227*
Turner, R. G., 43, *63*

U
Udelf, M. S., 56, *63*
Uexkull, J. von, 163, *172*
Umilta, C., 218, *224*

V
Vanderwolf, C. H., 196, 205, *224*
van Deventer, A. D., 205, *225*
Van Nes, F. L., 58, *67*
Venables, P. H., 183, *224*
Verhaart, W. J. C., 202, *227*
Verrillo, R. T., 47, *68*
Very, P. S., 211, *227*
Vigorito, J., 37, 49, *63*, 121, *130*
Vohr, B., 107, *130*
von Noorden, G. K., 59, 60, *62*
von Uexküll, J., 81, 84, *99*
Vygotsky, L. S., 264, *267*

W
Wada, J. A., 186, *227*
Waddington, C. H., 135, *172*, 232, 236, *251*
Wagonfeld, S., 118, *130*
Walk, R. D., 50, *68*
Walker, A. E., 201, *227*
Walker, L. D., 246, *251*
Walter, D. O., 185, *223*
Warner, R. R., 92, *99*
Watson, J. B., 36, *68*, 79, *99*, 103, *133*
Wechsler, A. F., 190, *227*
Wedenberg, E., 36, *65*
Weidmann, R., 145, *172*
Weidmann, U., 145, *172*
Weis, P., 233, *251*
Weiskopf, S., 155, 156, *167*
Weiss, P., 232, *251*

Werner, H., 233, *251*
Werner, J. S., 36, 51, 52, 53, 54, 55, *68*, 120, *133*
Wertheimer, M., 48, *68*
Wertz, D. C., 114, *133*
Wertz, R. W., 114, *133*
Wesman, A. G., 211, *222*
Westin, B., 36, *65*
White, B. L., 56, *64*
Whorf, B., 154, *172*
Wiens, J. A., 238, *251*
Wier, C., 159, *170*
Wiesel, T. N., 59, *65*, *68*
Wikner, K., 39, *67*
Wilmer, H. A., 50, *67*
Wilson, E. O., 89, *99*
Wilson, J., 201, *223*
Wilson, R. S., 28, 29, *31*,
Wilson, W. R., 161, *167*
Winick, M., 240, *251*
Witelson, S. F., 186, 210, 211, *228*
Witkin, H. A., 213, *228*
Wittgenstein, L., 262, *267*
Wood, C., 154, *172*
Wolff, P. H., 123, *133*
Wolter, J. R., 51, 53, *62*
Wooten, B. R., 51, 54, 55, *68*
Wright, A., 155, *172*

Y
Yakovlev, P. I., 202, *228*
Yang, K., 192, *227*
Yodlowski, M. L., 163, *172*
Yonas, A., 39, 50, *67*, *68*
Young, P. T., 102, *133*

Z
Zoccolotti, P., 215, *228*
Zoloth, S., 151, *172*
Zoloth, S. R., 151, 152, 153, *166*, *170*, *172*

Subject Index

A
Adaptation, 71, 75, 78, 136, 232
  perceptual, 137-138
Agraphia, 198
Anomia, 188
Aphasia, 185
Asymmetry, 173-228
  genetic basis, 175
  in invertebrates, 176-177
  origins, 175-179
  sex hormones, 210
Audition, 47-49
  auditory engram, 149
  temporal integration, 48

B
Babkin reflex, 46
Binocular organization, 59-61
  and visual experience, 59-61
Birdsong, 146-150
  learning, 146-147
  state dependence, 136
  universals, 146
Brain growth spurt, 244
Broca's aphasia, 188, 196

C
Canalization, 89, 96, 135, 147, 157, 165
Categorical perception, 155, 159, 160
Causality, 231
Central process theory, 243
Cerebral hemispheres
  bifunctional competence, 187
  early loss of language zones, 187
  equipotentiality, 185
  hyperdevelopment, 191
  infancy, 185-186
  speech localization, 198
Cerebral reorganization, 184-185, 186
Change, 232, 233
Chemical senses, 39-45
Chicks, 138-143
  ecological relevance of stimuli, 137
Chreod, 236
Chronological time, 244
Cognitive development, 242-243
Coherence, 232
Color vision, 154-157
  opponent processes, 155
Concept acquisition, 254
Conditioning, 77, 94-97, 101
  classical, 77-79, 115-118
  neonatal, 103
  operant, 77-79, 119-122
Contrast sensitivity, 55-59
  developmental changes, 58
Crib death, 127
Cultural evolution, 257
Cutaneous sensitivity, 45-47
  and age, 46

279

SUBJECT INDEX

D
Dandy-Walker syndrome, 201
Darwin, Charles, 12
Decussation, 196, 201
Developmental perspectives, 233-235
  epistemology, 255
  philosophy, 255
Developmental time, 244, 245
Dishabituation, 40, 48
Dizygotic twins, 192, 193
DNA, 19, 176, 240
Dysgraphia, 188, 196
Dyslexia, 188, 196

E
Early experience, 103
Ecological release, 89
Ecology, 137-140
Egocentric behavior, 240
Embryogenesis, 181
Enculturation, 257-262
Epistemology and developmenal psychology, 253-267
Ethological theory, 71, 81-87
Ethology of perceptual development, 162-166
Eurytopic species, 88-90
Evolution, 1-31
  of associative processes, 77-81
  and developmental plasticity, 88-93
  of polymorphism, 91-92
Experience, roles of, 237-239

F
Fitness, 233
Fixed action pattern (FAP) 85, 88, 89

G
Genes, 12, 15, 16-19, see also Asymmetry; Lateralization
Genetic constraints, 135
Genetic epistemology, 256
Growth gradients, 178, 199
Gustation, 36, 42-45, 105-112

H
Habituation, 40, 122-124
Hand-brain relationship, 196-209
Handedness 194, 196-209
  and aphasia, 195
  and culture, 198

Hardy-Weinberg-Castle equilibrium, 15
Heart rate, 110-112
  and conspecific song, 150
Herring gulls, 143-146
Hermaphroditism, 91
Homeorhesis, 236
Hominids, 9
Honeybees, 137-138
  ecological relevance of stimuli, 137
Hyperplasia, 240
Hypertrophy, 240

I
Imprinting, 140-143
Infancy
  anatomical asymmetry, 186, 210
  cerebral hemisphere differentiation, 185
  color classification, 155
  growth rate, 103
  initative behavior, 112-113
  learning processes, 101-133
  memory, 110
  sensory processes, 33-67
Innate releasing mechanism (IRM), 86, 96
Instinctive behavior, 75, 81, 82-84
  stimulus control of, 84-86
Interactionism, 231
Invertebrates, 7

L
Lamarck, 12
Lateralization, 180-196
  differentiation, 180
  environmental determinants, 181-192
  genetic determinants, 180, 192-196
  handedness, 194
  in twins, 192-193
Law of effect, 104
Learning Theory, 71, 77-81, 90, 149-150
  field expectancies, 80
  role of hedonism, 102

M
Macaques, 150-154
Maturation, 101
Mendel, Gregor Johann, 12, 16
Monozygotic twins, 192, 193, 195
Multimodal-polyphasic development, 239-243
Mutation, 19

SUBJECT INDEX

N
Neurogenesis, 35, 50, 199-203
  abnormal timing, 202
  agenesis of corpus callosum, 201, 203
  maturation rates, 199
  midline dysgenesis, 201
Nucleotide bases, 19

O
Olfaction, 39-42
  aversive qualities, 39-40
  hedonic qualities, 39
Ontogeny, 72
  and instinct, 82-84
  speech perception, 157-162
Operon model, 26-27
Organization, 231-232

P
Pain, 248
Perceptual development, see also Wavelength sensitivity
  asymmetry, 186
  color naming, 156-157
  color vision, 154-157
  natural categories, 153, see also Categorical perception
  perceptual learning, 143
  range of strategies, 164-166
  salience of stimuli, 154
Piaget, Jean, 233, 235, 255, 256, 257, 259
Plasticity
  asymmetry, 186
  CNS, 73
  constraints, 238
  in development, 72-73
  genetic expression, 178
  instinct theory, 87-88
  learning theory, 87-88
  quantitative variations, 237
  reorganizational capacity, 184-185
  sensory systems, 35-36
Pleasure principle, 102
Polyethisms, 88, 93
Polygenic systems, 21-26
Polymorphism, 91-93, 176
  obligate, 91
  facultative, 91
Population genetics, 15
Primates, 150-154
  graded auditory signals, 151
  species specific processing of auditory signals, 150-154
Pyramidal tracts, 196

R
Reinforcement, 105
Reptiles, 9
Retroactive causation, 74-77
RNA, 176
Rooting reflex, 46

S
Secondary simplifications, 233-234
Sensitive periods, 138-140, 149, 240, 243
Sensorimotor maturation, 199
Sensory development, prenatal origins, 36-37
Sensory precocity, 35
Serum growth levels, 248
Sex differences, 209-221
  behavioral measures, 211
  bilaterality, 213-214
  field-dependence, 214-216
  maturation, 212
  neuroanatomical measures, 211
  reaction time, 218
Sign stimulus, 85, 136, 143-146
Situs inversus, 179, 181
Social behavior, 247
Sound localization, 48
Spatial summation, 47
Speech perception, 157-162
Stage Theory, 246
Stenoptic species, 88-90
Stimulation
  minimal, 137
  optimal, 137

T
Teleology, 73-74, 75, 76
  and goal-directedness of behavior, 73
  mental purpose, 73-74, 76
Temporal summation, 47
Twinning, 182-183
Tympanic membrane, 47

V
Variability, 1-29, 87, 89, 193, 220-222, 232

Variability (*Cont.*)
  between species, 3-4,
  and evolution, 5-12
  and molecules, 19-21
  sources of, 12-17
  within organism, 26-29
  within species, 18-19
Vestibular system, 37-39
Visual evoked potential (VEP), 190
Vision, 49-61
  and ontogeny, 50
Vocal stimulation, 150-154
  species-specific processing, 150-154
Voice onset time (VOT), 158, 159

W

Wavelenth sensitivity, 51-55
  changes during ontogeny, 51
Werner, Heinz, 233
Word association, 247

LIBRARY OF DAVIDSON COLLEGE

ooks on regular loan may   for tw
presented at the Cir